Dr. Joachim Wuttke

Practical Guide
on
Transboundary
Waste Movements

Dr. Joachim Wuttke

Practical Guide
on
Transboundary
Waste Movements

Bibliografische Information der Deutschen Nationalbibliothek:
Die Deutsche Nationalbibliothek verzeichnet diese Publikation in der Deutschen
Nationalbibliografie; detaillierte bibliografische Daten sind im Internet über
http://dnb.dnb.de abrufbar.

Titelgestaltung: © Sabine Wuttke

Herstellung und Verlag: BoD – Books on Demand, Norderstedt

ISBN: 978-3-7557-6012-2

Preface

In view of the risk posed by the improper handling of hazardous waste, the extensive international, European and national regulations on the transfrontier shipment of waste aim to prevent the shipment of hazardous waste in particular to countries that do not have the appropriate capacities for environmentally sound disposal or recovery and to dispose of non-recyclable waste as close as possible to the place of origin.

The regulatory framework for transfrontier waste shipments is provided internationally by the Basel Convention on the Control of Transboundary Movements of Hazardous Wastes and other Wastes and their Disposal and the OECD Council Decision C(2001)107/Final on the Control of Transboundary Movements of Wastes destined for Recovery Operations, as well as at European level by Regulation (EC) No 1013/2006 on shipments of waste, which is directly applicable law in all Member States of the European Union.

This book was written from a European perspective. It is based in particular on experience gained in the EU Member State of Germany. One focus is thus on the European Waste Shipment Regulation. Nevertheless, it is also helpful for the implementation of the Basel Convention and the OECD Council Decision. The explanations and instructions are marked to indicate whether they are specific to one of the three instruments mentioned.

Due to numerous cross-references and interdependencies between the various legal provisions and the notification procedure with different case scenarios depending on the disposal method used, the export, transit and import countries involved and the waste to be shipped, it is not easy to get started with the subject matter and can be a challenge even for experienced actors.

The handbook is therefore intended to support all interested parties and stakeholders from authorities, industry and society in the application and implementation of the extensive national and international regulations on transboundary waste shipments. In addition to all relevant legal provisions, it contains a series of practical aids, especially on waste classification. A preceding explanatory text explains the interrelationships and most important legal regulations in a practical manner. This compilation of information provides waste producers, exporters and disposers with a good knowledge base and can contribute to the successful implementation of the legal regulations, to the avoidance of illegal waste shipments and to the support of enforcement.

I would like to thank Mr Ross Bartley for his critical review of the manuscript.

Dr. Joachim Wuttke

Content

Content

Content

Content

List of Abbreviations and Glossary

Bamako Convention	Bamako Convention on the Ban of the Import into Africa and the Control of Transboundary Movements and Management of Hazardous Wastes within Africa, adopted in Bamako, Mali on 30 January 1991, entered into force on 22. April 1998
Basel Protocol	Protocol on Liability and Compensation for Damage resulting from Transboundary Movements of Hazardous Wastes and their Disposal
BC Basel Convention	Basel Convention on the Control of Transboundary Movements of Hazardous Wastes and their Disposal of 22. March 1989 in the current version (State: 31 March 2020)
BT-Drs.	Printed matter of the German Bundestag
CA	Competent authority
cf.	compare
Chapt.	Chapter
CJEU	Court of Justice of the European Union
CLP CLP-Regulation	Regulation (EC) No 1272/2008 of the European Parliament and of the Council of 16 December 2008 on classification, labelling and packaging of substances and mixtures, amending and repealing Directives 67/548/EEC and 1999/45/EC, and amending Regulation (EC) No 1907/2006, OJ, L 353, 31.12.2008, p. 1–1355, as amended by Commission Delegated Regulation (EU) 2020/1677 of 31 August 2020 (C/2020/5758), OJ, L 379, 13.11.2020, p. 3–23
COP	Conference of the Parties
Community	see EU
DDT	Dichlorodiphenyltrichloroethane
EC Waste Shipment Regulation (1993)	Council Regulation (EEC) No 259/93 of 1 February 1993 on the supervision and control of shipments of waste within, into and out of the European Community, OJ, L 30, 6.2.1993, p. 1–28, last amended by Commission Regulation (EC) No 2557/2001 of 28 December 2001, OJ, L 349, p. 1 of 31.12.01
EEE	electrical and electronic equipment (non-waste)
EFTA	European Free Trade Association
Enforcement Guidance	LAGA Notice 25: "Enforcement Guidance for Regulation (EC) No. 1013/2006 of the European Parliament and of the Council of 14 June 2006 on shipments of waste and for the Waste

	Shipment Act (Ger.) of 19. July 2007 (WSA)", State May 2017 available in German only at: http://www.laga-online.de
EPR	Extended producer responsibility
ESM	Environmentally Sound Management
et seq.	and the following (et sequentes)
EU	European Union
EU POP-Regulation	Regulation (EU) 2019/1021 of the European Parliament and of the Council of 20 June 2019 on persistent organic pollutants, OJ. L 169 25.6.2019, p. 45 as amended by Commission Delegated Regulation (EU) 2021/115 of 27 November 2020 (C/2020/7980), OJ, L 36, 2.2.2021, p. 7–9
EU Regulation 1418/2007	Commission Regulation (EC) No 1418/2007 of 29 November 2007 concerning the export for recovery of certain waste listed in Annex III or IIIA to Regulation (EC) No 1013/2006 of the European Parliament and of the Council to certain countries to which the OECD Decision on the control of transboundary movements of wastes does not apply, OJ, L 316, 04.12.2007, p. 6, as amended
EU-WFD Waste Framework Directive	Directive 2008/98/EC of the European Parliament and of the Council of 19 November 2008 on waste and repealing certain Directives, OJ L 312, 22.11.2008, p. 3–30 as amended by Directive (EU) 2018/851 of the European Parliament and of the Council of 30 May 2018, OJ, L 150, 14.06.2018, p. 109–140
EU-WSR European Waste Shipment Regulation	Regulation (EC) No 1013/2006 of the European Parliament and of the Council of 14 June 2006 on shipments of waste, OJ, L 190, 12.7.2006, p. 1–98 as amended by Commission Delegated Regulation (EU) 2020/2174 of 19 October 2020 (C/2020/7091) OJ, L 433, 22.12.2020, p. 11–19
EWL European Waste List	Commission Decision of 18 December 2014 amending Decision 2000/532/EC on the list of waste pursuant to Directive 2008/98/EC of the European Parliament and of the Council (2014/955/EU), OJ, L 370, p 44 of 30.12. 2014
Federal Immission Control Act	Act on the Prevention of Harmful Effects on the Environment Caused by Air Pollution, Noise, Vibration and Similar Phenomena (Federal Immission Control Act),
HS	Harmonized Commodity Description and Coding System ("Harmonized System" for short) (developed by WCO)
i.c.w.	in connection with
i.e.	id est (that is)

IPPC Directive	Directive 2010/75/EU of the European Parliament and of the Council of 24 November 2010 on industrial emissions (integrated pollution prevention and control), OJ, L 334, 17.12.2010, p. 17–119, as amended
ISO	International Organization for Standardization
kg	kilogram
LAGA	German Federation/Federal States Working Party on Waste
LAGA Technical Notes	Technical notes on the classification of wastes according to their hazardousness
Liability Protocol	Protocol on Liability and Compensation for Damage Resulting from Transboundary Movements of Hazardous Wastes and Their Disposal, not yet in force
m^3	cubic meters
MFSU	manufacture, formulation, supply and use
mg	Milligram
MPPI	Mobile Phone Partnership Initiative
n.o.s.	not otherwise specified
OECD	Organisation for Economic Co-operation and Development
OECD Council Decision C(92)39	OECD Decision C(92)39/FINAL on the Control of Transfrontier Movements of Wastes Destined for Recovery Operations
OECD-Decision	Decision of the Council C(2001)107/Final of 22. May 2001 on the Revision of Council Decision C(92)39/FINAL on the Control of Transboundary Movements of Wastes Destined for Recovery Operations in the current version (State: 1. January 2021)
OJ	Official Journal of the European Union
PACE	Partnership for Action on Computing Equipment
PBDE	Polybrominated diphenyl ethers
PCB Directive	Council Directive 96/59/EC of 16 September 1996 on the disposal of polychlorinated biphenyls and polychlorinated terphenyls (PCB/PCT), OJ, L 243, 24.9.1996, p. 31).
PCBs	polychlorinated biphenyls
PCDD	polychlorinated dibenzo–p–dioxins
PCDD/PCDF	polychlorinated dibenzo-p-dioxins and dibenzofurans

PFOS	perfluorooctane sulfonic acid
PVC	polyvinyl chloride
REACH-Regulation	Regulation (EC) No 1907/2006 of the European Parliament and of the Council of 18 December 2006 concerning the Registration, Evaluation, Authorisation and Restriction of Chemicals (REACH), establishing a European Chemicals Agency, amending Directive 1999/45/EC and repealing Council Regulation (EEC) No 793/93 and Commission Regulation (EC) No 1488/94 as well as Council Directive 76/769/EEC and Commission Directives 91/155/EEC, 93/67/EEC, 93/105/EC and 2000/21/EC, OJ, L 396, 30.12.2006, p. 1, as amended
sect.	section
subst.	substance
TEQ	toxic equivalent
TFEU	Treaty on the Functioning of the European Union (TFEU), consolidated version of 26.10.2012, OJ,C 326, p.47
TFS	Transfrontier Shipment of Waste
UN	United Nations
UNEP	United Nations Environment Program
US	United States of America
Waigani Convention	Convention to Ban the Importation into Forum Island Countries of Hazardous and Radioactive Wastes and to Control the Transboundary Movement and Management of Hazardous Wastes within the South Pacific Region (the Waigani Convention), entered into force in 2001
WCO	World Custom Organisation
WEEE	waste electrical and electronic equipment
WEEE Directive	Directive 2012/19/EU of the European Parliament and of the Council of 4 July 2012 on waste electrical and electronic equipment (WEEE), OJ, L 197, 24.07.2012, p. 38, as amended by Directive (EU) 2018/849 of the European Parliament and of the Council of 30 May 2018, OJ, L 150, 14.06.2018, p. 93–99

I. Explanatory Notes on Waste Classification and Transfrontier Waste Shipments

In the explanatory notes, legal norms are indicated without legal sources. The full titles and references of the legal standards mentioned in the text as well as explanations of abbreviations are mainly listed in the glossary.

1. Introduction to the Legislation

The beneficial development of technology and the natural sciences over the last 150 years has been accompanied by an enormous exploitation of nature. There had already been systematic environmental destruction before, as the forests of Crete cleared for fuel in the copper furnaces for the production of bronze or the areas of the Lüneburg Heath cleared by farmers to obtain arable and pasture land as well as construction and firewood show. But they are insignificant compared to what has been achieved with the use of machines and chemicals: raw material deposits are mined to complete exhaustion, air, water and soil are polluted to the limit and the landscape is overgrown.

In such an environment, free human development is no longer possible. Not only our health, but also our economic success is endangered by such developments. In order to counter these dangers, environmental protection measures are necessary. Environmental protection is an independent, indispensable political task. Its objectives must be weighed against those of other policy areas, and wherever lives or serious threats to human health are at stake, environmental protection must claim priority.

Environmental policy must therefore take measures that are necessary to

- to secure an environment that is needed for the preservation of health and a dignified existence,

- to protect soil, air and water, plants and animals from adverse effects of human intervention, and

- to eliminate damage and disadvantages resulting from human intervention in the environment.

1.1 Protection Goals and Principles of Action

In order to protect humans and the environment from the harmful effects of all substances introduced into the atmosphere, waters or soils, a comprehensive and interdisciplinary concept is required. This means general protection goals and principles of action, which, together with principles for the use and further development of environmental policy instruments, determine medium and long-term environmental policy. Precaution against environmental pollution by substances is a primary concern of environmental policy for the following protection goals.

The primary objective of environmental policy is to protect human life and health, both now and in the future, from damage, particularly from substance inputs into the environment.

The existence of human beings is indispensably linked to the biotic communities of animals and plants via food chains and material cycles. It depends in many ways on maintaining or restoring the functioning of ecosystems. Not least for this reason, environmental policy aims to protect ecosystems in their entirety as well as to protect and conserve individual species of animals and plants. However, animals, plants and ecosystems are also protected for their own sake.

In the realisation of environmental policy goals, the principle of environmental precaution, the principle of cooperation and the polluter pays principle are followed. Environmental precaution is to be understood both as a principle of political action and as a legal principle, which, however, requires legal standardisation. As a principle of political action, environmental precaution encompasses all actions which

- serve to avert concrete environmental hazards,

- avoid or reduce risks to the environment in the run-up to danger prevention, and

- which, with foresight, serve to shape our future environment, in particular the protection and development of the natural foundations of life.

The cooperation principle is understood as a political procedural principle aimed at achieving environmental policy goals as amicably as possible. It requires a fair cooperation of all governmental and social forces in the decision-making process and in the realisation of environmental policy objectives. This improves the information situation of those involved as well as the acceptance and thus the effectiveness of environmental policy decisions. Unnecessary conflicts, administrative effort and costs are to be avoided or reduced. Of course, there are also limits to the cooperation between governmental and societal bodies. For example, the state cannot waive the competences assigned to it by the constitution or by law.

All sectors of the economy are potential polluters. Manufacturers and users of substances, companies and their associations are well aware of the possibilities and procedures for avoiding and reducing substance inputs. Therefore, it is expected that the economy uses creativity, scientific expertise and technical knowledge for the continuous further development of environmentally sound technologies.

The polluter pays principle in environmental protection is understood as a cost allocation principle and an economic efficiency criterion. The costs of preventing or eliminating environmental pollution must be borne by the party responsible for causing it.

1.2   Legal Instruments

In recent decades, a large number of legal provisions have been enacted and amended in the field of environmental protection including waste management. The legal instruments for ensuring that social forces act in accordance with environmental protection concerns consist largely of tools that can be attributed to regulatory law.

Among the instruments of regulatory law, the rules and prohibitions or restrictions based directly on laws or ordinances are particularly numerous. The obligations based on rules are of a very different nature, such as management obligations, maintenance obligations, environmental compatibility tests, care obligations, storage and packaging obligations, recycling obligations, precautionary measures, monitoring and security obligations.

The desired environmentally sound behaviour is usually achieved reliably and quickly with prohibitions and restrictions. The development of a comprehensive regulatory framework is therefore one of the central points of environmental policy. However, the effectiveness of the regulatory instruments is limited, especially ahead of prevention of hazards. In some cases, the state is unable to obtain the information required for bans and prohibitions, or can do so only at disproportionately high expense. The necessary state controls also tie up a large amount of human and financial resources.

The environmental impact assessment is an exemplary instrument of precautionary environmental policy. Environmental impact assessment involves the analysis and evaluation of the likely effects of technologies, programmes, plans or projects on the environment. With its help, these effects are to be described and secured in a transparent and comprehensible manner in a regulated procedure before the decision, in which other considerations can also be included, is made.

In order to avoid or reduce substance discharges into the environment, it is possible, in appropriate cases, to increasingly encourage polluters to enter into voluntary commitments. The state can refrain from issuing bans and prohibitions ahead of prevention of hazards, if the polluters voluntarily agree to the required environmentally sound behaviour and guarantee compliance with the commitment.

With voluntary commitments, polluters can demonstrate how acting on their own responsibility quickly and un-bureaucratically leads to environmental relief through cooperation with the state. This instrument will be particularly useful in cases where ongoing economic and technical developments are to be accelerated. However, a prerequisite for the realisation and success of voluntary commitments is that the group of polluters is manageable.

The experiences with the existing voluntary commitments in Germany[1] are ambivalent. One study found that only one third of the commitments set ambitious targets and the remaining two thirds only promised standards that were easy to meet. Industry sees the declarations primarily as a means of averting impending and probably stricter environmental regulations by the state.

The state instruments mentioned so far, economic incentives and agreements, but also bans and prohibitions are only effective to a limited extent if they do not meet with the insight of the polluters into the necessity of environmental protection measures. The decisive factor for the lasting success of environmental policy is that those affected accept the environmental policy goals and are prepared to actively contribute to achieving these goals. Environmental awareness is therefore a decisive prerequisite for the success of environmental policy measures. The further development of environmental awareness through rapid, comprehensive and precise information about possibilities of environmentally sound behaviour is therefore necessary.

1.3 European Waste Legislation *(EU-specific)*

In the EU, more than 2 billion tonnes of waste are generated every year, and their collection, storage, treatment and processing are regulated by a number of directives. These measures concern individual areas such as waste from titanium dioxide production, waste oil and waste containing PCBs. Transfrontier shipments of waste are to be controlled in accordance with the directly applicable Regulation (EC) No 1013/2006 on shipments of waste (EU-WSR).

Community policy on the environment shall contribute to the pursuit of the following objectives

- preserving, protecting and improving the quality of the environment,

- protection of human health,

- prudent and rational utilisation of natural resources, and

---

[1] The Role of Environmental Agreements, Cameron & May, 1999

- promoting measures at international level to deal with regional or worldwide environmental problems.

Community policy on the environment aims at a "high level of protection". It is based on the principle that environmental damage should as a priority be rectified at source, on the principles of precaution and prevention and on the polluter pays principle.

The Community's environmental policy objectives are defined and described in action programmes. Community waste management policy gives priority to prevention and recovery. The focus is on the following key objectives

- preventing the generation of waste by increasing the use of environmentally sound technology and the production of environmentally sound products,

- promoting recycling and

- improving waste disposal with the help of strict European standards, according to which waste must be disposed of as close to the producer as possible.

The common environmental policy is committed to the precautionary and polluter-pays principles. Environmental protection requirements are an integral part of other Community policies. The subsidiarity principle applies, i.e. the Community will only take action in the field of environmental policy if its environmental objectives can be better achieved at Community level. Provided that they are compatible with the Treaty on the Functioning of the European Union (TFEU), Member States of the EU have the right under Article 193 TFEU to maintain and adopt enhanced environmental protection measures.

1.3.1 Delegated and Implementing Acts

The Lisbon Treaty introduced delegated acts (Art. 290 TFEU) and implementing acts (Art. 291 TFEU) in 2009. Before the EU Commission can adopt an implementing act, one of the following procedures applies:

- examination procedure – used particularly for

    (i)   measures with general scope and

    (ii)   measures with a potentially significant impact (in areas such as taxation or agricultural policy)

- advisory procedure – generally used for all other implementing acts.

Both procedures require that a committee composed of representatives from all EU Member States provide a formal opinion, usually in the form of a vote, on the Commission's proposed measures. This procedure is known in the EU jargon as "comitology" (committee procedure). Depending on the procedure, committee opinions can be more or less binding on the Commission.

If the EU Commission has been given the relevant power by a legal act, it can adopt delegated acts. It is subject to strict conditions:

- the delegated act must not change the essential elements of the legal act,

- the objectives, content, scope and duration of the delegation must be explicitly set out in the act, and

- Parliament and the Council may revoke the delegation of power or express reservations in respect of the delegated act.

Delegated acts are prepared by the Commission and adopted after consultation of expert groups composed of representatives of all EU Member States, which meet on a regular or ad hoc basis. Once the Commission has adopted the act, Parliament and Council have two months to raise objections. Otherwise, the delegated act enters into force.

1.3.2 Regulatory Instruments

Regulations, directives, decisions, recommendations or opinions can be used as regulatory instruments. Regulations are the most important instrument due to their direct validity in the EU Member States, whereas directives are addressed to the Member States and must be implemented by them. So far, mainly directives have been issued in the area of waste law.

The waste-relevant EU Directives mainly deal with specific regulations for certain types of waste, such as packaging, batteries, PCBs and PCTs, waste from titanium dioxide production, sewage sludge, etc., or with plant-related regulations, such as for landfills or waste incineration plants.

In 2018, the European Parliament adopted extensive amendments to the directives on waste management. Based on the European Circular Economy Package published in December 2015, the Council Directive on Waste (EU-WFD) was revised, among other things.

1.4   Waste Framework Directive *(EU-specific)*

The most important of the existing directives is the Council Directive on Waste (EU Waste Framework Directive - EU-WFD). The amended EU-WFD entered into force on 12 December 2008 and was again amended in 2018[2]. These amendments mainly include extended requirements to promote the prevention of waste, the setting of targets for recycling and preparation for re-use of municipal waste, minimum requirements for extended producer responsibility schemes, extended criteria for the assessment of the end-of-waste status and new requirements for separate collection.

The EU-WFD contains general provisions on waste, in particular provisions on

- the concepts of waste and waste disposal
- the definition of hazardous waste,
- the polluter pays principle,
- the reporting obligations to be fulfilled,
- the licensing and monitoring obligations,
- waste management planning,
- producer responsibility,
- by-products and the end-of-waste
- the waste hierarchy,
- waste prevention and
- the disposal of waste oils.

1.4.1 Waste Definition in the EU

The legal definition of waste has remained unchanged from the previous regulation. "Waste" is still any substance or object which the holder discards, intends to discard or is required to discard. The EU-WFD regulates all types of waste that meet this definition.

A substance or object only becomes waste when the holder discards it (de facto waste term), wants to discard it (subjective waste term) or has to discard it (objective waste term). The waste definition in the EU is not fulfilled if substances or objects are directly used again for their original purpose. This also includes products that can be repaired or are in need of repair.

Please note that the Waste Definition of Basel Convention (cf. sect. I.3.1.2) and of OECD Council Decision (cf. sect. I.3.2.1) are not the same as in EU-WFD.

---

[2] Amended by Directive (EU) 2018/851 of the European Parliament and of the Council of 30 May 2018

1.4.2 Distinction between Waste and Non-waste in the EU

Substances and objects that can be used directly for their original purpose as a product, commodity or raw material without further treatment in the sense of a recovery operation according to Annex II to the EU-WFD are generally not wastes (for recovery). The prerequisite is that the items in question are destined with the above-mentioned purpose and are actually put to a corresponding use or handed over for this purpose.

For electrical and electronic equipment, the WEEE Directive (Art. 23 in conjunction with Annex VI) introduced a reversal of the burden of proof. Whoever transports used electrical or electronic equipment shall ensure that this equipment is accompanied by documents proving that the equipment in question is not waste equipment and that it is adequately protected against damage during transport and loading and unloading, in particular by adequate packaging and suitable stacking of the load.

In the absence of appropriate documentation to prove that an item is used EEE and not waste, and in the absence of adequate protection against damage during transport and loading and unloading, the European competent authorities will consider such an item to be waste and assume that the shipment constitutes an illegal shipment. In these circumstances, the cargo will be dealt with in accordance with Art. 24 and 25 of the EU-WSR.

If, during controls, the above-mentioned evidence is not provided to the authorities involved in the controls within the time limit set by them, or if these authorities consider that the evidence and information at their disposal is not sufficient for an assessment, the shipment in question will be considered an illegal shipment in accordance with Art. 50(4) d of the EU-WSR.

Further guidance gives the Correspondents' Guidelines No. 1 (for WEEE) and No. 9 (for end-of-life vehicles).[3]

1.4.2.1 Differentiation between Waste and By-products in the EU

Art. 5 of the EU-WFD establishes a regulation on the distinction between waste and by-products. According to this, a substance or object that is the result of a production process without its production being the main purpose of the process is only considered a by-product and not a waste if all of the following conditions are met in the specific case:

- it is certain that the substance or object will continue to be used,

- the substance or object can be used directly, without further processing other than normal industrial practice,

---

[3] cf. Correspondents' Guidelines: https://waste-move.eu/?p=539&lang=en

- the substance or object is produced as an integral part of a production process, and

- the further use is lawful, i.e. the substance or object fulfils all relevant product, environmental and health protection requirements for the particular use and will not lead to overall adverse environmental or health impacts.

1.4.2.2   End of Waste in the EU

Art. 6 of the EU-WFD regulates whether and under what conditions a waste can be released from the waste regime. The end of waste status is systematically linked to the recovery and disposal obligations under waste legislation. Waste status ends for certain specified wastes when they have undergone a recovery process and also fulfil specific criteria. The following conditions must all be met at the same time in a specific case:

- the substance or object is to be used for specific purposes,

- a market or demand exists for such a substance or object,

- the substance or object fulfils the technical requirements for the specific purposes and meets the existing legislation and standards applicable to products; and

- the use of the substance or object will not lead to overall adverse environmental or human health impacts.

In addition, the substances or objects must comply with the relevant requirements of chemicals and product legislation. This includes compliance with appropriate quality criteria and, if necessary, pollutant limit values for the end-of-waste status of materials obtained through the recovery process.

1.4.3   End-of-Waste Regulations in the EU

The EU has issued three end-of-waste regulations, namely for iron, steel and aluminium scrap, for glass cullet and for copper scrap. In addition, some EU Member States have adopted national end-of-waste regulations, such as Austria for recycled wood, recycled construction materials, substitute fuel products and compost.

The three EU regulations are structured in the same way and, in addition to a few articles, contain annexes with precise details of the criteria that must be met for the respective materials. Certain material-specific criteria have to be fulfilled, certain treatment processes have to be carried out and certain requirements for the quality of the material after treatment have to be met.

It is important that the producer of the materials with end-of-waste status applies a quality management system and proves with a declaration of compliance (cf. annex to the respective regulation) for each consignment that all required criteria of the End-of-Waste regulation are met. The producer must pass on the declaration of compliance to the next holder of the material.

1.4.3.1 *End-of-Waste Regulation for Iron, Steel and Aluminium Scrap*

With this End-of-Waste Regulation, specific criteria for iron, steel and aluminium scrap were defined, which determine the end-of-waste. Essential conditions for the end of waste are requirements regarding

- the quality of the scrap,
- the input material to be recycled, and
- the treatment processes and techniques.

All treatment steps such as crushing, shredding, cleaning or decontamination that are necessary to prepare the scrap for use in steel or aluminium plants must be completed before the waste status is terminated.

For iron and steel scrap, a total foreign matter content of two percent by weight was specified, for aluminium scrap a foreign matter content of no more than five percent by weight or a metal yield of at least 90 %. A further condition is that the scrap must be free of visible oil, oil emulsions and lubricants, with the exception of small quantities, which must not leak. Furthermore, it must not exhibit any of the hazardous properties listed in Annex III to the EU-WFD.

Within the framework of the quality management system, compliance with all requirements of this regulation shall be demonstrated. The producer shall grant the competent authorities access to the quality management system on request.

1.4.3.2 *End-of-Waste Regulation for Glass Cullet*

The End-of-Waste Regulation for glass cullet sets specific criteria for the end-of-waste status of glass cullet intended for the production of glass. For this purpose, criteria have been defined for the following non-glass components:

- ferrous metals: ≤ 50 ppm,
- non-ferrous metals: ≤ 60 ppm,
- organic substances: $\leq 2,000$ ppm,
- inorganic non-metal non-glass substances:
    - < 100 ppm for glass cullet of size > 1mm,
    - $< 1,500$ ppm for cullet of size ≤ 1 mm.

Inorganic non-metal non-glass materials include, for example, ceramics, stones, porcelain or pyro ceramics and organic materials include paper, rubber, plastic, fabric or wood.

The glass cullet must neither exhibit any of the hazardous properties listed in Annex III to the EU-WFD nor exceed the limit values according to Annex IV to the EU-POP Regulation. Only waste from the collection of recoverable container glass, flat glass or lead-free tableware may be used as input. Glass-containing waste from mixed municipal solid waste or healthcare waste and hazardous wastes such as lead glass waste from cathode ray tubes shall not be used as an input.

1.4.3.3   End-of-Waste Regulation for Copper Scrap

The End-of-Waste Regulation for Copper Scrap establishes specific criteria for the end of waste status of copper scrap. In particular, copper scrap is no longer considered waste if the following criteria are met, namely if

- the total content of impurities does not exceed 2 per cent by weight,

- the scrap is free from excessive metal oxide in any form, except for typical amounts resulting from outdoor storage of processed scrap under normal atmospheric conditions, and

- the scrap is free from visible oil, oil emulsions, lubricants or grease, except for insignificant quantities which must not leach out.

Impurities are metals other than copper and copper alloys such as

- non-metallic substances (soil, dust, insulating material and glass),

- combustible non-metallic substances (rubber, plastic, fabric, wood),

- other chemical or organic substances, and

- slag, dross, skimmings, baghouse dust, grinder dust and sludge. The copper scrap must not exhibit any of the hazardous properties listed in Annex III to the EU-WFD or exceed the limit values according to Annex IV to the EU POP-Regulation and it must be free of PVC in the form of coatings, paints and residual plastics.

1.5   National Waste Legislation *(Example of Germany)*

The first legal foundations for waste management in Germany were developed at the beginning of the 19th century. However, the waste regulations at that time were limited to individual parts of Germany, where they were integrated into regulatory and police law, such as sanitary police regulations. It was only after the links between a lack of urban hygiene and widespread diseases such as cholera became increasingly clear that more emphasis was placed on orderly drainage and waste disposal. These fields of activity were carried out by the municipalities as sovereign tasks. Until the 1960s, the collected waste was almost exclusively taken to the existing many small waste dumps.

Waste management goals did not find their way into the environmental policy of Germany until around 1970. In 1971, the Federal Government analysed the waste management situation in its environmental programme[4]. At that time waste management was characterised by the operation of about 50,000 small, often uncontrolled dumps. Hazardous waste was also deposited at these sites along with household waste. Criticism of this state of affairs culminated above all in the demand to create a few central and orderly landfills. After the Federal Government obtained legislative competence for the area of waste law in 1972 by amending the Basic Law, it enacted the Waste Disposal Act.

1.5.1   The Waste Disposal Act of 1972

The Waste Disposal Act (WDA) was primarily an organisational and planning law, the aim of which was to steer the disorderly so-called "dumping economy" into orderly channels through organisational guidelines. This was primarily achieved by regulating the responsibilities for waste disposal. The law was clearly disposal-oriented. Binding obligations for waste recovery were not included in this law. Objectives for the prevention and recovery of waste were formulated for the first time in the Federal Government's Waste Management Programme[5] of October 1975.

1.5.2   The Waste Management Act of 1986 (WMA)

Within the framework of the fourth amendment to the Waste Disposal Act in 1986, the Waste Disposal Act became the "Act on the Prevention and Disposal of Waste – Waste Management Act". This law for the first time contains principles and obligations for the prevention and recycling of waste. Among other things, the WMA newly regulates the waste recovery requirement, the disposal of waste oils, the authorisation to issue technical instructions and the extension of waste law monitoring to contaminated sites.

The authorisation to issue technical instructions was used to regulate the disposal of hazardous waste through the "TI Waste" and the disposal of municipal waste through the "TI Municipal Waste". The bundle of empowerments contained in Art. 14 WMA to enforce prevention and recovery was used to issue product-related regulations in advance of the disposal

---

[4]   BT-Drs. 10/5656

[5]   BT-Drs. 7/4826

obligation.[6] The best-known example is the Packaging Ordinance, which was further developed into the Packaging Act after several amendments.

1.5.3 Recycling and Waste Management Act of 1994

The Recycling and Waste Management Act (RWMA) is the third comprehensive revision of waste legislation in Germany, following the Waste Disposal Act of 1972 and the Waste Management Act of 1986.

The amendment also aimed to implement European directives such as the EU-WFD. The primary objective of the RWMA was to further develop waste management into a circular economy. Important cornerstones of the RWMA were the implementation of a new concept of waste adapted to the EU law, the hierarchy of obligations, extended producer responsibility and a partial re-organisation of the disposal system with extended possibilities for privatisation of waste management.

What was new was that the scope of application of the RWMA in principle also covers substances which until 1986 were designated as so-called economic goods or residual materials within the meaning of Art. 5 (1) No. 3 of the Federal Immission Control Act in distinction to the waste concept of Art. 1 WDA. The scope of application of the RWMA was thus significantly expanded compared to the WDA of 1972.

1.5.4 Circular Economy Act of 2012

The Act to Promote Circular Economy and Safeguard the Environmentally Sound Management of Waste (Circular Economy Act – CEA) came into force in June 2012. The CEA replaced the Recycling and Waste Management Act (RWMA). The CEA transposes the requirements of the revised EU-WFD into national law.

The primary objective was to focus the circular economy even more strongly on resource, climate and environmental protection. With the CEA, the concept of waste was aligned with the EU-WFD and expanded. A new regulation on the distinction between waste and by-products not subject to waste law was introduced in Art. 4 and a new provision on the end of waste status in Art. 5 CEA.

1.5.5 Circular Economy Act of 2020

The aim of the revised CEA, is to improve resource management and re-

---

[6] Wuttke, Joachim: "Erklärtes Ziel: Mehr Produktverantwortung - Maßnahmen zur Vermeidung und Verwertung von Abfällen nach § 14 AbfG", Entsorgungs-Technik **5** (1993) Nr. 4, S. 41/45 und Nr. 5, S. 42/46.

source efficiency in Germany and, in particular, to strengthen waste prevention. The recycling rates of certain waste streams, especially paper, metal, plastic and glass, but also municipal waste, are increased and the obligations for separate collection are extended to bio-waste, hazardous household waste, textiles and bulky waste.

Public authorities will be obliged to give preference to ecologically advantageous products in future procurement. Recycled products are to be given priority in public procurement. With this law, the Federal Government is obliging itself to give preference in procurement to products that conserve raw materials, are low-waste, repairable, low-pollutant and recyclable, provided that no unreasonable additional costs are incurred.

With a newly created duty of care, the state will in future for the first time have legal recourse against the destruction of new goods or returns. For the first time, there is also a legal basis for manufacturers and retailers of single-use plastic products, such as "to-go cups" or cigarette butts, to participate in the cleaning costs of parks and streets by decree.

1.5.6　Waste Shipment Act, Germany

The Waste Shipment Act (WSA) establishes the necessary legal regulations for the implementation of the Basel Convention, among other things, and at the same time creates necessary supplements to the EU-WSR, for example supplementary regulations on re-importation obligations, on the financial guarantee, on the assignment of competent authorities, on data exchange, and on penalties and fines.

Furthermore, the operator of a waste disposal facility has an obligation to inspect and provide information, with regard to the conformity of the waste actually delivered (quantity, designation and composition, physical properties, waste identification) with the information in the accompanying movement document. The quantity of waste determined on arrival at the facility should not exceed the quantity indicated in block 5 of the movement document. The facility operator must also check that the waste delivered corresponds to the information provided with the movement document.

The amendment of the WSA in 2016 brought it into line with the EU law and created more differentiated criminal sanction regulations. The previous sanctions had shown that the sanction structure was not sufficiently differentiated and led to problems of interpretation. With the Amending Act, the sanctions of Art. 326 (2) No. 1 of the Criminal Code *(Strafgesetzbuch)* for violations of the EU-WSR was transferred to the German WSA and additional offences for fines were introduced for certain violations in the German WSA, thus creating a more differentiated sanction structure.

2. Waste Classification and Waste Lists

Waste classification systems can be developed and designed according to different principles. In general, waste classification can be based on material or origin, taking into account inherent hazard properties and/or risk considerations.

In practice, however, systematically pure systems are rarely encountered, but rather mixed systems. In Germany, for example, the first waste classification system was developed by employees in the Federal States who were involved in waste management enforcement for practical reasons. The waste streams arising in practice were classified according to the common characteristics of composition and origin for the respective waste type according to a numerical sorting system into upper group, group, subgroup and waste type, with an increasingly precise description from the upper group to the waste type.

In summary, a waste type was characterized by a waste code, a waste type designation and an indication of its origin. From this waste catalogue, the waste classification system known as the LAGA Waste Catalogue[7] was finally developed, which was structured neither strictly according to hazard criteria nor strictly according to origin, but in a mixed system according to the characteristics of composition and origin.

This system was used in Germany until the end of 1998 for the enforcement of waste law and waste technology and was then replaced by the European Waste List (EWL), which is predominantly based on origin.

Since the entry into force of the EC Waste Shipment Regulation in 1994, a waste classification system has been applied in the area of transboundary waste shipments that is predominantly based on substances. Such a system was also implemented worldwide with the amendment of Basel Convention in 1998 by introducing two waste lists (list A and list B)[8] which have been developed based on the former three-tier waste list system of the Organization for Economic Cooperation and Development (OECD).

The risk-based three-tier waste list system of the OECD with a classification of waste into the so-called green, amber and red waste lists was changed to a two-list system (green and amber list) in sense of harmonisation with the system of Basel Convention. This was done with some deletions and additions. With the amendment of the European Waste Shipment Regulation (EU-WSR) that came into force in 2007 this system was also adopted

---

[7]   LAGA-Informationsschrift Abfallarten, 4. durchgesehene Aufl. 1992, Erich Schmidt Verlag, Berlin.

[8]   COP Decision IV/9 "Amendment and adoption of annexes to the Convention"

for the EU with further additions.

This means that two waste list systems based on different principles are used in European waste management practice. These waste classifications and waste catalogues or lists, which are used internationally and nationally, will be explained in the following, as they are of fundamental importance both for understanding the regulations on transboundary waste shipments and for the enforcement of waste legislation.

2.1 European Waste List *(EU specific)*

The European Waste List (EWL) is decisive for waste designations in the EU. However, substances or objects do not automatically become waste if they can be assigned to one of the designations in the EWL or to the waste lists of the Basel Convention or in the OECD/EU to "Green" or "Amber" Lists. The only decisive factor for the classification as waste is the fulfilment of the waste definition according the EU-WFD. It is therefore decisive whether the owner of an object or substance discards it, wants to discard it or has to discard it. The EWL applies to all waste, regardless of whether it is destined for disposal or recovery.

The EWL must be regularly reviewed and amended to reflect both the dynamics of the state of the art in waste management and scientific and technical progress. The most recent amendment of the EWL was in 2014, mainly due to changes in the chemicals legislation used as a basis, in particular the adoption of the EU-CLP Regulation, as well as developments in waste management and waste legislation.

2.1.1 Structure of the European Waste List

The EWL is structured in 20 chapters, which in turn are subdivided into predominantly origin-specific groups. Each group can contain up to 99 specific waste codes. However, the principle of origin-based description in the EWL is not consistently applied, as the substance-based chapters 13, 14, 15 and 16 show. For further exceptions to the origin-specific classification, see sect. I.2.1.6.

From the chapters to the waste codes, the EWL contains an increasingly precise description of waste. The waste codes are represented by a six-digit numerical sequence, with the first two digits representing the chapter to which the waste belongs and the next two digits representing the assignment to the group. The last two digits are then used to finally determine the waste type. The final number 99 is used to designate catch-all codes under which wastes can be classified for which no origin-specific codes exist. The designations of these codes are supplemented in the EWL by the wording "not otherwise specified", but this refers only to the waste types contained in this group.

Although the designation of waste types in the EWL is largely based on substance/material, the waste types are predominantly grouped according to origin and can only be considered in connection with the origin-related chapters and group designations.

Table 1: Chapters of the European Waste List

Chapter	Origin
01 00 00	Wastes resulting from exploration, mining, quarrying, physical and chemical treatment of minerals
02 00 00	Wastes from agriculture, horticulture, aquaculture, forestry, hunting and fishing, food preparation and processing
03 00 00	Wastes from wood processing and the production of panels and furniture, pulp, paper and cardboard
04 00 00	Wastes from the leather, fur and textile industries
05 00 00	Wastes from petroleum refining, natural gas purification and pyrolytic treatment of coal
06 00 00	Wastes from inorganic chemical processes
07 00 00	Wastes from organic chemical processes
08 00 00	Wastes from the manufacture, formulation, supply and use (MFSU) of coatings (paints, varnishes and vitreous enamels), adhesives, sealants and printing inks
09 00 00	Wastes from the photographic industry
10 00 00	Wastes from thermal processes
11 00 00	Wastes from chemical surface treatment and coating of metals and other materials; non-ferrous hydro-metallurgy
12 00 00	Wastes from shaping and physical and mechanical surface treatment of metals and plastics
13 00 00	Oil wastes and wastes of liquid fuels (except edible oils, 05 and 12)
14 00 00	Waste organic solvents, refrigerants and propellants (except 07 and 08)
15 00 00	Waste packaging; absorbents, wiping cloths, filter materials and protective clothing not otherwise specified
16 00 00	Wastes not otherwise specified in the list
17 00 00	Construction and demolition wastes (including excavated soil from contaminated sites)
18 00 00	Wastes from human or animal health care and/or related research (except kitchen and restaurant wastes not arising from immediate health care)
19 00 00	Wastes from waste management facilities, off-site waste water treatment plants and the preparation of water intended for human consumption and water for industrial use
20 00 00	Municipal wastes (household waste and similar commercial, industrial and institutional wastes) including separately collected fractions

2.1.2 Amendment of the EWL and Annex III to the EU-WFD

With the amendment of the EU waste classification in 2014 the international developments in the chemicals[9] law as well as waste management developments were taken into account.

Whereas only a few changes were made with regard to the list of wastes, the classification basis was completely changed on the basis of the CLP-Regulation. Now, all hazard properties are allocated with limit values or other requirements. An exception is the hazard property "infectious" (HP 9), which is defined according to national regulations. Sect. II.5 contains the EU-WFD including Annex III with hazard property, hazard statements, limit values, cut-off values and other requirements.

The European Waste List is a reference nomenclature that defines a common terminology for the entire European Union. Explanations on this can be found in a Technical Guide[10] issued by the EU Commission.

2.1.3 Basic Definitions

The Annex to the EWL, which contains the wastes list as a core element, makes the following basic definitions in the sense and for the purposes of the EWL, which are significant for the application:

<u>'hazardous substance'</u>

means a substance classified as hazardous as a consequence of fulfilling the criteria laid down in parts 2 to 5 of Annex I to the CLP-Regulation;

<u>'heavy metal'</u>

means any compound of antimony, arsenic, cadmium, chromium (VI), copper, lead, mercury, nickel, selenium, tellurium, thallium and tin, as well as these materials in metallic form, as far as these are classified as hazardous substances;

<u>'polychlorinated biphenyls and polychlorinated terphenyls' ('PCBs')</u>

means PCBs as defined in Art. 2(a) of PCB-Directive;

<u>'transition metals'</u>

means any of the following metals: any compound of scandium, vanadium, manganese, cobalt, copper, yttrium, niobium, hafnium, tungsten, titanium, chromium, iron, nickel, zinc, zirconium, molybdenum and tantalum, as well

---

[9] Implementation of the Globally Harmonised Systems (GHS) of UN (2002) in the EU by CLP-Regulation

[10] Commission notice on technical guidance on the classification of waste of 09.04.2018, OJ, C 124, p. 1

as these materials in metallic form, as far as these are classified as hazardous substances;

'stabilisation'

means processes which change the hazardousness of the constituents in the waste and transform hazardous waste into non-hazardous waste;

'solidification'

means processes which only change the physical state of the waste by using additives without changing the chemical properties of the waste;

'partly stabilised wastes'

means wastes containing, after the stabilisation process, hazardous constituents which have not been changed completely into non-hazardous constituents and could be released into the environment in the short, middle or long term.

2.1.4 Classification Principles

The classification of hazardous waste in the EWL is based on the classification of substances and formulations under chemicals legislation, although this system is only directly applicable to waste to a limited extent. As part of the revision of the hazard properties in the EU, waste legislation has been partially harmonized with chemicals legislation (CLP-Regulation).

It is therefore entirely possible that a substance that is hazardous under chemicals legislation is to be classified as non-hazardous as waste[11], or that a substance that is not hazardous under chemicals legislation is to be classified as hazardous waste on the basis of the application of the "ecotoxic" hazard property (HP14) or the additional hazardous property HP15 under waste legislation[12].

The EWL contains a dynamic reference to the classification of hazardous substances according to CLP-Regulation. The reference to hazardous substances is formulated in such a way that future changes in the classification of hazardous substances under chemical law can be incorporated into the EWL without requiring the adoption of a new decision.

With regard to the concentration limits from Annex III to the EU-WFD, a different approach was taken. Amendments to the relevant provisions of CLP-Regulation must be incorporated in the form of an amendment to Annex III

---

[11] for example, due to the higher limit value for "sensitizing" (HP13) in Annex III of EU-WFD compared to CLP Regulation

[12] e.g. national determination of eluate properties and pollutant contents, in German only, see the LAGA Technical Notes on the classification of wastes according to their hazardousness" cf. [11]

to the EU-WFD. The reason for this is that changes to the concentration limits and changes to the properties that classify a waste as hazardous waste are the basis for classification and therefore these changes must be explicitly decided in a formal procedure.

2.1.4.1   Classification as Hazardous Waste

When assessing the hazard properties of waste, the criteria of Annex III to the EU-WFD apply. When assessing the hazard property HP 4, HP 6 and HP 8 (cf. Table 2) cut-off values for individual substances according to Annex III to the EU-WFD apply. If a substance is present in the waste in a concentration below the cut-off value, it is not taken into account in the calculation of a limit value.

A waste is classified as hazardous in the EWL if this waste contains relevant hazardous substances to exhibit one or more of the hazard properties HP 1 to HP 8 or HP 10 to HP 15 listed in Annex III to the EU-WFD. HP 9 is to be defined nationally. In Germany the presence of the hazard property HP 9 is assumed in the case of waste contaminated with hazardous pathogens in accordance with the Infection Protection Act or the Ordinance on Notifiable Epizootics[13].

A hazard property can be assessed based on the concentrations of substances in the waste in accordance with Annex III to the EU-WFD or, unless otherwise specified in the CLP-Regulation, based on a test in accordance with the REACH-Regulation or on the basis of other internationally recognised test methods and guidelines, taking into account Art. 7 of the CLP-Regulation with regard to animal testing and testing on humans. Where a hazardous property of a waste has been assessed by a test and by using the concentrations of hazardous substances as indicated in Annex III to the EU-WFD, the results of the test shall prevail.

In determining the hazard properties of waste, the following comments in Annex VI to the CLP-Regulation may be taken into account:

- the notes on identification, classification and labelling of substances mentioned in Annex VI, 1.1.3.1: Notes B, D, F, J, L, M, P, Q, R and U,

---

[13] Art. 17 of the Infection Protection Act of 20 July 2000 (Federal Law Gazette I p. 1045), last amended by Art. 6a of the Act of 10 December 2015, Federal Law Gazette I p. 2229, and in the case of waste containing pathogens (infectious agents) of animal diseases pursuant to the Ordinance on Notifiable Animal Diseases in the version published on 19 July 2011, Federal Law Gazette I p. 1404. I p. 1404, last amended by Art. 6 of the Ordinance of 29 December 2014, Federal Law Gazette I p. 2481, or the Annex to Art. 1 of the Ordinance on Notifiable Animal Diseases in the version published on 11 February 2011, Federal Law Gazette I, p. 252, last amended by Article 381 of the Ordinance of 31 August 2015, Federal Law Gazette I p. 147.

- the notes on classification and labelling of mixtures referred to in Annex VI, point 1.1.3.2: Notes 1, 2, 3 and 5.

After evaluating the hazard properties of a waste in accordance with the aforementioned procedural steps, the waste is assigned a suitable hazardous or non-hazardous entry from the EWL. For classification under HP 4 and HP 8, a pH value ≤ 2 or a pH value ≥ 11.5 is indicative. A compilation of HP properties, hazard statements and limit values of the EU-WFD is contained in Table 2.

Table 2: Hazard properties (HP) and hazard statements according to Annex III to the EU-WFD with concentrations limits and flash points

HP	Hazard Statement		Limit Values
HP1	H200	Unstable explosives	
	H201	Explosive; mass explosion hazard	
	H202	Explosive, severe projection hazard	
	H203	Explosive; fire, blast or projection hazard	
	H204	Fire or projection hazard	
	H240	Heating may cause an explosion.	
	H241	Heating may cause a fire or explosion.	
HP2	H270	May cause or intensify fire; oxidizer	
	H271	May cause fire or explosion; strong oxidizer	
	H272	May intensify fire; oxidizer	
HP3	H220	Extremely flammable gas.	
	H221	Flammable gas	
	H222	Extremely flammable aerosol.	
	H223	Flammable aerosol	
	H224	Extremely flammable liquid and vapour	flash point < 60° > 55° and ≤ 75° for waste gas oil, diesel and light heating oils
	H225	Highly flammable liquid and vapour	
	H226	Flammable liquid and vapour	
	H228	Flammable solid	
	H242	Heating may cause a fire	
	H250	Catches fire spontaneously if exposed to air	
	H251	Self-heating: may catch fire	
	H252	Self-heating in large quantities; may catch fire	
	H260	In contact with water releases flammable gases which may ignite spontaneously	
	H261	In contact with water releases flammable gases	

I. Explanatory Notes

HP	Hazard Statement		Limit Values
HP4	H314	Causes severe skin burns and eye damage	$\Sigma_{subst.} \geq 1\ \%$; CV* = 1 % [$\geq$ 5 % $\Rightarrow$ HP8]
	H318	Causes serious eye damage	$\Sigma_{subst.} \geq 10\ \%$; CV* = 1 %
	H315	Causes skin irritation	$\Sigma_{subst} \geq 20\ \%$; CV* = 1 %
	H319	Causes serious eye irritation	$\Sigma_{subst.} \geq 20\ \%$; CV* = 1 %
HP5	H304	May be fatal if swallowed and enters airways	$\geq$ 10 %
	H335	May cause respiratory irritation	$\geq$ 20 %
	H370	Causes damage to organs	$\geq$ 1 %
	H371	May cause damage to organs	$\geq$ 10 %
	H372	Causes damage to organs	$\geq$ 1 %
	H373	May cause damage to organs through pro-longed or repeated exposure	$\geq$ 10 %
HP6	H300	Fatal if swallowed	$\geq$ 0,1 % (acute 1) $\geq$ 0,25 % (acute 2); CV * = 0,1 %
	H301	Toxic if swallowed.	$\geq$ 5 %; CV* = 0,1 %
	H302	Harmful if swallowed	$\geq$ 25 %; CV* = 1 %
	H310	Fatal in contact with skin	$\geq$ 0,25 % (acute 1) $\geq$ 2,5 % (acute 2); CV* = 0,1 %
	H311	Toxic in contact with skin	$\geq$ 15 %; CV* = 0,1 %
	H312	Harmful in contact with skin	$\geq$ 55 %; CV* = 1 %
	H330	Fatal if inhaled	$\geq$ 0,1 %; CV* = 0,1 %
	H331	Toxic if inhaled	$\geq$ 3,5 %; CV* = 0,1 %
	H332	Harmful if inhaled	$\geq$ 22,5 %; CV* = 1 %
HP7	H351	Suspected of causing cancer	$\geq$ 1 %
	H350	May cause cancer	$\geq$ 0,1 %
HP8	H314	Causes severe skin burns and eye damage	$\Sigma_{subst.}$ H314 (1a, 1b, 1c) $\geq$ 5 %; CV* = 1 %
HP9	-----	national definition	
HP10	H360	May damage fertility or the unborn child	$\geq$ 0,3 %
	H361	Suspected of damaging fertility or the unborn child	$\geq$ 3 %
HP11	H340	May cause genetic defects	$\geq$ 0,1%
	H341	Suspected of causing genetic defects	$\geq$ 1 %
HP12	EUH029	Contact with water liberates toxic gas	
	EUH031	Contact with acids liberates toxic gas	
	EUH032	Contact with acids liberates very toxic gas	

HP		Hazard Statement	Limit Values
HP13	H317	May cause an allergic skin reaction	≥ 10 %
	H334	May cause allergy or asthma symptoms or breathing difficulties if inhaled	≥ 10%
HP14	H400	Very toxic to aquatic life	≥ 25 %
	H410	Very toxic to aquatic life with long-lasting effects	$100 \times \Sigma_c (H410)) + 10 \times \Sigma_c (H411) + \Sigma_c (H412) \geq 25\%$ for one or more aquatic chronic 1, 2 and 3 substances[*]
	H411	Toxic to aquatic life with long-lasting effects	
	H412	Harmful to aquatic life with long-lasting effects	
	H413	May cause long-lasting harmful effects to aquatic life	$\Sigma_c H410 + \Sigma_c H411 + \Sigma_c H412 + \Sigma_c H413 \geq 25\%$ for one or more aquatic chronic 1, 2, 3 and 4 substances[**]
	H420	Harms public health and the environment by destroying ozone in the upper atmosphere	≥ 0,1 %
HP15	H205	May mass explode in fire	
	EUH001	Explosive when dry	
	EUH019	May form explosive peroxides	
	EUH044	Risk of explosion if heated under confinement	

[*] CV = cut-off values; Σ = sum; c = concentration

[*] the cut-off value for substances classified as H410 is 0,1 % and for substances classified as H411 or H412 is 1 %

[**] the cut-off value for substances classified as H410 is 0,1% and for substances classified as H411, H412 or H413 is 1 %

2.1.4.2 Exemption Clause for Metal Alloys

The concentration limits specified in Annex III to the EU-WFD apply to pure metal alloys in solid form only if they are contaminated by hazardous substances. Those waste alloys that are considered as hazardous waste are specifically enumerated in the EWL and marked with an asterisk (*).

2.1.4.3 Up and down Grading Clause

Art. 7 (2) of the EU-WFD contains a clause with which – in exceptional cases – non-hazardous waste can be upgraded to hazardous waste, or according to Art. 7 (3) of the EU-WFD hazardous waste can be downgraded to non-hazardous waste.

In the event of a possible downgrading of hazardous waste, the producer or owner of the waste shall submit to the competent authority in the EU Member State appropriate documents on the basis of which a downgrading may be carried out by the authority.

In addition, EU Member States may upgrade non-hazardous waste to hazardous waste if one or more of the hazardous properties in Annex III to the EU-WFD Directive are met.

Decisions of the Member States shall be notified to the EU Commission, which shall examine whether it is necessary to revise the EWL.

2.1.5 Assignment of Hazardous Waste to the EWL

The vast majority of hazardous wastes always fulfil a hazard property and are therefore generally classified as hazardous waste (absolute hazardous entry) without the need for further testing. Examples are:

06 01 02* hydrochloric acid

14 06 01* chlorofluorocarbons, HCFC, HFC

16 01 06* oil filters

It is clear that the above classification principles are particularly important for wastes listed as "mirror entries". These wastes may be hazardous or non-hazardous, depending on the circumstances of their generation, and are identified in the list by at least two entries, one for hazardous and one for non-hazardous waste.

Some examples of mirror entries are compiled below:

Example of alternative entries where the assignment of the waste to the corresponding waste types depends on whether or not the waste contains "hazardous substances":

17 04 10* cables containing oil, coal tar and other hazardous substances

17 04 11   cables other than those mentioned in 17 04 10

Example of alternative entries where the assignment of the waste to the corresponding waste types depends on whether or not the waste contains certain hazardous constituents mentioned in the waste designation:

06 03 15* metallic oxides containing heavy metals

06 03 16   metallic oxides other than those mentioned in 06 03 15

The examples show that the above-mentioned general definitions play an essential role. However, specific ingredients such as cyanides are also mentioned.

Example of entries with multiple references to several corresponding waste types. The classification depends on the specific waste origin or certain waste properties as well as the hazardous substances contained:

06 03 11* solid salts and solutions containing cyanides

06 03 13* solid salts and solutions containing heavy metals

06 03 14   solid salts and solutions other than those mentioned in 06 03 11
           and 06 03 13

2.1.6 Assigning a Waste to the Waste Type

The correct assignment of a specific waste to a waste code of the EWL, i.e.,
to one of the 842 waste types, 408 of which are considered hazardous, is of
great importance. When assigning a waste to a waste type, the industry or
manufacturing process in which the waste was generated is decisive. The
stepwise classification procedure is described in the introduction to the EWL
and shown in Figure 1.

The procedure is as follows. First, the waste must be assigned to a six-digit
waste code of chapters 01 to 12 or 17 to 20 according to its origin (excluding
the codes of these chapters ending in 99). A waste producer may have to
assign his wastes to waste codes of several chapters depending on the activity. For example, an automotive manufacturer may assign its wastes to
chapters 12 (wastes from shaping and physical and mechanical surface
treatment of metals and plastic), 11 (wastes from chemical surface treatment and coating of metals and other materials), and 08 (wastes from the
manufacture, formulation, supply and use of coatings), depending on the
manufacturing stage.

Separately collected packaging waste (including mixed packaging of different materials) is not classified in group 20 01, but in group 15 01, even if the
packaging waste is collected separately from municipal waste (e.g. glass
packaging collected in bottle containers or systems for the collection of
mixed packaging waste). This has also been made clear in the headings of
Chapters 15 01 and 20 01.

However, the list of wastes also contains deviations from this strict requirement. For 19 groups of the waste list in chapters 06, 07 and 08 which have
the abbreviation MFSU (manufacture, formulation, supply and use) in the
group heading, wastes from manufacture, formulation, supply and use can
also be assigned to the waste codes of the corresponding groups. Furthermore, wastes can be assigned to the waste codes of the two groups 1908
and 1912 that have the wording "wastes not otherwise specified" in the
group heading, irrespective of their origin.

Fig. 1: General waste classification scheme

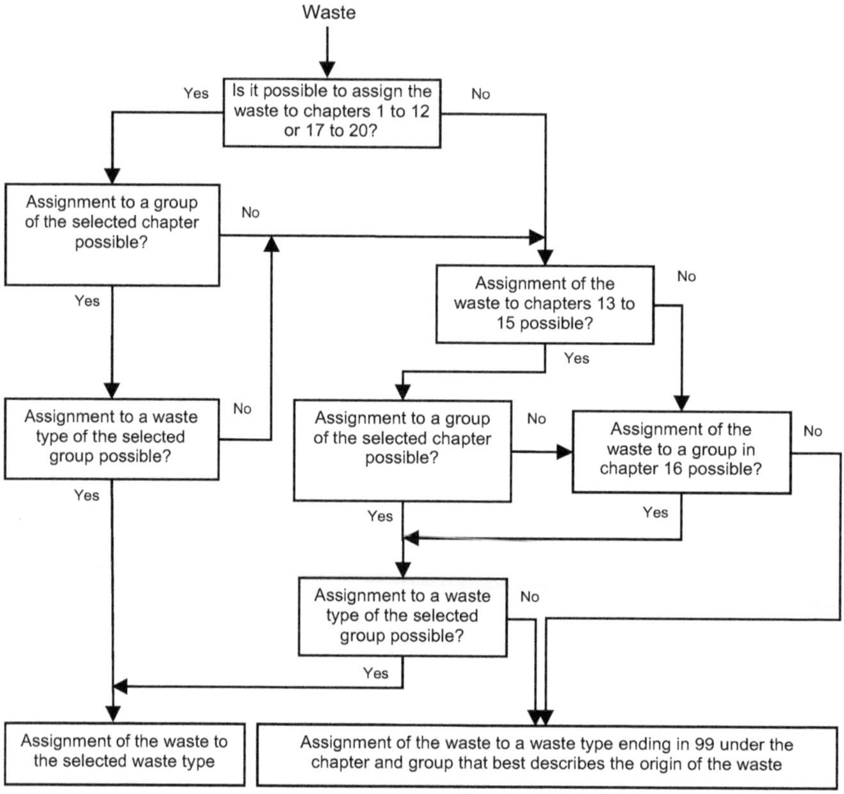

2.1.6.1 Assignment of Wastes to Mirror Entries

The EWL contain 178 hazardous mirror entries, which can lead to problems due to different interpretations with regard to classification, because there is neither a uniform national nor a uniform European enforcement practice. In principle, the definition for hazardous waste is to be used as a basis for the allocation of a waste to the hazardous or non-hazardous waste type of a mirror entry. The EWL refers back to the definition for hazardous waste according to the EU-WFD with the hazardous properties specified therein in Annex III. This means:

- A waste is to be assigned to the hazardous waste type as soon as a hazard property is met.

- A waste can only be assigned to the type of waste not designated as hazardous if it can be plausibly demonstrated that all the hazard properties are not met.

The waste has to be assigned to the waste types listed in the EWL by the producer, owner, dealer, broker, transporter and disposer. If one of the above-mentioned persons wants to make a classification to the waste code of the mirror entry that is not marked as hazardous, he has to substantiate this by submitting suitable evidence, unless corresponding information is already available from enforcement experience. This presumption is based on the general environmental precautionary principle and is reflected in many environmental regulations.

According to the abovementioned a waste is assigned to one of the two waste types of a mirror entry according to a stepwise procedure. First, the waste is assigned to a mirror entry and then it is examined step by step to what extent the available findings on the hazardous substance classification and labelling of the waste, according to waste-specific regulatory assumptions, from the performance of a declaration analysis − without performing additional, costly chemical analyses − allows a conclusive assignment of the waste. The following procedure can be followed:

1. allocation of the waste on the basis of empirical data,

2. allocation of the waste on the basis of available experience (assumption of the rule),

3. allocation of the waste on the basis of knowledge of the production process and the origin of the waste,

4. allocation of the waste on the basis of plausibility considerations with regard to the content of hazardous substances (concentration values),

5. allocation of the waste on the basis of chemical analytical examination.

The chemical analytical examination of the presence or absence of hazardous substances in the almost predominantly heterogeneous mixture of waste requires extensive and complicated investigations, which are not without problems. One problem is the representative sampling, which in the chain of sampling, sample preparation and analysis is subject to the greatest error, which is also difficult to quantify, and which dominates the total error according to the law of fault propagation. In order to take a sample whose properties correspond as far as possible to the average properties of the waste, the respective sampling methods have to be followed. In Germany the sampling regulations of LAGA PN 98[14] have to be observed, for which a supplementary LAGA Manual[15] has been published. The European Committee for Standardisation (CEN) has developed several standards, e.g. CEN-EN 14899.[16]

Of the methods for determining hazardous properties, only a few, such as the determination of flash point, gas formation rate or pH value, can be used directly. For example, it is almost impossible to determine the CMR criteria (carcinogenic, mutagenic and toxic for reproduction) by chemical analysis in daily waste management.

These verifications are often dispensable due to the level of knowledge. In most cases, sufficient information is available to determine the presence of hazardous substances based on the knowledge of the origin or the former purpose of a material or object that has become waste, and thus to assign the waste to a waste type of a mirror entry.

2.2 Waste List System according to the BC, the OECD Decision and the EU-WSR

The waste classification system, which is applied in transfrontier waste shipments, is a material/substance orientated system. Based on the waste lists of Basel Convention (list A, Annex VIII to the BC and list B Annex IX to the BC) the OECD has established Green and Amber Waste Lists (cf. Appendix 1 to the explanatory notes).

The background to the adoption of the OECD waste lists based on Annexes VIII and IX to the Basel Convention and into the corresponding annexes to the EU-WSR was the general desire for worldwide harmonisation of the waste list systems used in transfrontier waste shipments. The OECD therefore reverted to the classification system of the Basel Convention and adopted it with some additions or modifications. The EU then adopted the OECD system and supplemented it further. It should be noted that the hazard characteristics of both the Basel Convention and the OECD Council Decision are based on the hazard criteria for the transport of dangerous goods, unlike the properties from Annex III to the EU-WFD, which are based on chemical legislation.

The wastes in these waste list systems are grouped according to their material composition into

---

[14] LAGA PN 98 "Richtlinie für das Vorgehen bei physikalischen, chemischen und biologischen Untersuchungen im Zusammenhang mit der Verwertung/Beseitigung von Abfällen", Erich Schmidt Verlag, Berlin 2002.

[15] Handlungshilfe zur Anwendung der LAGA Mitteilung 32 (LAGA PN 98), State: 5. May 2019: https://www.laga-online.de/documents/hinweise_pn98_stand_2019_mai_1564665128.pdf

[16] cf. Annex 4 "Sampling and chemical analysis of waste" of [12]

- metal wastes and wastes containing metals,

- predominantly inorganic waste, which may contain metals and organic substances,

- predominantly organic wastes that may contain metals and inorganic materials, and

- waste that may contain both inorganic and organic substances.

(EU-specific)

In implementing the OECD system by the EU further possibilities of listing wastes or mixtures of wastes that have not been listed so far are offered by establishing three further annexes in addition to the Green List (Annex III to the EU-WSR) and the Amber List (Annex IV to the EU-WSR):

- Annex III A for mixtures of green wastes,

- Annex III B for additional green waste, and

- Annex IV A for green waste subject to notification.

These options have been used to list mixtures in Annex III A to the EU-WSR and some waste streams in Annex III B; Annex IV A is empty. A regularly updated and consolidated version of the waste lists of the EU-WSR is available on the webpage of the German Federal Environment Agency and a commented version of this list is contained in Appendix 1 to these explanatory notes.

Since the classification of waste in the waste lists of the EU-WSR is based on substance, individual descriptions of origin are only indicative. An example of this is plastic aluminium composite material from packaging (e.g. packaging of dry powder) that do not originate from the pre-treatment of liquid packaging. These can be assigned to waste code B3026, specifically to the second indent. This is supported by the substance/material reference, the risk based approach of the EU-WSR waste listing, and the principle of equal treatment.

As already described the basic system of the EU-WSR differs fundamentally from that of the EWL. Furthermore, the assignment of wastes to entries of the Green or Amber Waste Lists differs from that in the EWL, because in these lists the classification is based on a risk assessment, whereas in the EWL the classification as hazardous is based exclusively on the substance-inherent hazard property based on European chemicals law without carrying out a risk assessment.

Table 3: Waste code B3026 of the Green List

Code	Waste description
B3026	The following wastes from the pre-treatment of composite packaging for liquids, not containing Annex I materials in concentrations sufficient to exhibit Annex III characteristics:
	— Non-separable plastic fraction
	— Non-separable plastic-aluminium fraction

2.2.1 Mixtures of Waste and Composite Material

(Chapt. II, B.(8) OECD Decision; Art. 2 No 3 i.c.w. Annex IIIA EU-WSR)

Some specific mixtures of waste are already listed in the waste lists of the BC, the OECD Decision and the EU-WSR, for example, entry A1170 "Unsorted waste batteries excluding mixtures of only list B batteries ...", entry B1050 "Mixed non-ferrous metal, heavy fraction, scrap ...", entry Y46 "Wastes collected from households "or entry A3120 "FLUFF — light fraction from shredding", which are controlled according to their listing on the respective list.

Because more mixtures of waste occur in waste management the OECD Decision and the EU-WSR contain provisions in order to clarify the status of mixtures of wastes.

According to the <u>OECD Decision </u>mixtures of wastes are in general defined as a waste that results from an intentional or unintentional mixing of two or more different wastes. To the mixtures of wastes for which no individual entry in the OECD Decision exists, the following rule shall apply:

- A mixture of two or more Green wastes shall be subject to the Green control procedure;

- A mixture of a Green waste and more than a de minimis amount of an Amber waste, or a mixture of two or more Amber wastes, shall be subject to the Amber control procedure. The interpretation of the term "a de minimis amount", in absence of internationally accepted criteria, is to be defined according to national regulations and procedures.

According to the <u>EU-WSR</u> only specific mixtures of wastes, which can be shipped as green wastes without notification are listed in Annex III A (cf. appendix 1), for example

- mixtures of waste classified under Basel entries B1010 and B1050,

- mixtures of waste classified under Basel entries B1010 and B1070,

- mixtures of wastes classified under Basel entries B3040 and B3080.

The waste lists of the BC, the OECD Decision and the EU-WSR also contain some specific waste entries of composite material, for example B1115 "Waste metal cables coated or insulated with plastics, not included in list A1190, excluding …" or B3026 "The following wastes from the pre-treatment of composite packaging for liquids …". But neither the BC and the OECD Decision nor the EU-WSR contain general provisions concerning further composite materials.

This is surprising because, in contrast to the mixtures mentioned above with changing percentage composition, composite materials occur in a defined composition, such as bottle caps or window frames. Since a large number of products made of composite materials exist, it would be desirable to have a general mirror entry with composite materials listed as examples on the waste lists.

From the point of view of the author such a green entry should be restricted to composite materials, which can be easily separated into its materials and which are recyclable.

2.2.2 Classification of Wastes in the System of the OECD/EU-WSR *(OECD/EU specific)*

Before classifying a waste in the OECD or the EU-WSR list system, a first step is taken to check whether the waste has a hazard property – comparable to the procedure in the EWL. In a second step, a risk assessment is carried out in deviation from the EWL. The essential basis for this is Annex 6 to the OECD Decision, the so-called "Criteria for the OECD Risk Approach", which was taken over unchanged from the predecessor OECD regulation.

Based on this, it is examined whether the waste in question presents a hazard property in its existing physical state. In other words, it must be estimated whether the hazardous property associated with the substance/material poses a risk. For example, for metal in solid form, the hazard criterion "toxic by inhalation or ingestion" is not applicable.

On the other hand, waste that does not have a hazard property can also be placed on the Amber List based on waste management criteria. The Chapeau of the Green List (introductory sentence above the Green List) and the explanatory footnote on the "dispersion risk" (cf. sect. I.2.2.3 below) provide further guidance on the application of waste classification in the waste list system of the OECD/EU-WSR. According to this, waste may not be classified as "green" if, due to contamination by other materials, the risks associated with the waste are increased to such an extent that, taking into account the hazard properties listed in Annex III to the EU-WFD[17], the application of the prior written notification and consent procedure appears appropriate or the environmentally sound recovery of the waste is prevented.

As a consequence of the different classification bases, this leads both to green listed wastes that have inherent hazardous properties and to non-hazardous wastes being listed on the Amber List and thus being subject to prior written notification and consent in the case of transfrontier shipment (for recovery). Examples of the different classification are listed Table 4.

Table 4: Differences in waste classification based on the risk approach (examples)

Waste designation	EU-WSR-Code		EWL-Code	
	Amber	**Green**	**hazardous**	**Non hazard.**
Sewage sludge	AC 270			19 08 09
Municipal waste	Y46			20 01 03
Mixtures of plastic wastes	Y48			20 01 39 19 12 04 17 02 03 16 01 19 15 01 02
Pig manure, faeces	AC260			02 01 06 20 03 04
Wastes from treated cork and treated wood	AC170[#)]		03 01 04* 15 01 10* 17 02 04* 19 12 06*	03 03 01 03 01 05
Waste blasting abrasives	AB130			12 01 17
Spent catalysts		B1120	16 08 02*	
Auto catalysts in metal casing		B1130	16 07 08*	
Casting cores and moulds	AB070		10 10 05* 10 10 07*	10 10 06 10 10 08
Waste blasting material	AB130		12 01 16*	12 01 17

#) The OECD/EU-WSR classifies all - more than mechanically treated wood - as amber waste

2.2.3 Dispersion Risk

For some entries of list B (BC) or the Green List (OECD/EU), it is pointed

---

[17] In the OECD Decision it reads: "…increases the risks associated with the wastes sufficiently to render them appropriate for submission to the amber control procedure, when taking into account the criteria in Appendix 6" of the OECD-Decision

out that material can only be assigned to the entry if they are present in massive non-dispersible form. The associated footnote states that this does not include any wastes in the form of powder, sludge, dust or solid items containing encased hazardous waste liquids.

The OECD had commissioned a study[18] on the question of dispersibility. The conclusion of the study is that there is no risk of exceeding the air concentration limits for most toxic metals, provided that a maximum of 0.1 % fine particles are present. Fine particles are particles that are small enough to be blown away by wind (particle size $\leq$ 100 µm).

For example, a metal waste is classified as "scrap" if the majority of the waste is in metallic form (i.e. no metal oxides or other compounds) with no risk of dispersion and only a small proportion of the waste has a particle size below 100 µm (the above-mentioned guideline value for dispersibility).

As there is no worldwide harmonised interpretation of the term dispersible, the interpretation by individual Countries can lead to different, sometimes contradictory results.

2.2.4 Permissible Impurities in Waste

The question of the extent to which impurities may be contained in waste has also been the subject of controversial discussion for years and is handled differently in the countries.

It is clear that impurities are a characteristic property of waste. This can also be seen from the EU's End-of-Waste Regulations (cf. sect. I.1.4.3), which allow small proportions of foreign matter even in by-products, whereby the proportions are higher for metals than for cullet, and a quality management system must be applied to the by-products and proven with a declaration of conformity for each consignment. Furthermore, the criteria of the End-of-Waste Regulation must be complied with.

What can be concluded from this for impurities in wastes is - first of all - the fact that impurities differ depending on the waste type. In practice, therefore, starting with the rule of thumb, which assumes 5 % impurities, there are further waste-specific interpretations in place, which are predominantly 5 to 10 % for metals and 2 % for paper and plastics.[19]

---

[18] Study prepared for OECD: J.A. Garland: "A Criterion for Non-Dispersibility of Metal and Metal Containing Material in Waste Classification", AEA Technology Consulting Services, February 1994

[19] Threshold values for contaminants in "green"-listed wastes established by Member States – A Compilation Document, February 2016, webpage accessed in November 2021:
https://ec.europa.eu/environment/waste/shipments

(EU-specific and national)

In the Netherlands[20], rules have been established in this respect. In most EU Member States, however, decisions are mainly made on a case-by-case basis. In the meantime, the European Court of Justice (CJEU) has also dealt with the issue of impurities, taking entry B3020 as an example, and has reached a decision[21] on whether and to what extent a waste shipped across borders may contain impurities. In the light of this judgement, the above-mentioned impurities must be assessed on a case-by-case basis and not as a general rule. According to the CJEU ruling, the decisive factor for the admissibility of impurities is whether these impurities increases the hazard potential of the waste or impedes its environmentally sound recovery.

2.2.5 Decision amending the Entry for Plastic Waste

The export of plastic waste, among other things in the form of packaging waste, has been critically viewed and discussed for a long time. These global discussions were concluded in 2019 with Decision 12 of COP 14. According to this decision, only unmixed waste and mixtures of the plastics polypropylene (PP), polyethylene (PE) and polyethylene terephthalate (PET) that are destined for material recycling can be freely traded with other countries.

Hazardous plastic waste and waste that is unlikely to be recycled is now subject to the Basel Convention. This means that they may only be exported with the consent of the authorities of the exporting and importing countries and must be recycled in an environmentally sound manner. With the decision of COP 14, a multiple mirror entry on plastic waste was introduced. In addition to the new entry B3011 in Annex IX, which replaces the old entry B3010, there is a new entry A3210 in Annex VIII for hazardous plastic waste, as well as a new entry Y48 in Annex II for plastic waste that is not hazardous but requires monitoring and control.

(OECD-specific)

By decision of 7 September 2020, Annexes 3 and 4 to the OECD Council Decision were amended, but there was no consensus among OECD member countries to adopt the entries of Basel Convention as they have been adopted by COP 14.

The OECD Decision stipulates that each OECD Member country retains its right to control non-hazardous plastic waste (Basel entries Y48, B3011) in

---

[20]   webpage accessed in May 2020:
https://zoek.officielebekendmakingen.nl/stcrt-2015-43221.html?zoekcriteria=%3fzkt%3dUitgebreid%26pst%3dStaatscourant%26dpr%3dAlle%26spd%3d20151218%26epd%3d20151218%26jgp%3d2015%26nrp%3d43221%26sdt%3dDatumPublicatie%26planId%3d%26pnr%3d1%26rpp%3d10&resultIndex=0&sorttype=1&sortorder=4
[21]   Legal case C-654/18 of 28. May 2020

conformity with its national legislation and international law. OECD Member countries agreed to inform the OECD Secretariat about their controls applied for non-hazardous plastic waste, as well as about any future changes of such controls. The respective information is available on the OECD webpage.[22]

(EU-specific)

With the Delegated Regulation (EU) 2020/2174 of 19 October 2020, adjusted entries were adopted in the EU-WSR, each with different entries for intra-EU, OECD and third country shipments (cf. sect. I.3.1.3.7).

For shipments for recovery within the Community, the new regulation in the EU includes an adapted entry EU3011 for non-hazardous plastic waste, destined for material recycling and energy recovery, an entry AC300 for hazardous plastic waste originating from the OECD Decision and an entry EU48 for non-hazardous plastic waste requiring monitoring and control.

Since January 2021, exports of plastic waste under entries AC300 and Y48 from the EU to countries to which the OECD Decision applies and imports of such plastic waste from OECD countries into the Union are subject to the procedure of prior written notification and consent.

According to Article 36(1)(a) and (b) and Annex V to the EU-WSR, the export of plastic waste of entries A3210 and Y48 to third countries to which the OECD Decision does not apply is prohibited. The import of such waste into the EU is subject to prior written notification and consent.

The draft Correspondents' Guidelines No.12[23] contains limit values on permissible impurities for plastic wastes. Limit values for impurities are provided for the first ident of the entry B3011 and the first three idents of the entry EU3011, namely 2 % by weight for the entry B3011 and 6 % by weight for the entry EU3011, in order to fill out the terms "almost free of impurities and other types of waste" and "almost exclusively consisting of". Member States may justifiably apply the stricter limit also to EU3011. The same applies to the other idents or mixtures from paragraph 4 of Annex III A of the WSR.

The maximum levels above for the interpretation of the terms "almost free from contamination and other types of wastes"and "almost exclusively consisting of" should be measured on the gross weight of the waste in question, after emptying moisture, not taking into account residual moisture and easy separable usual secondary components of products that become waste, such as caps and labels.

---

[22]   accessed in March 2021:
http://www.oecd.org/environment/waste/Reporting-of-controls-non-hazardous-waste.pdf
[23]   *Draft* Correspondents' Guideline No.12 of 23 September 2021

3.   Regulations on Transboundary Waste Movements

In the 1970s initial bottlenecks appeared in the disposal of hazardous waste in industrialised countries. So called "toxic waste scandals" occurred because no sufficient waste disposal capacities for hazardous waste were available. This led to substantial cost increases in the disposal of such waste and also resulted in illegal disposal at home and abroad.

1976 a chemical accident occurred in Italy, which had a substantial influence on the development of the regulatory framework for transfrontier shipments of waste especially in Europe. The malfunction of a chemical reactor for the manufacture of hexachlorophene led to the release of a huge amount of polychlorinated dibenzo–p–dioxins (PCDD). For the highly contaminated waste from this accident no adequate disposal possibility could be found in subsequent years.

In 1982 the hazardous waste resulting from the accident was packed into 41 barrels, transported off without any controls and "disappeared", until the barrels were found in France after more than six months of intensive search. The media interest this case triggered, and the search for the waste in different countries, made the risks and problems of an insufficiently controlled transfrontier shipment of waste crystal clear to the interested public. The arising demands for regulation of transfrontier shipments of waste were reinforced, and obtained a global dimension, by cases of export from industrial countries into developing countries which occurred in the 1980s.

The failure to prohibit shipment of hazardous waste to less developed countries from developed countries was shown by prominent cases. The so called "Khian Sea incident" occurred in 1986 when the ship Khian Sea sailed from US to the Bahamas carrying around 14,000 tons of incinerator ash from a municipal waste incinerator in Philadelphia with the aim to dispose of the ash in the Bahamas. When this disposal was refused in the Bahamas they tried to dispose of the ash in Honduras, Panama and Guinea-Bissau without success. The permit of Haiti to dispose of the cargo was rescinded after some thousand tons had been unloaded. The ship tried to return to US but was stopped by the Coast Guard, which ordered to remain anchored. But in the middle of the night the ship disappeared and tried to get rid of the cargo in several countries including Yugoslavia, Sri Lanka, Senegal and Indonesia. None of these countries accepted the cargo. The ash was illegal spilled beginning in the area of the Suez Canal. When the ship arrived in Singapore no ash was on-board of the Khian Sea.[24]

Another important case, which occurred in 1987, concerned the importation

---

[24]   Daniel Jaffe: "The international effort to control the transboundary movement of hazardous waste: The Basel and Bamako Conventions'", ILSA Journal of Int'l & Comparative Law, Vol. 2, p 123, 1995

into Nigeria of 18,000 barrels of hazardous waste by Italian companies. The barrels, contained hazardous waste including polychlorinated biphenyls. Even in this century illegal shipments took place for example the so called "Probo Koala Incident". In 2006, a Panama-registered cargo tanker, chartered by Trafigura, a commodities' trading multinational, dumped over 500 m^3 of highly toxic waste in Abidjan, killing 17 people and poisoning thousands.

At the end of the 1980s and the beginning of the 1990s, after many years of global efforts to control transfrontier shipments of waste, especially waste exports from industrialised countries to developing countries, a comprehensive set of regulations was created. This consists of international, regional and national laws that are constantly being revised and modified. When looking at the details or dealing with individual cases, this set of regulations turns out to be a very complex and difficult to understand set of regulations. The most important regulations are in detail:

- Basel Convention of March 1989 on the Control of Transboundary Movements of Hazardous Wastes and their Disposal,

- OECD Council Decision C(2001)107 on the Control of Transboundary Movements of Wastes Destined for Recovery Operations,

- European Regulation (EC) No 1013/2006 on Shipments of Waste (EU-WSR), and

- bilateral, multilateral and regional agreements or arrangements on transfrontier waste shipments, such as the Bamako and Waigani Conventions or the Canada-US and Mexico-US agreements (cf. sect. I.3.1.4).

The instruments of the Basel Convention, the OECD Decision and the EU Waste Shipment Regulation will be further described in detail in the following.

3.1 Basel Convention

The Basel Convention is based upon the Cairo Guidelines[25], issued on 1 December 1985 by UNEP (United Nations Environment Program). They were the first milestone in the establishment of a regulatory system for world-wide supervision and control of transfrontier waste shipments. In the subsequent development of the Basel Convention, preliminary work done by the OECD was utilised, for example concerning disposal methods or categories of wastes and waste constituents.

Each State Party to the Basel Convention is bound by all the obligations

---

[25] Cairo Guidelines and Principles for the Environmental Sound Management of Hazardous Waste, UNEP WG 122/3, UNEP GC 14/17 Annex II

under the Convention. A State that is a Party to the Basel Convention has to have national legislation implementing the requirements of the Convention. Any person within the national jurisdiction of a State that is a Party to the Basel Convention, who is involved in transfrontier shipment of hazardous wastes or other wastes, is therefore legally bound to comply with the relevant national laws and regulations governing the transfrontier shipments of wastes that implement the Basel Convention provisions.

3.1.1 Regulations of the Basel Convention

The Basel Convention, which was adopted on 22 March 1989 in Basel, contains the first outlines of a global "waste management convention", such as the principle of waste disposal at the site of generation, giving priority to measures aimed at reducing the volumes of waste and the task of formulating general principles for an environmentally sound system of waste disposal which is applicable globally.

The following provisions form the core of the Convention:

- the import, export and transit of hazardous waste for the purpose of disposal[26] is only permitted providing all the participating states have been informed beforehand and have consented to the shipment.

- shipment to "non-party states" is not permitted unless bi- or multilateral regulations which do not derogate from the requirements of the Basel Convention are in place.

- the exporter or, alternatively, the state from which the waste originates is responsible for compliance with the Convention and, if necessary, obligated to take back the waste. This obligation applies in particular to waste which is "illegally shipped".

The adoption of these provisions did not end the discussions as to which waste types should be allowed to be exported, particularly from developed countries. This debate continued to play an important role at the 2nd Conferences of the Parties (COP 2) and initially led to a non-binding political decision (Decision BC 2/12 on the "total ban") to extend the export ban to certain hazardous wastes destined for recovery.

Decision BC 3/1 of COP 3 confirmed this extended export ban and contained at the same time an amendment to the Basel Convention requiring ratification. This amendment comprised the inclusion of a preamble, a new Art. 4a as well as a new Annex VII, which introduced a general ban on the shipment of hazardous wastes from states listed in Annex VII to the Basel

---

[26] including recycling and recovery

Convention*) to states not listed therein. Decision BC 3/1 applies to wastes destined for disposal as well as to wastes destined for recovery. After ratification by the necessary number of Parties to the Basel Convention in 2019, the ban took effect under international law at the beginning of 2020.

Irrespective of this, all EU Member States have been subject to an export ban on hazardous waste for recovery in countries to which the OECD Decision does not apply in accordance with Art. 36 EU-WSR. The hazardous wastes affected by the export ban and wastes not subject to the export ban are specified in Annex V to the EU-WSR. The accession of further countries to the EU, which has since been completed, has in fact added these countries to the above-mentioned Annex VII to the Basel Convention.

3.1.2 Definition of Wastes and Hazardous within the Basel Convention

In the Basel Convention, waste is defined as "substances or objects which are disposed of or are intended to be disposed of or are required to be disposed of by the provisions of national law". With this definition the BC "adopts the concept of the (never finalised) 1985 OECD draft agreement on hazardous wastes which defines the wastes to be covered by references to a set of technical annexes, supplemented by a provision that allows every party to determine by national legislation, additional hazardous wastes".[27]

Wastes are defined as substances or objects which are subject to disposal. Disposal is defined by reference to Annex IV to the BC. Therefore a substance or object undergoing one of the operations listed in Annex IV is waste. These are operations which lead to (final) disposal (e.g. landfill, ...) or may lead to resource recovery, recycling, reclamation, direct re-use or alternative use. "However some of the operations describe activities (e.g. 'use as a fuel') that may also be applied to non-waste.

When assessing whether a substance or object is intended to be disposed of, all the circumstances need to be taken into account on a case by case basis. The origin and destination of the substance or object may be relevant. In addition, factors such as appearance, obsolescence, insufficient functionality and insufficient protection against damage during transport, loading and unloading may be relevant. These factors may suggest an intent to dispose of the object or substance, which would make it a waste.

A good may be a waste if the waste definition applies. A good is a substance or object that has economic value and which is capable, as such, of forming the subject of commercial transactions. 'Good' is a wider term than 'product'.

---

*)   which includes Lichtenstein as well as the OECD and the EU Member States
27   Literature [4], sect. I.C.4., p. 48

I. Explanatory Notes

A used good is one that is or has been used, either by its first or subsequent owner. A used good may or may not be a waste. 'Use' means the utilization of a good, except in a recovery operation, whether by its first or a subsequent owner."[28]

In contrast to regulations, for example, of the EU (cf. sect. I.1.4.1/I.1.4.2), the BC does not clarify when a waste ceases to be a waste (end of waste).

Concerning the definition of the term 'hazardous waste' the BC offers only a framework. According to Art. 1 of the BC the following wastes are hazardous wastes:

(a) wastes that belong to any category contained in Annex I unless they do not possess any of the characteristics contained in Annex III, and

(b) wastes that are not covered under paragraph (a) but are defined as, or are considered to be, hazardous wastes by the domestic legislation of the Party of export, import or transit.

This general definition based on the waste streams and constituents specified in Annex I to the BC, in combination with the hazardous characteristics in Annex III to the BC, contributed to the emergence of different interpretations of the definition of hazardous waste. The implementation of the total export ban on hazardous wastes made an exact and globally valid definition of the term hazardous waste absolutely necessary.

Therefore, COP 4 decided to incorporate new waste lists as new annexes

- Annex VIII (List A - waste classified as hazardous) and

- Annex IX (List B - waste not classified as hazardous)

to the Basel Convention. Since these were included via a modification of Annex I to the BC, a formal ratification by the Parties was not necessary.

Further amendments to Annexes VIII and IX were adopted in 2002, 2004 and 2020 respectively at COP 6, COP 7 and COP 14.

(OECD/EU-specific)

These Basel waste lists each complemented by some waste types from the old OECD waste list system[29] which are not listed in the Basel waste lists, and with some deviations regarding the application of certain waste codes, were incorporated as waste lists into the revised OECD Decision from the year 2001. The OECD waste lists (Green and Amber) in turn were incorporated as waste lists into the EU-WSR (as Annexes III and IV).

---

[28] Basel Convention: Glossary of terms", Secretariat of the Basel Convention, December 2017, p. 8/11
[29] from the Green, Amber and Red Waste Lists of the OECD Decision C(92)39/final of 1992

3.1.3 Selected Work Priorities under the Basel Convention

The most important issues discussed at COP 5 in December 1999 in Basel were the adoption of a political declaration ("Basel Declaration on Environmentally Sound Management") and the decision regarding the liability protocol. The Basel Declaration comprised a 10-year agenda and placed emphasis on practical environmental protection, e.g. promoting the development of environmentally sound production and disposal plants, particularly in developing countries. In the following years, work focused on the development of partnership programmes with non-governmental organisations (NGO's) as well as the development of technical guidelines.

During COP 5 a Mobile Phone Partnership Initiative (MPPI) for environmentally sound management of used and end-of-life mobile phones was suggested, the first concrete world-wide Public Private Partnership. In December 2002, during COP 6, the world's main mobile phone manufacturers signed a declaration concerning the partnership for environmentally sound management of used and end-of-life mobile phones. Guidelines concerning the design, collection, transfrontier shipment, repair, refurbishment, recovery and disposal of mobile phones[30] have been developed from that time up to COP 8 by working groups consisting of government representatives and NGO's.

During COP 7 and COP 8 technical guidelines on the environmentally sound management of waste containing persistent organic pollutants (POP) were developed and agreed. Furthermore at COP 8 in November 2006 in Nairobi an emphasis was put on the development of an initiative for environmentally sound management of electrical and electronic scrap.

COP 10 adopted a Strategic Framework for the implementation of the Basel Convention and COP 14 new waste entries for plastic waste.

3.1.3.1 *Basel Protocol – Liability Protocol*

One significant outcome of COP 5 is the adoption of the "Basel Protocol" following eight years of difficult negotiations. The objective of the Protocol is to provide for a comprehensive regime for liability as well as adequate and prompt compensation for damage resulting from the transboundary movement of hazardous wastes and other wastes, including incidents occurring because of illegal traffic in those wastes.

The Protocol addresses who is financially responsible in the event of an incident. Each phase of a transfrontier shipment, from the point at which the

---

[30] cf. Basel webpage:
http://www.basel.int/Implementation/TechnicalAssistance/Partnerships/MPPI/Overview/tabid/3268/Default.aspx

wastes are loaded on the means of transport to their export, international transit, import, and final disposal, is considered.

OECD countries are exempted from applying the Protocol of Liability for environmental damages in their countries caused during transfrontier waste shipments between OECD countries.

The Basel Protocol is the first liability regime in the environment sector worldwide and can be acknowledged as an immense success in international environmental protection. The Protocol shall enter into force on the ninetieth day after the date of deposit of the twentieth instrument of ratification, acceptance, formal confirmation, approval or accession[31].

3.1.3.2 Strategic Framework for the Implementation of the BC

At COP 10 the Conference of the Parties, by decision BC-10/2, adopted a Strategic Framework for the implementation of the Basel Convention for 2012-2021, setting out the following strategic goals and objectives:

Goal 1: Effective implementation of parties' obligations on transfrontier shipments of hazardous and other wastes;

Goal 2: Strengthening the environmentally sound management of hazardous and other wastes; and

Goal 3: Promoting the implementation of the environmentally sound management of hazardous and other wastes as an essential contribution to the attainment of sustainable livelihoods, the Millennium Development Goals and the protection of human health and the environment.

3.1.3.3 Initiative for ESM of Electrical and Electronic Waste

Globally, 20 to 50 million tonnes of electrical and electronic scrap are generated each year. Many electronic devices contain hazardous pollutants such as lead, cadmium and brominated flame retardants. The export of electrical and electronic scrap under the label of reuse in the importing country, which often is not possible, presents a large problem. The most serious problem is the processing of electrical and electronic scrap in ways that are not environmentally sound and that pose a risk to human health.

Against this background, a "World Forum on E-waste" took place on 30 November 2006 in Nairobi as part of COP 8 to discuss innovative solutions for the environmentally sound management of used and end-of-life electrical and electronic devices. As a result, decisions and a Ministerial Declaration were adopted, which stresses inter alia that the prohibition on exports of

---

[31] status of ratifications see: http://www.basel.int/ratif/protocol.htm

hazardous electrical and electronic waste from industrialised countries into developing countries has to be implemented more effectively. World-wide, devices free of hazardous substances shall be developed, and electrical and electronic waste is to be collected separately and disposed of in an environmentally sound way. In addition the take-back of old devices by the manufacturers shall be expanded world-wide.

3.1.3.4 Partnership Acting on Computing Equipment

The Partnership for Action on Computing Equipment (PACE) was launched in 2008 by COP 9. PACE was developed as a multi-stakeholder public-private partnership that provides a forum for representatives of personal computer manufacturers, recyclers, international organizations, associations, academia, environmental groups and governments to tackle environmentally sound refurbishment, repair, material recovery, recycling and disposal of used and end-of-life computing equipment.

In the framework of PACE guidelines on re-use, refurbishment and repair, environmentally sound material recovery and recycling of used computing equipment have been developed[32]. The 'Guideline on environmentally sound testing, refurbishment and repair of used computing equipment' and the 'Guideline on environmentally sound material recovery and recycling of end-of-life computing equipment' were field tested to take into account practical experiences of private companies that agreed to evaluate these two guidelines and to provide recommendations for revisions.

The guidance on procedures for transboundary movement of used and end-of-life computing equipment was not approved. It shall be adjusted in light of technical guidelines on transboundary movements of used electronic and electrical equipment and e-waste.

3.1.3.5 Development of Technical Guidelines on e-waste

A technical guidelines on transboundary movements of electrical and electronic waste and used electrical and electronic equipment, in particular regarding the distinction between waste and non-waste under the Basel Convention, was adopted, on an interim basis, at COP 14 by decision BC-14/5. The Parties acknowledged the need to look further into some sub-paragraphs, in particular concerning the distinction between waste and non-waste.

---

[32] cf. Basel webpage:
http://www.basel.int/Implementation/TechnicalAssistance/Partnerships/PACE/Overview/tabid/3243/Default.aspx

I. Explanatory Notes

The guidelines focus on clarifying aspects related to transboundary movements of e-waste and used equipment that may or may not be waste. Only the transfrontier transport of whole used equipment and components that can be removed from equipment, be tested for functionality and subsequently be directly reused, sent for failure analysis or reused after repair or refurbishment is considered in the present guidelines. For the purpose of these guidelines, the term "equipment" also covers such components. Transfrontier shipments of materials that have been removed or that derive from the dismantling or recycling of e-waste and are waste, such as metals, plastics, PVC coated cables or activated glass, are not addressed in the present guidelines, regardless of whether or not they fall under the provisions of the Convention.

The guidelines provide:

- Information on the relevant provisions of the Convention applicable to transboundary movements of e-waste;

- Guidance on the distinction between waste and non-waste when used equipment is shipped across borders;

- Guidance on the distinction between hazardous waste and non-hazardous waste when used equipment is shipped across borders; and

- General guidance on transboundary movements of hazardous e-waste and used equipment and enforcement of the control provisions of the Convention.

The guidelines are intended for government agencies, including enforcement agencies, that wish to implement, control and enforce legislation and provide training regarding transfrontier shipments. They are also intended to inform all actors involved in the management of e-waste and used equipment so they can be aware of the application of the Basel Convention and other considerations when preparing or arranging for transboundary movements of such items.

Their application should help reduce transfrontier shipments of e-waste in the scope of the Convention to the minimum consistent with the environmentally sound and efficient management of such waste and reduce the environmental burden of e-waste that currently may be exported to countries and facilities that cannot handle it in an environmentally sound manner.

The guidelines do not address other aspects of environmentally sound management (ESM) of e-wastes, such as collection, treatment, disposal or extended producer responsibility (EPR). These aspects are covered in other guidance documents developed under the Basel Convention. There are documents covering ESM generally, including the ESM toolkit, for example a practical manual on EPR. There is also a series of guidelines developed in the context of the following two public-private partnership initiatives under the Basel Convention (MPPI and PACE).

3.1.3.6 Guidelines on POP-containing Waste

COP 7 adopted two technical guidelines on the environmentally sound management of waste containing POPs (a general guideline concerning all POPs as well as a special guideline dealing with PCB, PCT and PBB). At COP8 these guidelines were updated and three further guidelines on waste containing POPs such as pesticides, hexachlorobenzene, DDT as well as polychlorinated dioxins and furans were adopted.

The general guideline contains limit values for POP content above which the POPs in waste have to be destroyed in principle (low POP content):

- polychlorinated biphenyls (PCB): 50 mg/kg

- polychlorinated dibenzo-p-dioxins and dibenzofurans (PCDD/PCDF): 15 µg TEQ/kg

- other POPs (aldrin, chlordane, DDT, dieldrin, endrin, heptachlor, hexachlorobenzene, mirex and toxaphen): 50 mg/kg each

The list of POPs has been enlarged by including new POPs in the Annexes to the Stockholm Convention by decisions of COP 4 and COP 5 of Stockholm Convention 2011 as follows:

- the pesticides chlordecone, alpha hexachlorocyclohexane, beta hexachlorocyclohexane, lindane, and pentachlorobenzene;

- the industrial chemicals hexabromobiphenyl, tetra-, penta-, hexa- and heptabromodiphenyl ether, pentachlorobenzene, perfluorooctane sulfonic acid (PFOS), its salts, perfluorooctane sulfonyl fluoride;

- the by-products alpha hexachlorocyclohexane, beta hexachlorocyclohexane and pentachlorobenzene; and

- endosulfane

The follow-up conferences under the Basel Convention updated the existing guidelines and added new technical guidelines, for example for PFOS and waste containing PBDE.

3.1.3.7 Decision on Shipments of Plastic Waste

In accordance with the decisions of COP 14 of the Basel Convention only single-variety waste and mixtures of plastics such as polypropylene, polyethylene and PET that are intended for recycling can be freely traded with other countries. Since these plastics are in demand worldwide and have a market value, it is unlikely that they will be landfilled. Hazardous plastic waste and those that are difficult to recycle are now subject to the requirements of the Basel Convention. This means that they may only be exported and disposed of in an environmentally sound manner with the consent of the authorities of the exporting and importing countries. Annexes II, VIII and IX to the Basel Convention have been amended accordingly.

Implementation in OECD and EU

By decision[33] of 7 September 2020, Annexes 3 and 4 to the OECD Council Decision were amended, but there was no consensus among OECD member countries to adopt the new Basel entries.

The Delegated Regulation[34] (EU) 2020/2174 of 19 October 2020 adopted adapted entries in the EU-WSR, different entries for intra-EU, OECD and third country shipments. The new regulation in the EU includes an adapted entry EU3011 for non-hazardous plastic waste, an entry AC300 for hazardous plastic waste originating from the OECD Decision and an entry EU48 for non-hazardous but controlled plastic waste for shipments for material and energy recovery within the Community.

The import of plastic waste under entries AC300 and Y48 into the EU from countries to which the OECD Decision does not apply, as well as the import of such plastic waste from OECD countries into the EU, is subject to the procedure of prior written notification and consent. In accordance with Art. 36(1)(a) and (b) and Annex V to the EU-WSR, the export of plastic waste of entries A3210 and Y48 to third countries to which the OECD Decision does not apply is prohibited.

3.1.4 Bilateral, Multilateral and Regional Agreements

According to Art. 11 of Basel Convention Parties to the Basel Convention may enter into bilateral, multilateral, or regional agreements or arrangements regarding transboundary movement of hazardous wastes or other wastes with Parties or non-Parties provided that such agreements or arrangements do not derogate from the environmentally sound management of hazardous wastes and other wastes as required by the Basel Convention.

---

[33] See: http://www.oecd.org/environment/waste/theoecdcontrolsystemforwasterecovery.htm

[34] Commission Delegated Regulation (EU) 2020/2174 of 19 October 2020 amending Annexes IC, III, IIIA, IV, V, VII and VIII to Regulation (EC) No 1013/2006 of the European Parliament and of the Council on shipments of waste, OJ L 433, p. 11

The agreements or arrangements which have been concluded by parties can be downloaded from the webpage[35] of the Basel Convention. Some of these agreements or arrangements are shortly described in the following.

3.1.4.1   Bamako Convention

Established in 1991, the Bamako Convention came into force in 1998. The Bamako Convention is a response to Art. 11 of the Basel Convention, which encourages parties to enter into bilateral, multilateral and regional agreements on hazardous waste to help achieve the objectives of the BC. Just under half of the African countries have ratified this multilateral agreement prohibiting the import into Africa of any hazardous waste. The Bamako Convention uses a format and language similar to that of the Basel Convention, but:

- it prohibits all imports of hazardous waste, and

- it does not make exceptions on certain hazardous wastes (like those for radioactive materials, otherwise covered by IAEA Code of practice[36]) made by the Basel Convention.

The purpose of the Convention is inter alia

- to prohibit the import of all hazardous and radioactive wastes into the African continent for any reason;

- to minimize and control transboundary movements of hazardous wastes within the African continent;

- to ensure that disposal of wastes is conducted in an environmentally sound manner"; and

- to establish the precautionary principle.

Furthermore the Bamako Convention has a broader scope than the Basel Convention, as it covers waste streams and constituents as well as waste that meets the hazard criteria.

Countries should ban the import of hazardous and radioactive wastes as well as all forms of ocean disposal. For Intra-African waste shipments, parties must minimize the transfrontier shipments of wastes and only conduct it with consent of the importing and transit states among other controls. They should minimize the production of hazardous wastes and cooperate to ensure that wastes are treated and disposed of in an

---

[35]   http://www.basel.int/Countries/Agreements/MultilateralAgreements/tabid/1518/Default.aspx
http://www.basel.int/Countries/Agreements/BilateralAgreements/tabid/1517/Default.aspx
[36]   https://www.iaea.org/publications/documents/infcircs/code-practice-international-transboundary-movement-radioactive-waste

environmentally sound manner.

3.1.4.2   OECD Council Decisions

The OECD Decision is a multilateral agreement based on Art. 11 of the BC through which the OECD regulates transfrontier shipment of wastes destined for recovery operations to or from OECD countries. This OECD Decision, together with other OECD Council decisions, constitutes a multilateral agreement pursuant to Art. 11 (2) of the Basel Convention.

Unlike the Basel Convention or the EU-WSR, the OECD Decision only contains stipulations relating to shipments of wastes destined for recovery, not for disposal. On the other hand, the OECD Decision applies not only to hazardous wastes, but to all types of wastes. In addition the OECD Decision contains more concrete provisions, including in particular concerning the deadlines of the notification procedure.

3.1.4.3   Waigani Convention

Based on Art. 11 the BC the purpose of the Waigani Convention is to ban the importation into Forum Island countries of hazardous and radioactive wastes and to control the transboundary movement and management of hazardous wastes within the South Pacific Region. There are however some differences between Waigani and Basel Convention. The Waigani Convention also covers radioactive wastes; and its territorial coverage includes each Party's Exclusive Economic Zone (200 nautical miles) (rather than extending only to outer boundary of each Party's territorial sea (12 nautical miles) as under Basel).

The Waigani Convention is designed to:

- reduce or eliminate transboundary movements of hazardous and radioactive wastes into and within the Pacific Forum region;

- minimize the production of hazardous and toxic wastes in the Pacific Forum region;

- ensure that disposal of wastes is done in an environmentally sound manner and as close to the source as possible; and

- assist Pacific island countries that are Parties to the Convention in the environmentally sound management of hazardous and other wastes they generate.

The major benefit of the Waigani Convention is the establishment of a system to prevent hazardous and radioactive waste entering the region or being dumped there.

3.1.4.4 Canada-US Agreement

Canada and the United States have entered into a comprehensive agreement on the transfrontier shipment of hazardous waste. The Canada-US Agreement, together with its supporting regulatory framework, is compatible with the control procedures under the Basel Convention and constitutes a bilateral agreements according to Art. 11 of the BC. It sets out specific administrative conditions for the export, import, and transportation of hazardous waste between the two countries.

Wastes are considered hazardous if defined as such by the legislation of the exporting country. In addition to regulating hazardous waste transported domestically and between Canada, the United States and other countries, both Canada and the United States support the OECD Decision. The Canada-US Agreement takes into account the OECD Decision, as well as other international programs regarding hazardous waste.

Hazardous waste generators and parties wishing to transport hazardous waste across the border must first submit, together with other relevant documents, a notice that contains a variety of detailed information about the proposed shipment. The notice is submitted to the designated authority in the exporting country, which notifies the designated authority in the importing country of the proposed shipment.

After acknowledging receipt of the Notice, the importing country has 30 days to review the request and indicate its objection or consent to the proposed shipment. The importing country has the right to alter the conditions of transport as described in the Notice. Responses to the Notice are then provided to the designated authority in the exporting country. If no response is received within the 30-day period, the importing country is considered to have no objections to the shipment (tacit consent).

3.1.4.5 Mexico-US Agreement

On 12 November 1986 Mexico and US signed the "Agreement of Cooperation between the United States of America and the United Mexican States Regarding the Transboundary Shipments of Hazardous Wastes and Hazardous Substances" which entered into force on 29 January 1987. It constituted Annex III to the "Agreement between the United States of America and the United Mexican States on Cooperation for the Protection and Improvement of the Environment in the Border Area" and addresses the transfrontier shipment of hazardous wastes and hazardous substances between the two countries. This means that the agreement also covers hazardous substances in addition to waste.

Concerning waste the bilateral agreement between Mexico and the United

States addresses the transfrontier shipment of hazardous wastes from Mexico to the United States for recycling or disposal, and the transfrontier shipment of hazardous wastes from the United States to Mexico for recycling only.

3.2   OECD Council Decision C(2001)107/FINAL

Since March 1992, transboundary movements of wastes destined for recovery operations between member countries of the OECD have been supervised and controlled according to the OECD Council Decision C(92)39.

The developments under the Basel Convention, in particular the adoption of two waste lists in 1998, made it necessary to revise the OECD Council Decision C(92)39 in order to harmonise procedures and requirements and to avoid duplication with the Basel Convention. Result of the revision process was the adoption of Council Decision C(2001)107/FINAL (OECD Decision). Provisions of the revised OECD Decision have been harmonised with those of the Basel Convention. However, certain procedural elements of the OECD Council Decision C(92)39, which do not exist in the Basel Convention, such as time limits for approval process, tacit consent and pre-consent procedures have been retained.

The OECD Decision from 2001 is compatible with the environmentally sound management of hazardous wastes and other wastes pursuant to Art. (2) of the BC. Consequently, it is the OECD Decision which applies when transboundary movements of wastes destined for recovery operations take place from one OECD Member country to another. As OECD Council Decisions are legally binding for member countries, the OECD Decision has to be implemented in member countries through the enactment of national legislation.

For example, in the EU Member States, the OECD Decision is implemented through the EU-WSR. In Canada, the "Export and Import of Hazardous Waste and Hazardous Recyclable Material Regulations" fully implement the requirements of the OECD Decision, the Basel Convention, and the Canada-US Agreement on Transboundary Movement of Hazardous Waste. In Switzerland, the OECD Decision is translated into national legislation through the "Swiss Ordinance on the Movements of Waste". In Japan, the "Ordinance Designating Materials to be controlled by the OECD Decision C(2001)107/FINAL Concerning the Control of Transboundary Movements of Wastes Destined for Recovery Operations" has been adopted. In Korea, the requirements of the OECD Decision have been transposed into the "Act on the Control of Transboundary Movements of Hazardous Wastes and their Disposal".

3.2.1 Waste Definition

For the purposes of the OECD Decision wastes are substances or objects, other than radioactive materials, which:

(i) are disposed of or are being recovered; or

(ii) are intended to be disposed of or recovered; or

(iii) are required, by the provisions of national law, to be disposed of or recovered.

Like the EU-WFD the OECD Decision distinguishes between the terms "disposal" and "recovery", whereas, in the Basel Convention, the term "disposal" covers both disposal and recovery. The disposal operations are specified in Appendix 5.A and recovery operations in Appendix 5.B to the OECD Decision.

Although the OECD definition of waste covers wastes destined for disposal and recovery, the OECD Decision only applies to wastes destined for recovery. Wastes destined for disposal are subject to different legal control, in particular those established by the Basel Convention and any applicable national law.

OECD Member countries have different interpretations "of the definition of waste[37]. Consequently, different decisions may be made in different countries about the status of the same material and therefore, the same material may be regarded as waste in one country but as a commodity or raw material in another country"[38].

3.2.2 OECD Waste Lists

The waste lists determining wastes subject to control are harmonised with the Basel Convention. The former OECD green, amber and red lists of wastes have been replaced by Annexes II, VIII and IX to the Basel Convention. The OECD Green waste list contains Annex IX wastes as well as some additional wastes, which were included in earlier OECD green waste list. The OECD Amber waste list contains Basel Annexes II and VIII wastes as well as some additional wastes which were included in earlier OECD amber and red waste lists. However, some adjustments have been made with respect to certain entries in Annexes VIII and IX to the Basel Convention for the purposes of the OECD Decision (cf. sect. I.2.2 and appendix 1 for further details).

---

[37] For further information on the definition of waste, see: "Guidance Document for distinguishing waste from non-waste" [ENV/EPOC/WMP(98)1/REV1], published in 1998.

[38] [2]: https://www.oecd.org/env/waste/guidance-manual-control-transboundary-movements-recoverable-wastes.pdf

3.3   EU Regulations on Waste Shipments *(EU specific)*

In the EU, the barrels with dioxin containing waste "roving around" led to the issue of Directive 84/631/EEC, based on the internal market competence of the EC Treaty in force at that time. This Directive remained however relatively ineffective. Therefore, the EC Waste Shipment Regulation (1993) was adopted which already based on the environmental protection competence of Art. 192 TFEU.

The EC Waste Shipment Regulation (1993) already had a certain "pooling" function, since it transposed in particular the regulations of the Basel Convention and the OECD Decision into directly applicable EU law. After many years of discussions[39] the EC Waste Shipment Regulation (1993) was replaced by the Regulation of the European Parliament and of the Council (EC) No 1013/2006 on Shipments of Waste (EU-WSR). The EU-WSR fully transposes the procedural provisions of the Basel Convention and the OECD Decision into directly applicable Community law. Its provisions are intended to ensure the proper transfrontier shipment and the environmentally sound disposal of waste. Environmental and health risks are to be prevented and developing countries are to be protected from waste imports.

The EU-WSR retained the basic idea of EC Waste Shipment Regulation (1993) that certain procedures have to be followed in the transfrontier shipment of waste, depending on which type of waste is shipped into which destination country and how the waste is managed in the country of destination (recovery or disposal).

In the interest of precautionary environmental protection, the EU-WSR is designed as preventive prohibition with reservation of authorisation. This can be seen clearly in the two overview tables 5 and 6 in sect. I.4.4.4. The procedure of prior written notification and consent is regulated in Art. 4 et seq. (Title II of the EU-WSR). This is described in more detail in sect. I.4.6. An exemption from the notification requirement applies to Green listed wastes (cf. sect. I.4.5), whereby special provisions are laid down for third countries in the EU Regulation (EC) No 1418/2007 on shipments of 'green wastes' to countries to which the OECD Decision does not apply.

The EU-WSR is based solely on the competence of environmental protection, which in the first recital of the EU-WSR is expressed as follows:

"The main and predominant objective and component of this Regulation is the protection of the environment, its effects on international trade being

---

[39]   Wuttke, Joachim.: "Grenzüberschreitende Abfallverbringungen von Beseitigungs- und Verwertungsabfällen – Was ändert sich nach der novellierten Abfallverbringungsverordnung", in K. Wiemer, M. Kern (Hrsg.): Bio- und Sekundärstoffverwertung, stofflich – energetisch, Witzenhausen-Institut für Abfall, Umwelt und Energie GmbH, Witzenhausen (2006) S. 279/293

only incidental."

Furthermore, the following regulations as particularly pertaining to waste definition and classification are applied in the EU:

- EU Waste Framework Directive (EU-WFD) and
- Decision of the Commission on the List of Waste (EWL).

Concerning EU waste definition and classification please cf. sect. I.1.4.1 and I.2.1.

3.3.1 Proposal for a new regulation

On 17 November 2021, the Commission adopted a proposal on a revised Regulation on waste shipments[40]. According to Commission the proposed new Waste Shipment Regulation has three goals:

- ensuring that the EU does not export its waste challenges to third countries;
- making it easier to transport waste for recycling and reuse in the EU; and
- better tackling illegal waste shipments.

Waste exports to non-OECD countries will be restricted and be made conditional on an official request from the country to import non-hazardous waste from the EU and demonstration that it can recover it in a sound manner. A list of countries authorised to import waste from the EU will be set up.

Waste shipments to OECD countries will be monitored and can be suspended if they generate serious environmental problems in the country of destination. Under the proposal, all EU companies that export waste outside the EU should ensure that the facilities receiving their waste are subject to an independent audit showing that they manage this waste in an environmentally sound manner.

Within the EU, the Commission is proposing to simplify the established procedures considerably, facilitating waste to re-enter the circular economy, without lowering the necessary level of control. The new rules are also bringing waste shipments to the digital era by introducing electronic exchange of documentation.

To address waste being illegally presented as "used goods", specific binding criteria shall be developed to differentiate between waste and used goods

---

[40] European Commission: "Proposal for a new regulation on waste shipments", 17 November 2021, see: https://ec.europa.eu/environment/publications/proposal-new-regulation-waste-shipments_en

I. Explanatory Notes

for specific commodities of a particular concern, such as used vehicles and batteries.

Concerning shipments within the EU it is proposed

- to fully digitalise all procedures governing the shipments of waste between EU Member States, in particular for "green-listed" waste;

- to support the use of fast-track procedures for shipments of waste destined to recovery, when they are destined to facilities certified by the EU Member States ("pre-consented facilities");

- to harmonise the classification of waste at the EU level, to help overcome the current fragmentation of the EU market, where a shipment of waste can be subject to different interpretations and procedures when crossing borders in the EU;

- to make it possible to streamline at the EU level the calculation of financial guarantees that operators have to establish before shipping "notified" waste abroad;

- to include new and stricter conditions for shipments of waste for incineration or landfilling, so that they are only authorised in limited and well-justified cases, as they are the least preferred options for the management of waste.

The proposal covers all types of waste. The proposal would make it easier to recycle and re-use hazardous and other non-hazardous waste subject to notification, like mixed municipal waste and unsorted plastic waste within the EU. There will be no major changes to the current rules on EU exports of such waste.

The proposal contains important changes on the export of "green-listed" waste. The proposal does not establish new rules applying to the shipments of green-listed waste within the EU. However, it proposes that this waste should be more easily traceable; notably the documentation accompanying their circulation should be digitalised.

3.3.1.1   *Export of Green listed Wastes to Non-OECD countries*

With the proposed amendment of Regulation 1013/2006, the export of Green List wastes to non-OECD countries will also receive a completely new basis. According to the amendment, exports will only be possible to countries on a list drawn up by the Commission. This changes the regulatory system decisively. Under the current law, all exports to non-OECD countries that do not respond to a corresponding request from the Commission are subject to the notification requirement. According to the new regulation, an export ban then applies.

If one analyses the previous statements or non-statements of the third countries, this means that about 50 states with export bans will be added to the 25 states with export bans so far.

In addition, the states that want to be listed must fulfil new requirements, among other things they must have a comprehensive waste management system and a legal framework for waste management, as well as be party to various conventions (cf. Annex VIII in conjunction with Art. 39 of the new regulation). Furthermore, according to Art. 40, the Commission shall assess the responses of the third countries and only list them if the requirements according to Art. 39 are fulfilled. If during the assessment the Commission comes to the conclusion that the reply is not complete or sufficient, the third country has three months to complete it. The time limit can be extended by three months on the basis of a reasoned request by the third country.

This review procedure is likely to lead to an export ban for further states in the future.

This new regulation is to take effect 30 months after the entry into force of the amendment to Regulation 1013/2006. For this purpose, the Commission is to start a survey of third countries 3 months after the amendment enters into force on the basis of the questionnaire from Annex VIII.

3.3.1.2 *Empowerments for the Commission*

In addition to amendments to the provisions of the Regulation, the proposal contains a whole series of empowerments for the Commission to adopt delegated acts, inter alia, to

- establish criteria to distinguish between used goods and waste (Art 28 para 4),

- setting of demarcation criteria between green waste and amber waste,

- establish contamination thresholds (Art 28 para 4),

- establish a harmonised calculation method for the financial guarantee (2 years after entry into force),

- establish a list of non-OECD countries (30 months after entry into force, provided that countries have applied for inclusion by then and meet the requirements) to which the export of non-hazardous waste is permitted (Art. 38 (1)); for inclusion in the list, demanding require-ments and documentation obligations apply (cf. Art. 39 (3) in conjun-ction with Annex VIII). Annex VIII),

- amend Annexes IA, IB, IC, II, III, IIIA, IIIB, IV, V, VI, and VII, as well as Annexes IX and X.

Furthermore, the Commission shall be empowered to adopt implementing acts in order to

- to determine the conditions for electronic data exchange (1 year after entry into force), and to

- to draw up a correlation list between customs codes and waste codes (green and amber lists of waste and European Waste List).

This is supplemented by the provisions on actions to be carried out by the Commission (Art. 64 to Art. 68) for the enforcement of the regulations.

3.3.1.3   *Entry into force*

The proposed new Regulation shall enter into force on the twentieth day following that of its publication in the Official Journal of the European Union and it shall apply from two months after the entry into force. However, Articles 5, 8 and 9, Article 14(14) and (15), Articles 15, 16, 18, Article 26(1), (2) and (3), and Articles 35, 41, 47, 48, 49, 50, 51, 54 and 55 shall apply from two years after the date of entry into force and Articles 37, 38, 39, 40, 43 and 44 shall apply from three years after the date of entry into force (compare Article 82).

3.3.1.4   *Possible date of full implementation*

The Commission's proposal is now before the Council and Parliament for discussion. Experience shows that this will take at least one year. However, the provisions for delegated acts of the Commission could meet with resistance in the Council and thus extend the period until publication in the Official Journal.

Since Article 82 provides for transition periods of two or three years, full entry into force is not expected before 2025/26.

For the current status of the revision, see the website: www.waste-move.eu.

4. Practical Guidance on Waste Shipments

Several international instruments have been established to control export and import of wastes, which may pose a risk or a hazard to human health and the environment. These instruments are the Basel Convention, the OECD Decision and the EU-WSR and/or other regional agreements or arrangement (cf. sect. I.3.1.4).

4.1 Scope of Application

(Art. 1 (1) and (2) BC; Chapt. I (1) OECD-Decision; Art. 1 (2) EU-WSR)

The Basel Convention and the EU-WSR concern international shipments of waste, whether destined for disposal or recovery, whereas the OECD Decision only concerns shipments of wastes destined for recovery operations within the OECD area.

The Basel Convention covers shipments of hazardous and other wastes[41]. The OECD Decision regulates shipments of all wastes for recovery. As the EU-WSR implements both the Basel Convention and the OECD Decision all waste for recovery and disposal are covered.

All of the instruments operate a range of administrative controls by the countries/parties/states implementing them.

4.2 Definition of Waste

(Art. 2 (1) BC; Chapt. II A.(1) OECD-decision; Art. 3 (1) EU-WFD)

According to Basel Convention waste are defined as "*substances or objects which are disposed of or are intended to be disposed of or are required to be disposed of* by the provisions of national law". According to Art. 1 (1) b and Art. 3 of the BC additional wastes may be defined as hazardous in national legislation over and above those listed in the BC.

The waste definition of the OECD Decision is harmonised with that of the Basel Convention. According to the OECD, wastes are "*substances or objects which are disposed of or are being recovered; or are intended to be disposed of or recovered; or are required*", by the provisions of national law, "*to be disposed of or recovered*".

The EU defines waste as any "*substance or object, which the holder discards, intends to discard or is required to discard*". The notion of discard is not legally defined, but it can be interpreted as to dispose of a substance or

---

[41] other waste are wastes listed in Annex II of the Basel Convention

object.

Regarding waste classification of the BC, the OECD Decision and the EU-WSR see sect. 2.2 and appendix 1.

4.3   Illegal Shipments

(Art. 9 (1) BC; Chapt. II D.(3) and (4) OECD-Decision; Art. 2 No. 35 EU-WSR)

According to the Basel Convention, any transfrontier shipment of hazardous wastes or other wastes is deemed to be illegal traffic if:

- it is carried out without notification pursuant to the provisions of the Basel Convention;

- it is carried out without consent in accordance with the Basel Convention;

- the consent for it is obtained from the Courtiers concerned through falsification, misinterpretation or fraud;

- it does not conform in a material way with the relevant documents; or

- it results in deliberate disposal (e.g. dumping) of hazardous wastes or other wastes in contravention of the Basel Convention and of general principles of international law.

The EU-WSR adopts this definition of illegal shipment and adds further EU-specific points to it.

4.3.1   Take-back Obligations

(Art. 9 BC; Art. 22 and 24 EU-WSR)

In the case where a transfrontier shipment is deemed to be illegal shipment as the result of conduct on the part of the notifier/exporter or generator the Country of export shall ensure that the wastes in question are taken back by the notifier/exporter or generator or, if necessary, by itself into the State of export; or if this is impracticable, are otherwise disposed of within a time period specified in the respective instrument or a period of time as the Countries concerned may agree.

If the illegal traffic is the result of conduct on the part of the importer/consignee or disposer, the Country import, within a time period specified in the respective instrument or a period of time as the Countries concerned may agree, shall ensure that the wastes in question are disposed of in an environmentally sound manner by the importer/consignee or disposer or, if necessary, by itself.

In cases where the responsibility for the illegal traffic cannot be assigned, the Parties concerned or other Parties, as appropriate, ensure, through co-operation, that the wastes in question are disposed of as soon as possible in an environmentally sound manner.

Illegal shipments or legal shipments that could not be completed as intended must be taken back or otherwise disposed of in an environmentally sound manner. The person obliged to take back the waste shall bear the costs for the return and environmentally sound disposal of the waste.

(EU-specific)

If the responsible person cannot be held responsible, the country of export has subsidiary liability for the take-back obligation. Take-back obligations for waste according to Art. 22 to 25 of the EU-WSR can be considered

- in the case of legal transboundary shipments of waste on the basis of official permits, if the shipment cannot be completed, including the disposal operation planned in the country of destination, or

- in the case of illegal transboundary shipments of waste within the meaning of Art. 2 No. 35 WSR.

According to Art. 24 Para. 3 WSR, the consignee is solely responsible if the consignee carries out the disposal in deviation from the information in the notification and the notifier was not aware of this.

According to Art. 24 (5), there are also cases in which both the notifier and the consignee have jointly caused the illegal shipment. These cases, which occur frequently in practice, require a joint assessment of the respective case. This becomes particularly difficult when there is no notification and thus no financial guarantee. This can lead to a dispute about the extent to which the recipient is partly to blame. Griesbach[42] assumes that at least in the case of an illegal shipment to an importer/consignee without a facility permit, both parties involved can be responsible.

The take-back of illegally shipped waste or shipments that could not be completed as intended requires a reasoned request by the competent authority of destination or the competent authority of transit and a consensual decision by all competent authorities concerned. Further explanations on take-back obligations can be found in the German Enforcement Guidance [7] as well as by Backes in [8], pp. 589 to 644.

---

[42] Griesbach, Angela: "Die Rücknahme illegal grenzüberschreitend verbrachter Abfälle", Verlag Dr. Kovač, Hamburg 2011.

4.4 Requirements for Waste Shipments

Depending on the intended disposal method and the classification of the waste, a transfrontier shipment of waste is either subject to the procedure of prior notification and consent by the authority (cf. sect. I.4.6) or not.

While the Basel Convention has no provisions for the shipment of non-hazardous wastes, both the OECD Decision and the EU-WSR contain rules for the shipment of such wastes (green listed wastes). The OECD applies a Green control procedure and the EU has implemented general information requirements.

(EU-specific)

Concerning shipments of waste out of or into the EU Art. 34 to 45 of the EU-WSR regulate the conditions under which EU Member States may ship waste to or from OECD Countries and to or from third countries. In these regulations, the export bans according to Art. 36 of the EU-WSR and the import bans according to Art. 41 of the EU-WSR must be observed first and foremost.

The countries addressed, to which the OECD Decision cited in Art. 2 No. 17 of the EU-WSR applies, currently include twelve OECD countries[43]. In the case of newly acceded OECD Countries, it must be examined whether they have already implemented the OECD Decision nationally, or whether there is still a transitional period and they are thus to be considered third states within the meaning of the EU-WSR, i.e. states to which the OECD Decision does not apply; as is currently the case for Colombia.[44]

4.4.1 Green Control Procedure *(OECD-specific)*

(Chapt. II B.(2) a) OECD-Decision)

Wastes subject to the Green control procedure are deemed to pose negligible risks for human health and the environment during their transfrontier shipment for recovery within the OECD area.

Therefore they are not controlled, but the OECD Decision imposes a general requirement that Green wastes shall be destined for

- recovery operations within a recovery facility and

- recovered in an environmentally sound manner according to national legislation.

---

[43] State May 2021: Australia, Chile, Costa Rica, Israel, Japan, Canada, Korea, Mexico, New Zeeland, Turkey, United Kingdom, United States

[44] State: May 2021

Furthermore, it is required that all persons involved in any contracts or arrangements for such transfrontier shipment should have the appropriate legal status, in accordance with national legislation and regulations. Those shipment shall also be subject to applicable international transport agreement and other existing controls normally applied in commercial transactions.

4.4.2 Information Requirements *(EU-specific)*

(Art. 3 (2) and (4) EU-WSR)

The EU by implementing the OECD Decision requires that certain information, signed by the person arranging the shipment of green wastes, accompanies each shipment of such waste, in order to assist the tracking of these shipments. This procedure applied is called "information requirements". They must be met in the case of shipments of

- waste listed in Annex III or IIIB to the EU-WSR for recovery ("green" waste list), as well as non-hazardous waste mixtures listed in Annex IIIA to the EU-WSR of more than 20 kg or of

- waste for laboratory analysis up to a maximum of 25 kg

for which a document pursuant to Annex VII to the EU-WSR (see Fig. 3) is to be used. The information requirements are described in more detail in sect. I.4.5.

4.4.3 Obligation to Notify

(Art. 6 (1) and (2) BC; Chapt. II B.(2) b) OECD-Decision; Art. 4 EU-WSR)

Normally the exporter (directly or via/by the competent authority of the exporting country) has the responsibility to notify. It depends on the applicable legislation whether the exporter has to provide the written notification to the competent authorities of the countries of export, import and transit or the competent authority of if the country of export has to transmit the notification to other countries concerned. Such a requirement is considered useful by some, for example by the EU and its Member States in order to ensure that the information provided in the notification is complete.

The main purpose of notification is to inform authorities in advance of the nature, transport and disposal of the waste and to enable them to take all necessary environmental and health protection measures. It also gives the authority the opportunity to raise objections to the shipment. In substance, the notification procedure is a transnational authorisation procedure.

According to the BC the state of export shall not allow the export of waste until the state of import has consented to the shipment.

I. Explanatory Notes

According to the OECD Decision and the EU-WSR the notification procedure is no single administrative act, but all the authorities involved take a decision independently of each other. The notifier/exporter is obliged to start the shipment only when he has received all consents from all authorities involved including transit states. The procedure is described in more detail in sect. I.4.6.

4.4.3.1 Unilateral Notification Procedure

(Art. 6 (5) BC; Chapt. II, B.(4) d) OECD Decision)

Due to differing opinions on the classification of waste by the countries involved, a so-called unilateral notification (also called consent procedure) may result. Both the Basel Convention and the OECD Decision contain provisions on this. The EU-WSR, on the other hand, conclusively regulates in Art. 28 that in the case of shipments with the exclusive participation of EU Member States, the stricter requirement applies in each case.

However, if a shipment is to take place between a Basel Party or an OECD country that is not an EU Member State and an EU Member State, this is a case in which unilateral notification may be required.

For example if a Green List waste requires notification only in another country outside the EU, the so-called unilateral notification procedure is applied. This is the case for Green List wastes that are subject to control in transfrontier shipment in the exporting country, e.g. Switzerland.[45]

4.4.4 Export and Import of Waste for Disposal

As already mentioned, both the Basel Convention and the EU-WSR contain provisions on shipments of waste for disposal. The most important provision is the export ban on waste for disposal under Art. 4a (1) Basel Convention, which is further extended by the EU-WSR in Art. 34 (cf. sect. I.4.4.4.1).

The procedure of prior written notification and consent applies to permitted shipments.

---

[45] In Switzerland, this applies to the following wastes: B 3140 "Used tyres", B1250 "Drained and stripped end-of-life vehicles", B1115 "Used metal cables with plastic insulation", GC010 "Used electronic devices (e. g. CD players, telephones, printers, radio sets)", GC020 "Components from used electronic devices (such as de-polluted circuit boards)", B3065 "Used cooking oils", B2130 "Recovered asphalt with 250 to 1000 mg PAH per kg and less than 50 mg/kg benzopyrene".
cf: FOEN (ed.): Transboundary movement of waste, FOEN Information brochure for applicants, Published by the Federal Office for the Environment FOEN, Bern, 2017, Enforcement No. 1702: 62 pp.

4.4.4.1   Export from and Import to the EU (EU-specific)

(cf. specification of relevant Articles of the EU-WSR in table 5)

Export of waste for disposal from EU Member States is only allowed to other EU Member States and EFTA countries[46] that are Party to the Basel Convention.

Import of waste for disposal into the EU is only permitted from those countries that are Party to the Basel Convention, that are members of EFTA and Party to the Basel Convention, or with which the EU or individual Member States have concluded bilateral or multilateral agreements or arrangements.

Table 5: Regulatory Areas of the EU-WSR for Waste for Disposal

Transfrontier shipment	within the EU Art. 4 to 17	import into the EU, Art. 41 to 42	transit through the EU, Art. 47	export from the EU to EFTA countries, Art. 34, and 35	export out of the EU to non EFTA countries, Art. 34
Waste for disposal	permitted notification according to Art. 4 to 17	prohibited, with excep-tions[1] notification according to Art. 41 and 42	permitted notification according to Art. 47	permitted notification according to Art. 35	*prohibited*
1) The import from Parties to the Basel Convention and countries with which bilateral agreements exist is permitted					

4.4.5 Export and Import of Waste for Recovery

Shipments of waste for recovery are regulated by the Basel Convention, the OECD Decision as well as the EU-WSR. While the latter two instruments regulate all wastes for recovery, the provisions of the Basel Convention only cover hazardous wastes and other wastes as defined by the Basel Convention.

The most important provision is the export ban on hazardous waste for recovery under Art. 4a (2) Basel Convention, which is further extended by the EU-WSR in Art. 36 (cf. sect. I.4.4.5.1).

For permitted shipments of hazardous waste and other waste (*in the OECD/EU: amber listed waste*) the procedure of prior written notification and consent applies, which is not the case for non-hazardous waste (*in the OECD/EU: green listed waste*).

---

[46]   Iceland, Lichtenstein, Norway, Switzerland

4.4.5.1   *Export from and to the EU (EU-specific)*

(cf. specification relevant Articles of the EU-WSR in table 6)

Export and import of hazardous waste for recovery from and to the EU are also prohibited, depending on the third country involved. The export for recovery of hazardous waste listed in Annex V to the EU-WSR from the EU to countries that have not implemented the OECD Decision is prohibited.

The import of waste for recovery into the EU is permitted from countries to which the OECD Decision applies, which are parties to the Basel Convention and from countries with bilateral agreements (cf. Table 6).

A control system is applied that distinguishes between two waste categories: While wastes on the "amber" waste list and unlisted wastes are subject to prior notification and consent, wastes on the "green" waste list are exempt from the requirement to undergo a notification procedure. These wastes can be shipped within the EU and the countries to which the OECD Decision applies without notification, although within the EU, according to Art. 18 of the EU-WSR, information requirements (document according to Annex VII to the EU-WSR) must be provided and further obligations must be fulfilled (cf. sect. I 4.2).

4.4.5.2   *Export Ban to Countries where the OECD Decision does not apply (EU-specific)*

(Art. 36 EU-WSR)

Annex V to the EU-WSR lists the wastes covered by the ban on export of hazardous wastes to countries where the OECD Decision does not apply. Cf. also the decision scheme in Fig. 2.

Annex V has three parts:

- Part 1 is divided into List A and List B.

If a particular waste is listed in List A, its export to countries where the OECD Decision does not apply is prohibited.

If a waste is listed on List B, its export to a country where the OECD Decision does not apply is potentially permitted. If a waste listed on List B is classified as hazardous by reference to EU-criteria in an EU Member State in accordance with Art. 36(4) and (5) of the EU-WSR, its export to a country where the OECD Decision does not apply is prohibited.

Table 6: Regulatory Areas of the EU-WSR for waste for recovery

Transfrontier shipment	within the EU; Art. 3 to 17	import into the EU; Art. 43 to 46	transit through the EU; Art. 48	export from EU to OECD-countries; Art. 18, and 38	export out of EU to non OECD countries; Art. 18, 36 and 37
Waste for recovery; Annexes III, IIIA and IIIB ≤ 20 kg	free movement Art. 3 (2)	free movement Art. 3 (2)	free movement Art. 3 (2)	Applicable only to waste listed in Annexes III and IIIA, and without a limit on quantity as in EU:	
Waste for recovery Annexes III, IIIA and IIIB > 20 kg	information requirements Art. 18	information requirements Art. 18	information requirements Art. 18	information requirements within EU Art. 18; Art. 38	information requirements within EU, Art. 18, but special provisions according to Art. 37[1]
Waste for recovery Annexes IV und IVA	permitted notification accordingto Art. 4 to 17	prohibited, with exceptions[2] notification according to Art. 43 to 46	permitted notification according to Art. 48	permitted notification according to Art. 38	not applicable
Waste for laboratory analysis ≤ 25 kg	information requirements Art. 3 (4), Art. 18	information requirements Art. 3 (4), Art. 18	information requirements Art. 3 (4), Art.18	not applicable	
Hazardous waste for recovery according to Annex V	not applicable				prohibited

1)   according to Commission Regulation No 1418/2007 including issued corrections and amendments

2)   the import from countries to which the OECD Decision applies, Parties to the Basel Convention and countries with which bilateral agreement exist is permitted

- Parts 2 and 3 of Annex V apply only if a waste is not listed in either List A or List B of Part 1.

  If a waste is identified as hazardous (by an asterisk) in Part 2 of Annex V or is listed in Part 3 of Annex V, its export to countries where the OECD Decision does not apply is prohibited.

  If a waste is not marked with an asterisk in Part 2 of Annex V, its export to countries where the OECD Decision does not apply is potentially permitted.

  If a waste is not listed in Part 2 or 3 of Annex V, its export to countries where the OECD Decision does not apply is potentially permitted and is subject to the procedure of prior written notification and consent.

This means that exports of waste to countries where the OECD Decision does not apply are potentially permitted if the waste is listed in Part 1, List B, or its export is not otherwise prohibited by listing in Annex V.

Fig. 2: Decision Tree for the Export Prohibition for Hazardous Wastes Listed in Annex V to the EU-WSR (EU-specific)

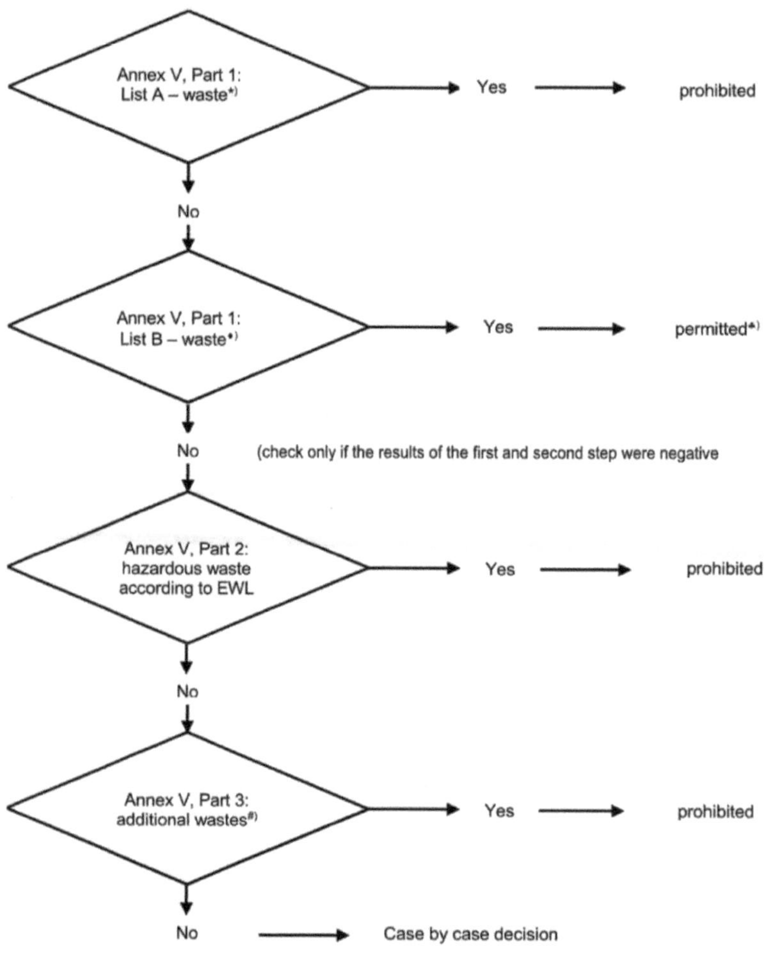

*)   hazardous waste listed in Annex VIII to the Basel Convention
♦)   non-hazardous waste listed in Annex IX to the Basel Convention
♣)   taking into account export prohibitions and authorisation requirements laid down in Commission Regulation (EC) No 1418/2007, including corrections and amendments
#)   Y46, Y47, Y48, AA010, AA060, AA190, AB030, AB070, AB120, AB150, AC060, AC070, AC080, AC150, AC160, AC170, AD090, AD100, AD120, AD150, RB020
no export ban for: AB 130, AC250, AC260, AC270

An export ban also applies in the two cases mentioned above if the waste is exceptionally classified as hazardous in the EU Member State of dispatch (export) with reference to EU-criteria in accordance with Art. 36 (4) and (5) of the EU-WSR.

4.4.5.3   *Special Regulations for Shipments of Green-listed Waste for Recovery to third Countries (EU specific)*

(Art. 37 EU-WSR in conjunction with EU Regulation 1418/2007)

In the case of shipments of green listed wastes for recovery (Annex III, IIIA to the EU-WSR) from the EU to a country to which the OECD Decision does not apply, special regulations must be observed. The regulations based on a survey of third countries by the Commission are laid down in the EU Regulation (EC) No. 1418/2007, which is adapted from time to time. It specifies in tabular form, depending on the country of destination and type of waste, whether the export of a "green" waste from the EU to the respective third country

- is prohibited (column a),

- can be shipped with prior written notification and consent as described in Art. 35 of EU-WSR (column b),

- can be shipped without control in the country of destination (column c) or

- other control procedures will be followed in the country of destination under applicable national law (column d).

For those third countries that have not responded, i.e. are not listed in the EU Regulation 1418/2007, the procedure of prior written notification and consent applies (Art. 37 (2) of the EU-WSR).

With regard to these EU Regulation, an overview is available which lists the provisions for each individual state in alphabetical order. This list of states that includes the name, ISO country code and includes information on when a country became a party to the Basel Convention and whether it is a member country of the EU, the EFTA or the OECD.

With the proposed amendment of Regulation 1013/2006, the export of Green List wastes to non-OECD countries will receive a completely new basis. According to the amendment, exports will only be possible to countries on a list drawn up by the Commission (cf. sect. I.3.3.1.1.).

4.4.6 Transit of Waste

(Art. 6 (4) BC; Art. 47 and 48 EU-WSR)

According to Art. 5 (1) of the BC and Art. 53 of the EU-WSR every country/state has to designate one competent authority to receive the notification in case of transit of waste. If a State of transit is not Party to the Basel Convention, the competent authority of such State must nevertheless be notified of the proposed transit of waste in the same way as if it were a Party to the Convention (Art. 7 of the BC). Although not explicitly required by the Basel Convention, many countries require that the transit of waste shall not be allowed to proceed, until the competent authority of such State of transit has given written consent to the movement.

Some Countries not Party to the Basel Convention have provided the Secretariat of the Basel Convention with information on focal points and/or competent authorities who should be contacted in case of intended transit of waste through their territory. Information on the competent authorities responsible for the control of transfrontier shipments of wastes under the Basel Convention is available from the secretariat of the Basel Convention and from their website.[47] With regard to Countries that have not provided such a contact point, a relevant government authority to be contacted is normally the Ministry of the Environment of these Contrives or the Ministry of Foreign Affairs.

In cases where waste for recovery is shipped from one OECD Member Country to another and where a transit third country is passed through, the shipment is to be treated in accordance with the relevant international law and any applicable national legislation. For example, if the country of transit is a Party to the Basel Convention, the transit of waste must be notified and authorised in accordance with the provisions of the Basel Convention. Non-parties to the Basel Convention should also be notified of transit of waste. According to the OECD Decision Recommendation C(86)64(Final), transit through a non-OECD country is prohibited without prior notification of the planned shipment. It is recommended that transit should not be allowed until the competent authority of that country has given its consent to the shipment.

In the case of a shipment of waste between EU Member States with transit through a third country, a copy of the notification must be sent to the competent transit authority of the third country (e.g. Switzerland, in the case of a shipment from Italy to Germany) in accordance with Art. 48 of the EU-WSR. If the third country is a Party to the Basel Convention, it must be taken into account that the decision deadlines are extended from 30 days to 60 days. If the transit country is not a party to the Basel Convention, the authorities have to agree on a time limit.

---

[47]   http://www.basel.int/Countries/CountryContacts/tabid/1342/Default.aspx

4.4.6.1 *Shipments of Waste destined for Disposal between EU Member States with Transit through a third Country (EU-specific)*

(Art. 47 EU-WSR)

Concerning waste destined for disposal shipped through EU Member States from and to third countries all consents from competent authorities of transit outside the EU must in principle be given in writing[48] according to Art. 35 (4) a) of the EU-WSR. For such authorities in the EU, however, the possibility of tacit consent pursuant to Art. 9 (1) sub-para 2 of the EU-WSR remains unaffected.

4.4.6.2 *Shipments of Waste for Recovery between EU Member States with Transit through a third Country (EU-specific)*

(Art. 48 EU-WSR)

Concerning the transit of waste destined for recovery, the following cases must be distinguished:

1. In the case of transit of waste through EU Member States from and to countries to which the OECD Decision does not apply, the procedure of prior written notification and consent applies to amber waste (Annex IV and IVA to the EU-WSR) and waste not on the green list (III, IIIA and IIIB of the EU-WSR). For green waste, the general Art. 18 information requirements apply.

2. In the case of transit through Member States from and to countries to which the OECD Decision applies, the provisions of Title II of the EU-WSR shall apply mutatis mutandis. In the case of competent authorities of transit in the EU, their tacit consent shall be sufficient. For competent authorities of transit in third countries to which the OECD Decision applies, tacit consent may be sufficient.

Where waste destined for recovery is shipped through EU Member States from a country to which the OECD Decision does not apply to a country to which the OECD Decision applies or vice versa, No. 1 shall apply as regards the country to which the OECD Decision does not apply and No. 2 shall apply as regards the country to which the OECD Decision applies.

---

[48] Until to 2020, no Basel Party has specified a deviating regulation to the Secretariat of the Basel Convention.

4.4.7 Pre-consented Recovery Facilities *(OECD/EU-specific)*

(Chapt. II D.(2) case 2 OECD-Decision; Art. 14 EU-WSR)

In order to simplify and accelerate the notification procedure, provisions have been included in the OECD Decision and the EU-WSR to identify pre-consented recovery facilities.

In the case of regular transfrontier shipments of certain types of waste to such a recovery facility, the competent authorities do not raise objections. The approval period for shipments to a pre-consented recovery facility can be extended from one calendar year to three calendar years. Furthermore the decision-making periods of the authorities in the notification procedure involving a pre-consented recovery facility are reduced from 30 days to 7 days. However, according to Art. 14 (5) of the EU-WSR, this is not an exclusion period. The transit authority can extend the deadline to a maximum of 30 days if it requires additional information and documents for the examination.

The OECD Decision and the EU-WSR does not specify the term "specific recovery facility" (pre-consented recovery facility) to which the pre-consent is linked. According to the interpretation of the German Enforcement Guidance [7], such specific recovery facilities do not include, in particular, facilities in which interim recovery (Procedures R12 and R13) takes place. In applying Art. 14 of the EU-WSR, it is important whether the waste is destined for recovery and not whether a facility may have a status as a "recovery facility".

The informal application for pre-consent according to Art. 14 of the EU-WSR has to be sent by the recovery facility in the respective OECD/EU country to the responsible competent authority. The pre-consents may be given for a specific period of time and can be revoked at any time. Competent authorities must inform, normally through the focal point of the member country concerned, the OECD Secretariat of any pre-consent they grant to their recovery facilities, and of any revocations of pre-consents. The OECD Secretariat makes this information available on its website[49] in the form of an Excel table.

Even if a pre-consent for a recovery facility is submitted, the shipment of waste to this facility is still subject to the procedure of prior written notification and consent.

---

[49] see OECD webpage:
http://www.oecd.org/environment/waste/theoecdcontrolsystemforwasterecovery.htm

4.5 Procedural Rules on Information Requirements *(EU-specific)*

(Art. 18 and Annex VII EU-WSR)

The following provides the necessary explanations for completing the Annex VII document. The Annex VII document is intended to accompany a shipment of waste at all times[50] from the moment the shipment starts in the country of dispatch (export) to its arrival at a recovery facility or laboratory in another country. The instructions are applicable to shipments of waste as long as the waste shipped is located within the EU, i.e. the Annex VII document relates to waste shipments within the EU, and from, into or through the EU. Relevant parties involved in the shipment (the person who arranges the shipment, carriers, the consignee (importer) where applicable, and the relevant recovery facility or laboratory) are to sign the document either upon delivery or receipt of the waste concerned.

A planned shipment subject to Art. 18 of the EU-WSR may take place only after the Annex VII document has been fully completed pursuant to the EU-WSR[51].

The Annex VII document should be completed either in typed format or by using capital letters in permanent ink throughout. Signatures should always be written in permanent ink and the name of the authorised representative should accompany the signature in capital letters. In the event of a minor mistake, a correction can be made by the person who arranges the shipment. The correction should be marked and signed or stamped, and the date of the modification noted. For major changes or corrections, a new document should be completed.

The Annex VII document has also been designed to be easily completed electronically. In such cases, appropriate security measures should be taken to prevent any misuse of the document, such as converting into an unalterable electronic format. Any changes made to a completed document should be visible. An Annex VII document may accompany the transport in an electronic form with a digital signature, if it can be read at any time during the transport and if this is acceptable to the competent authorities concerned.

To simplify translation, a code rather than text is required for the completion of some blocks. Where text is required, however, it should be in a language

---

[50] The person arranging the shipment is to ensure that the shipment is accompanied by the Annex VII document.

[51] It is noted that, according to paragraph 13 of Correspondents' guidelines No. 10, there may be an illegal shipment pursuant to Article 2(35)(g)(iii) of the EU-WSR if the Annex VII document is missing or if important information in this document is missing, including the signature in block 12, or is otherwise not correct.

acceptable to the authorities of all countries involved[52].

A six-digit format should be used to indicate the date. For example, 29 January 2021 should be shown as 29.01.21 (Day.Month.Year).

Where annexes providing additional information are attached to the Annex VII document, these should include clear references (e.g. enumerated references) pointing to the relevant parts of the Annex VII document as well as a citation of the block(s) to which they relate.

4.5.1 Instructions for Filling in the individual Blocks of Annex VII[53]

Before the actual start of the shipment, the person who arranges the shipment or its representative[54] is to complete blocks 1 to 12 and sign block 12, with the exception of block 5. At the time of taking possession of the consignment, the respective carrier or its representative[55] is to complete block 5. The consignee (importer) is to complete and sign block 13 in the event that this is not the recovery facility or the laboratory and when it takes charge of a shipment of waste after it arrives in the country of destination. The recovery facility or the laboratory is to complete block 14 after receipt of the waste.

Block 1: Provide the name, address and all other required details of the person who arranges the shipment. The address should include the name of the country and telephone and fax numbers including the country code. Provide the phone and fax numbers and the e-mail address, which should facilitate the contact of all relevant persons regarding an incident during the shipment. If a fax number cannot be provided write N/A. Where the person who arranges the shipment is a legal person such as a company, the name of the company should be filled-in under "Name" and information of an authorised person who can give additional information if needed should be added under "Contact person". The contact person should be responsible for the shipment including any incidents that may occur during the shipment. In case the person who arranges the shipment is a natural person, no contact person may need to be added.

Block 2: Provide the required information. Normally, the consignee would be the recovery facility or laboratory given in block 7. In some cases, however, the consignee may be another person, for example a dealer, a broker, or a corporate body, such as the headquarters or a mailing address of the receiving recovery facility in block 7. In order to act as a consignee, a dealer, broker or corporate body must be under the jurisdiction of the country of destination and possess or have some other form of legal control over the waste at the moment the

---

[52] cf. Art. 27 of the EU-WSR.

[53] These instructions have been taken from the Correspondents" Guidelines No. 10

[54] A "representative" is a person employed and authorised in writing by the person who arranges the shipment.

[55] A "representative" is a person employed and authorised in writing by the carrier.

shipment arrives in the country of destination[56]. The country appearing in the address of this block should be the same as that of block 7.

Block 3: Give the actual weight of the waste in tonnes (1 tonne equals 1 megagram (Mg) or 1000 kg.

Block 4: Enter the date when the shipment actually starts.

Block 5 (a-c): The information and signature required in block 5 should be provided by each carrier or carrier's representative when taking possession of the consignment. The address should include the name of the country, and telephone and fax numbers should include the country code. If a fax number cannot be provided write N/A. When more than three carriers are involved, appropriate information on each carrier should be attached to the Annex VII document.

Block 6: Provide the name, address and all other required information of the "waste generator"[57]. If the waste generator is the same as the person who arranges the shipment, then write "Same as block 1". If the waste has been produced by more than one generator, write "See attached list" and append a list providing the requested information for each generator (see footnote 3 of the Annex VII document). A "waste generator" may include the original producer, a new producer or a licenced waste collector. In case the generator is unknown, the name of the person in possession or control of such waste (holder) should be provided.

Block 7: Provide the name, address and all other required information (give destination of the shipment by ticking either recovery facility or laboratory). The address should be the actual address (i.e. no P.O. Box). The country appearing in the address of this block should be the same as that of block 2. If the recovery facility or laboratory is also the consignee, state here "Same as block 2".

Block 8: In case "Recovery facility" is ticked in block 7, indicate the type of recovery operation by using R-codes of Annex II to Directive 2008/98/EC on waste or, in case "Laboratory" is ticked in block 7, the type of recovery or disposal operation by using R-codes or D-codes of Annexes I and II to Directive 2008/98/EC on waste.

Block 9: Give the name or names by which the material is commonly known or the commercial name. In the case of a mixture of wastes listed in Annex IIIA to the EU-WSR, provide the same information for the different fractions.

Block 10: Fill in the code or codes that identify the waste according to Annexes III, IIIA or IIIB to the EU-WSR in the subheadings as indicated below. A code corresponding to one of the four following categories is to be specified in block 10:

---

[56] Some Member States have a stricter approach as to when such other person may act as consignee, for example, that the consignee must have physical control over the waste at the moment the shipment arrives in the country of destination. See also the reply to question 5.2 in the Frequently Asked Questions (FAQs) on Regulation (EC) No 1013/2006 on shipments of waste (available at http://ec.europa.eu/environment/waste/shipments/pdf/faq.pdf

[57] As regards the completion of this block, see the Court judgement on Case C-1/11 (see: http://curia.europa.eu/juris/document/document.jsf?text=&docid=121166&pageIndex=0&doclang=EN& mode=lst&dir=&occ=first&part=1&cid=839361).

I. Explanatory Notes

Subheading (i):

Basel code(s) from Basel Convention Annex IX which are listed in Part I of Annex III to the EU-WSR should be provided (see also List B in Part 1 of Annex V to the EU-WSR).

Subheading (ii):

OECD codes should be used for wastes listed in Part II of Annex III to the EU-WSR, i.e. wastes that have no equivalent listing in Annex IX to the Basel Convention or that have a different level of control under the WSR from the one required by the Basel Convention.

Subheading (iii):

In case of mixtures of wastes listed in Annex IIIA to the EU-WSR, provide the relevant code(s) as indicated in Annex IIIA as appropriate in sequence (cf. footnote 4 of Annex VII). Certain Basel entries such as B1100, B3010 and B3020 are restricted to particular waste streams only, as indicated in Annex IIIA.

Subheading (iv):

In case of waste listed in Annex IIIB to the EU-WSR, provide the BEU codes listed in Annex IIIB (cf. footnote 5 of Annex VII). These codes are only valid in the EU and are to be used for shipments within, into or through the EU.

In addition, a code corresponding to the following three categories should be specified in block 10:

Subheading (v):

Provide the codes included in the European Waste List (see Commission Decision 2000/532/EC as amended)[58].

Subheadings (vi):

Where applicable, national identification codes (other than the codes of the European list of waste) used in the country of dispatch and, if known, in the country of destination should be provided.

Subheadings (vii):

In case plastic waste (EU 3011) is shipped destined for recovery provide the code EU 3011. The code is only valid in the EU and is to be used for shipments within, into or through the EU.

Block 11: Provide the name of the countries of dispatch, transit and destination or the codes for each country by using the ISO standard 3166 abbreviations[59].

Block 12: At the time of shipment, the person who arranges the shipment or its representative[60] should fill in his/her name and sign and date the Annex VII document. In case the person who arranges the shipment is a legal person, its representative that signs block 12 may differ from its representative that signs the contract but they should sign on behalf of the same legal entity.

---

[58] See http://ec.europa.eu/environment/waste/framework/list.htm and http://ec.europa.eu/environment/waste/legislation/a.htm.

[59] See https://www.iso.org/obp/ui/#search/code/.

[60] A "representative" is a person employed and authorised in writing by the person who arranges the shipment.

Fig. 3: **Information to be supplied for Shipments of Waste referred to in Art. 3 (2) and (4) - Document according to Annex VII of the EU-WSR**

CONSIGNMENT INFORMATION (¹)

1. Person who arranges the shipment:	2. Importeur / Consignee:
Name:	Name:
Adress:	Adress:
Contact person:	Contact person:
Tel.: Fax:	Tel.: Fax:
E-Mail:	E-Mail:

3. Actual quantity: Tonnes (Mg): m³:	4. Actual date of shipment:

5.(a) First Carrier (²)	5.(b): Second Carrier	5.(c): 3. Third Carrier
Name:	Name:	Name:
Adress:	Adress:	Adress:
Contact person:	Contact person:	Contact person:
Tel.:	Tel.:	Tel.:
Fax:	Fax:	Fax:
E-Mail:	E-Mail:	E-Mail:
Means of transport:	Means of transport:	Means of transport:
Date of transfer:	Date of transfer:	Date of transfer:
Signature:	Signature:	Signature:

6. Waste generator (³) Original producer(s), new producer(s) or collector:	8. Recovery operation (or if appropriate disposal operation in case of waste referred to in Article 3(4)):
Name:	R-code / D-code:
Adress:	
Contact person:	9. Usual description of the waste:
Tel.: Fax:	
E-Mail:	

7. Recovery facility ☐ Laboratory ☐	10. Waste identification (fill in relevant codes):
Name:	i) Basel Annex IX :
Adress:	ii) OECD (if different from (i)):
Contact person:	iii) Annex IIIA (⁴):
Tel.: Fax:	iv) Annex IIIB (⁵):
E-Mail:	v) EU list of wastes:
	vi) National code:
	vii) Other (please specify):

11. Countries / State(s) concerned:

Export / Dispatch	Transit	Import / Destination

12. Declaration of the person who arranges the shipment: I certify that the above information is complete and correct to my best knowledge. I also certify that legally-binding written contractual obligations have been entered into with the consignee (not required in case of waste referred to in Article 3(4)):

Name: Date: Signature:

13. Signature upon receipt of the waste by the consignee:

Name: Date: Signature:

TO BE COMPLETED BY THE RECOVERY FACILITY OR BY THE LABORATORY:

14. Shipment received at recovery facility: ☐ or laboratory: ☐	Quantity received: Tonnes (Mg): m³:
Name: Date:	Signature:

(¹) Information accompanying shipments of green listed waste and destined for recovery of waste destined for laboratory analysis pursuant to Regulation (EC) No 1013/2006. For completing this document, see also the corresponding specific instructions as contained in Annex IC of Regulation (EC) No 1013/2006.
(²) If more than 3 carriers, attach information as required in blocks 5 (a), (b), (c).
(³) When the person who arranges the shipment is not the producer or collector, information about the producer or collector shall be provided.
(⁴) The relevant code(s) as indicated in Annex IIIA to Regulation (EC) No 1013/2006 are to be used, as appropriate in sequence. Certain Basel entries such as B1100, B3010 and B3020 are restricted to particular waste streams only, as indicated in Annex IIIA.
(⁵) The BEU codes listed in Annex IIIB to Regulation (EC) No 1013/2006 are to be used.

Block 13: This block is to be completed and signed by the consignee, shown in block 2 in case the consignee is neither the recovery facility not the laboratory and in case the consignee takes charge of the waste after the shipment arrives in the country of destination.

Block 14: This block is to be completed and signed by a representative[61] of the recovery facility or the laboratory upon receipt of the waste consignment. Tick the box for either recovery facility or laboratory. Give the quantity of the waste received in tonnes (1 tonne equals 1 megagram (Mg) or 1000 kg).

4.5.2 Disclosure of the Waste Producer

According to the ECJ ruling[62] of 29 March 2012, article 18(4) of WSR must be interpreted as not permitting an intermediary dealer[63] arranging a shipment of waste not to disclose the name of the waste producer to the consignee of the shipment, as provided for in Article 18(1) of WSR in conjunction with Annex VII to that regulation, even though such non-disclosure might be necessary in order to protect the business secrets of that intermediary dealer.

Article 18(1) of WSR must be interpreted as requiring an intermediary dealer, in the context of a shipment of waste covered by that provision, to complete Field 6 of the document contained in Annex VII to WSR, and transmit it to the consignee, without any possibility of the scope of that requirement being restricted by a right to protection of business secrets.

4.5.3 Jurisdiction and Residence Requirement

According to the Correspondents Guidelines' No. 10,[64] there are different views in the EU Member States on the question of who is subject to the jurisdiction of the country of export/dispatch. The decision on this has been left to the individual EU Member States. It is important that the decision of the country of export/dispatch is recognised by all other countries concerned.

For example in Germany's federal system, the enforcement authorities have differing views on how this should be interpreted. The spectrum ranges from "all are subject to jurisdiction" to the view that the initiator must have a place of business (residence requirement) in the country of export/dispatch.

Historical, one finds that there was no regulation or special definition for the term "notifier" in the EC Waste Shipment Regulation (1993). According to

---

[61] A "representative" is a person employed and authorised in writing by the recovery facility or the laboratory.

[62] Case C – 1/11 - Interseroh Scrap and Metals Trading GmbH

[63] In German: Streckenhändler

[64] cf. Correspondents' Guidelines: https://waste-move.eu/?p=539&lang=en

this definition, a "notifier" was any person who was obliged to notify. Requirements for this person did not exist and the right to notify was left open.[65]

The new definition of notifier was introduced with the entry into force of the EU-WSR in 2007 and should be seen against the background of the Basel Convention, which already contains a definition for "area under the national jurisdiction of a state". Furthermore, the definition states that it means "any land, marine area or airspace within which a State exercises administrative and regulatory responsibility in accordance with international law in regard to the protection of human health or the environment".

The use of the concept of sovereignty in the Basel Convention was intended to avoid any problems that might have arisen, for example, with the use of the concept of territory, and to allow free navigation and air traffic for transit. If the term territory had been used, controls on shipping and air traffic would have been necessary [4].

The use of the concept of sovereignty and in particular the inclusion of Art. 4 (12) in the Basel Convention took account of this:

"Nothing in this Convention shall affect in any way the sovereignty of States over their territorial sea established in accordance with international law, and the sovereign rights and the jurisdiction which States have in their exclusive economic zones and their continental shelves in accordance with international law, and the exercise by ships and aircraft of all States of navigational rights and freedoms as provided for in international law and as reflected in relevant international instruments".

The EU-WSR contains only the term "under the jurisdiction of" and not the term "sovereignty". The question therefore arises as to why the EU-WSR does not also speak of the sovereignty of a country as in the Basel Convention, but instead uses the term "jurisdiction" and above all, what meaning this has for transfrontier shipments in according with Art. 18 of the EU-WSR.

The term jurisdiction is understood to mean the exercise of the administration of justice, in particular the administration of justice. In terms of subject matter, jurisdiction is limited to the extent that only those matters fall within its remit for which legal recourse is open. Any person who is subject to e. g. German state jurisdiction is therefore subject to German jurisdiction, which is therefore geographically limited to the national territory. This does not exclude a conviction for performance or omission abroad or the enforcement of an act abroad through the use of coercion in Germany.[66]

In principle, all shipments from the country of export/dispatch are subject to

---

[65] Bert-Axel Szelinski, Schneider, S.: Grenzüberschreitende Abfallverbringungen, Behr's Verlag, Hamburg 1995, S. 29.

[66] Schoch/Schneider/Bier, VwGO § 40 Rn. 30.

the jurisdiction of that country within the meaning of Art. 18 (1) of the Basel Convention. If in one country, persons who do not have a residence or place of business in that country, but who carry out or wish to carry out activities there, such as arranging a transfrontier shipment, are also subject to the jurisdiction of the respective competent authority of the exporting country. This means that authorities or that country can also take sovereign action vis-à-vis such persons, in particular by issuing favourable or onerous administrative acts, such as the granting of permits for the transport of hazardous waste to waste transporters who are only resident abroad.

4.5.3.1   Residence Requirement

Whether only the sovereign competence or, a residence and place of business requirement can be derived from the term "jurisdiction" is a matter of dispute.[67/68]

The European Commission is of the opinion that there is no need for the notifier/initiator to have a place of business in the country of export/dispatch. According to the understanding of the territoriality principle, the concept of "natural or legal person under the jurisdiction of a Member Country" gives the respective Member Country the power to exercise its jurisdiction also where a shipment has taken place. It is not necessary for the notifier/initiator to be established in the country of export/dispatch. From the definition of the notifier/initiator as a natural or legal person under the jurisdiction of a Member Country, all the necessary powers for a Member Country can be derived. A residence requirement is not necessary for this exercise, for example the issuing of an administrative act.

The legislative history and the Freedom to Provide Service (Art. 56 to 58 TFEU) within the EU speak for the fact that the definition of notifier/initiator does not imply a requirement for a residence, because when interpreting the term jurisdiction from an EU regulation, it is not possible to fall back on national legal provisions. For example, according to Rogusch-Sießmayer, the German Administrative Procedure Act cannot be used, as it only applies in Germany.[69]

However, before initiating a transfrontier shipment, it is advisable to make enquiries as to what view the competent authority at the place of export/dispatch takes, be it the residence requirement or the existence of a broker's licence in the country of export/dispatch. This is mainly to avoid

---

[67]   For residence requirement: Kropp, AbfallR 2012, p. 12 f.
[68]   Against residence requirement: Rogusch-Sießmayr, AbfallR 2012, p. 58.
[69]   See footnote above

problems during the transfrontier shipment, for example with the control authorities.

4.6 Procedural Rules of Prior Notification and Consent

(Art. 8 BC; Chapt. II D. OECD Decision; Title II, Chapt. 1 EU-WSR)

The prior written notification and consent procedure applies to all transfrontier shipments of waste that are not excluded from notification. This procedure is subject to pre-shipment requirements (prior to the shipment of waste) and destination control requirements (for each shipment of waste).

Furthermore, in the EU additional requirements are In place for shipments of waste destined for interim disposal or recovery (disposal operations D 12 to D 15 and R 12 to R 13, according to Art. 15 of the EU-WSR) and partly different requirements for shipments of waste to a recovery facility with pre-consent, according to Art. 14 of the EU-WSR (cf. sect. I.4.4.7).

4.6.1 Who can notify? *(EU-specific)*

The notifier must notify the competent authority in his country of origin of the planned shipment of waste by means of a notification document and other required documents. Pursuant to Art. 2 No. 15 of the EU-WSR, the notifier may be

- the initial producer,
- the authorised new producer
- the authorised collector
- a registered dealer or registered broker, or
- the holder of the waste.

The Enforcement Guidance issued by LAGA assumes that first producers, new producers and collectors are equally eligible. This takes account of the consideration in recital 18 of the EU-WSR that traders and brokers can only act as notifier if they are registered and commissioned or authorised.

4.6.2 Notification

Individual and general notifications are possible. The general notification may be used if waste destined for disposal or recovery with essentially the same physical and chemical properties is to be shipped regularly by the same route to the same importer/consignee. A general notification may cover a shipment period of up to one year.

Each notification may only contain one waste code according to the waste lists of the BC, the OECD Decision or the EU-WSR. If none of these codes is applicable, the waste is considered unlisted. In this case, the type of waste shall be adequately described in block 12 or by attaching an annex.

Transfrontier shipments of waste are only permitted if the competent authorities of export/dispatch and import/destination have given their consent in writing, and any competent authorities of transit have given their consent at least tacitly. The consent of all authorities must be given cumulatively and prior to the shipment. If, for example in the OECD/EU, the written consent of the authority of export/dispatch is not available, the shipment of waste is not permitted, even if the written consents of the authority of import/destination and all authorities of transit are at least tacitly available.

Detailed questions concerning a specific notification should be discussed by the notifier with the competent authority before the written application is submitted. This also applies to all documents required by competent authorities.

4.7 Notification Documentation to be submitted

(Annex VA and VB of BC; Appendix 8 of OECD Dec.; Annex II Part 1, 2 and 3 of EU-WSR)

The necessary notification documentation shall be submitted by the exporter/notifier according to the need of the competent authority of export/dispatch. This documentation includes above all

- notification and movement document

- contract between exporter/notifier and importer/consignee, this may include

  - a brokerage agreement - if a broker organises the shipment on behalf of the waste holder.

- financial guarantee (cf. sect. I.4.9.4), this may include

  - information on the calculation of the guarantee

- description of transport route; this may include information on

  - transport duration and transport distance

  - measures required to ensure transport safety

  - the port's authorisation for the transhipment of waste (if the waste is transhipped in a port), and

- in case of combined transport, indication of the place where the transhipment takes place from the point of generation to the exporter/disposer.

- proof of liability and environmental liability insurance for the means of transport used for the movement,

- description of the waste generation process

Competent authorities in the <u>OECD and the EU in case of shipments of waste for interim disposal or recovery</u> will ask for

- an indication of all facilities where subsequent recovery or disposal takes place, and a

- contract between the interim and non-interim facility, if applicable.

If a <u>waste is destined for recovery competent authorities in the EU</u>, according to Art. 12 (1) g of the EU-WSR will ask for

- Planned method of disposal of the non-recoverable fraction after recovery,

- quantity of recovered materials in relation to the non-recoverable waste,

- estimated value of the recovered materials,

- cost of recovery and disposal of the non-recoverable fraction.

Furthermore documentation may be requested from the competent authorities, especially in the EU on

- chemical analysis and/or composition of the waste

- copy of the IPPC Facility Permit (Directive 96/61/EC, IPPC Directive)

- Permit documents (type and period of validity) of the disposal facility

- Description of the treatment process in the facility receiving the waste

- any other information relevant to the assessment of the notification under the EU-WSR and national legislation.

4.7.1 Special Features of Interim Disposal Procedures

(Chapt. II D.(6) OECD Decision; Art. 2, Nos. 5 and 15 of EU-WSR)

Some of the listed operations in Annex IVA and IVB of the BC, Appendix 5.A and 5.B to the OECD Decision or Annex I and II to the EU-WFD are to be considered as "intermediate or temporary operations", that is, after these operations wastes still need to undergo further treatment before the last disposal operation takes place. These operations are:

- R12   exchange of wastes for submission to any of the operations numbered R1-R11, and

- R13   accumulation of material intended for any recovery operations,

and <u>in the EU</u> in addition:

- D13   blending and mixing prior to submission to any of the disposal operations,

- D14   repackaging prior to submission to any of the disposal operations,

- D15   storage pending any of the disposal operations.

In the case where the transfrontier shipment of wastes takes place in order to undertake recovery operations R12-R13 or disposal operations D13-D15, the competent authorities may require that the subsequent intended disposal or recovery operation(s) should be specified on the notification as additional information. The competent authority may decide not to authorize the proposed movement of waste if it is not convinced that the waste will be disposed of in an environmentally sound manner at its final destination.

<u>*(EU specific)*</u>

The regulations of the EU-WSR on interim recovery (R12 and R13) originate from the OECD Decision. The EU has enlarged the stipulation to disposal facilities. According to the EU-WSR, interim disposal operations are D13 to D15 and the interim recovery operations are R12 and R13 as defined in the EU-WFD. Typical interim processes used in practice, unless another D or R code is relevant, may include:

- dismantling/sorting/classification/separation corresponds to D13 /R12,

- shredding/shredding/crushing corresponds to D13/R12,

- palletising/densification corresponds to D13/R12,

- disassembling/mixing/blending corresponds to D13/R12,

- repackaging corresponds to D14/R12,

- storage corresponds to D15/R13.

All operations, i.e. also the subsequent interim and non-interim recovery or disposal operations are part of the review by the competent authorities. This means that the entire disposal route, including the subsequent interim and non-interim recovery or disposal, must be presented in the respective notifycation (pursuant to Art. 15 lit. a, and Art. 4 No.2 sub-para 1 and 2 of the EU-WSR) and examined in the procedure. In the case of recovery or disposal operations, e.g. sorting, all wastes generated and their further disposal routes must be presented. This additional work should be taken into account when preparing the application.

In particular, processes D8 (biological treatment), D9 (chemical-physical treatment), R3 (recycling of organic materials) and R5 (metal recycling) are not interim processes within the meaning of the definition of the EU-WSR. These processes can also be followed by other processes mentioned in the lists without being classified as interim disposal or recovery.

The verification of individual shipments and the documentation of disposal also differ in interim processes from shipments in non-interim processes. This is because the facility carrying out the first interim recovery or disposal operation must obtain a certificate of final disposal of the waste pursuant to Art. 15 lit. e of the EU-WSR from the facility carrying out the non-interim recovery or disposal operation.

Since the movement document for the documentation only contains a block for the certification of the completion of the first interim recovery or disposal operation, the movement document remains with the operator of this facility. With regard to the certificates to be submitted by the facility carrying out the first interim recovery or disposal operation, reference is made to Correspondents' Guidelines No. 3.[70]

4.8 Competent Authorities

(Art. 5 of BC; Chapt. II A. No. 12 OECD Decision; Art. 53 and 54 of EU-WSR)

The countries must designate competent authorities responsible for the enforcement of the BC, the OECD-Decision or the EU-WSR. In addition each country according to the BC and the EU-WSR has to designate a focal point, which acts as an information and advice point for the economy, but in particular also as a contact between the Countries and in the EU also to the EU Commission. The competent authorities and the focal point are determined according to different aspects in the concerned countries.

As example in some EU Member States, responsibility is separated into export, import and transit, based on the political structure (counties, districts, regions, departments). In other EU Member States there is only one competent authority (e.g. in the Netherlands, Ireland or Poland). Addresses and

---

[70] https://ec.europa.eu/environment/pdf/waste/shipments/correspondents_guidelines3_en.pdf

contact persons can be downloaded from the webpage of the Basel Convention[71] or the EU Commission[72].

4.9 Application Procedure

The application for a transfrontier shipment of waste (notification) must be submitted by using the notification (cf. Figs. 4 and 5) and the movement documents (cf. Figs. 7 and 8) and further documentation (cf. I.4.7). The worldwide harmonised notification and movement documents also serve as a certificate of disposal/recovery.

Both documents contain blocks for official entries, such as acknowledgement of receipt, consent or stamp of the customs offices *(in the EU)*. They cover the legal instruments of the Basel Convention, the OECD-Decision and the EU-WSR.

Because the documents have been made broad enough to cover all three instruments, however, not all blocks in the document will be applicable to all of the instruments and it therefore may not be necessary to complete all of the blocks in a given case. Any specific requirements relating to only one control system have been indicated with the use of footnotes or directly in the instructions (cf. sect. 4.9.2). It is possible that national or multilateral implementing legislation may use terminology that differs from that adopted in the Basel Convention and the OECD Decision.

The terminology used in national implementing legislation may also differ from that used in the Basel Convention and the OECD Decision. For example, the EU-WSR, uses "shipment" instead of "movement". This is taken into account in the headings of the notification and movement document by placing both terms ("movement/shipment") next to each other.

The documents include both the term "disposal" and "recovery", because the terms are defined differently in the three instruments. The EU-WSR and the OECD Decision use the terms "disposal" and "recovery" for disposal or recovery operations listed in Annexes to the EU-WFD and the OECD Decision. In the Basel Convention, however, the term "disposal" is used to refer to both disposal and recovery operations.

The notification and movement documents are generally provided and issued (both in paper and electronic form, when capability exists and relevant

---

[71] http://www.basel.int/Procedures/CompetentAuthorities/tabid/1324/Default.aspx
[72] https://ec.europa.eu/environment/waste/shipments/pdf/Competent_Authorities_EN_2_Apr
_2020.pdf

legal requirements[73] are fulfilled) by the competent authorities of export/dispatch[74]. When doing so, they will use a numbering system, which allows a particular consignment of waste to be traced. The numbering system should be prefixed with the country code that can be found in the ISO standard 3166 abbreviation list.

Within the EU, the two-digit country code must be followed by a space. This may be followed by an optional code of up to four digits specified by the competent authority of export/dispatch followed by a space. The numbering system must end with a six-digit number, the so-called notification number. For illustration, if the country code is DE (for Germany) and the six-digit number 123456, the notification number would be DE 123456 if no optional code were specified. Where an optional code, for example 1005, is specified, the notification number would be XY 1005 123456. However, in case a notification or movement document is transmitted electronically and no optional code is specified, '0000' should be inserted instead of the optional code (e.g. DE 0000 123456); in case an optional code of less than four digits is specified, for example 12, the notification number would be DE 0012 123456.

Countries may wish to issue the documents in a paper size format that conforms to their national standards (normally ISO A4, as recommended by the United Nations). In order to facilitate their use internationally, however, and to take into account the difference between ISO A4 and the paper size used in North America, the frame size of the forms should not be greater than 183 X 262 mm with margins aligned at the top and the left side of the paper.

The notification document (block 1 to block 21 including footnotes) should be on one page and the list of abbreviations and codes used in the notification document should be on a second page. With regard to the movement document, block 1 to block 19 including footnotes should be on one page and block 20-22 and the list of abbreviations and codes used in the movement document should be on a second page.

Detailed instructions on how to fill in the documents are given further down.

(EU-specific)

In the EU like in the BC the application must be made via the competent authority at the place of export/dispatch (Art. 4 of the EU-WSR). The only exception is the import of waste for recovery from third countries to which the OECD Decision applies (Art. 44 (2) of the EU-WSR).

The application for a shipment of waste must be properly executed by the

---

[73] In the EU see for example Article 26, Section 4 of the Waste Shipment Regulation.
[74] In Germany, for example, the document can be obtained from the competent authorities or, among others, also from specialist publishers or specialised stationery shops.

notifier, i.e. submitted in full to the competent authority of export/dispatch. Completeness is defined in the form of so-called "mandatory information" in Art. 4 and Annex II Part 1 to the EU-WSR: These "mandatory information" include the fully completed notification document and the movement document completed as far as possible. The "mandatory information" also includes evidence of the existence of a disposal contract (cf. Art. 5 and Annex II Part 1 No. 20 to the EU-WSR) between the notifier and the consignee/importer, as well as evidence of financial guarantee covering the costs of any repatriation of the waste (Art. 6 of the EU-WSR).

It is stipulated that

- the competent authority of dispatch/export must set or approve the financial guarantee,

- only the competent authority of dispatch/export has access to the financial guarantee, and

- in cases of "illegal" or "not completed as intended" shipments, financial resources are to be made available to the other authorities concerned, if necessary.

Furthermore, according to Art. 6 (1) b) of the EU-WSR storage costs (for a maximum of 90 days) are to be charged in addition to the previously defined costs of repatriation and disposal (cf. sect. I.4.9.4).

4.9.1 General Guidance on Completing the Documents

The purpose of the notification document is to provide the competent authorities concerned with the information they need to assess the acceptability of notified shipments of waste. The document includes blocks to acknowledge receipt of the notification and to give written consent to the shipment concerned.

The movement document is intended to accompany a shipment of waste throughout its transport from the waste generator/producer to its arrival at a disposal or recovery facility in another country.

Each person who assumes responsibility for a shipment (the carrier and possibly the importer/consignee[75]) must sign the movement document either at the time of delivery or at the time of receipt of the waste in question. In addition, space is provided in the document for detailed information on all carriers of the consignment. There are also spaces in the movement document for recording passage of the consignment through the customs offices

---

[75] In the EU 'consignee' may be used instead of 'importer'.

of all countries concerned (while not strictly required by applicable international instruments, national legislation in some countries[76] requires such procedures, as well as information to ensure proper control over movement). Finally, the document is to be used by the relevant disposal or recovery facility to certify that the waste has been received and that the recovery or disposal operation has been completed.

A planned shipment subject to the procedure of prior written notification and consent may only take place after the notification and movement documents have been completed in accordance with the valid legal instruments[77] during the period of validity of the written or tacit consents of all competent authorities concerned.

For ease of translation, codes shall be entered in place of text in several blocks on the forms. Where textual information is required, a language acceptable to the competent authorities in the receiving country/state and, where necessary, to the other authorities concerned, shall be used.

Where forms are to be accompanied by annexes containing additional information, each annex shall bear the reference number of the form concerned and the number of the block to which it relates.

4.9.2 Instructions for Completing the Notification and Movement Documents[78]

Those filling out printed versions of the documents should use typescript or block capitals in permanent ink throughout. Signatures should always be written in permanent ink and the name of the authorised representative should accompany the signature in capital letters. In the event of a minor mistake, for example the use of the wrong code for a waste, a correction can be made with the approval of the competent authorities. The new text must be marked and signed or stamped, and the date of the modification must be noted. For major changes or corrections, a new form must be completed.

To simplify translation, the documents require a code, rather than text, for the completion of several blocks. Where text is required, however, it must be in a language acceptable to the competent authorities in the country of import/destination and, where required, to the other concerned authorities.

When submitting the notification, the exporter/notifier fills in blocks 1 to 18. In non-EU Member States or non-OECD Countries, the competent authority

---

[76] required in the EU
[77] Article 16(a) and (b) of the EU-WSR, Art. 6 (8) of the BC, Chapt. II D.(2) g) of the OECD-Decision
[78] These instruction are a compilation of instructions contained in [1], [2] and Annex IC of the EU-WSR.

of export may complete these boxes.

If - <u>in the EU</u> - the notifier is not the first generator/producer, the generator/producer or one of the persons referred to in Art. 2(15) (a) (ii) or (iii) of the EU-WSR, in accordance with Art. 4 (2) (1) and Annex II, Part 1, point 26 of the EU-WSR, shall also sign block 17, if practicable.

When applying for a transfrontier shipment of waste, the movement document containing the entries in boxes 1, 3, 4, 7, 8 (as far as possible at the time of application) and 9 to 14 shall be submitted in addition to the notification document. It is important to ensure that the notification number in block 3 of the notification document corresponds to that in block 1 of the movement document.

The forms have also been designed to be easily completed electronically. Where this is done, appropriate security measures should be taken against any misuse of the forms. Any changes made to a completed form with the approval of the competent authorities should be visible. When using electronic forms transmitted by e-mail, a digital signature is necessary.

4.9.2.1 *Electronic Data Exchange (EU specific)*

Concerning the electronic data exchange, the EU Correspondents adopted the "Correspondents' Guidelines on the specification of a data model for electronic data exchange under the WSR" on 19 July 2019. This Correspondents' Guidelines No. 11 shall apply in cases where the competent authorities concerned and the notifier agree to exchange information and documents by means of electronic data exchange in accordance with Art. 26 (4) sub-para 1 of the EU-WSR.

The same applies to persons and parties involved in the general information requirements under Art. 18 of the EU-WSR and to the authorities responsible for enforcing the EU-WSR, provided that the competent authorities concerned accept the exchange of information and documents by means of electronic data exchange.

The guidelines describe an EU-wide harmonised data model for the electronic data exchange of information and documents

- pursuant to Art. 26 (1) of the EU-WSR,

- according to Art. 18 of the EU-WSR,

which is to be applied to the electronic data exchange.

The EU Correspondents have agreed that the data model contained in the specification in the Annex[79] to the Correspondents Guidelines No. 11 should be used for the electronic data exchange of the above-mentioned information and documents.

4.9.2.2 Detailed Help for the Individual Boxes

A six-digit format should be used to indicate the date. For example, 29 January 2021 should be shown as 29.01.21 (Day.Month.Year).

Where it is necessary to add annexes or attachments to the documents providing additional information, each annex should include the reference number of the relevant document and cite the block to which it relates.

Blocks 1 and 2:	Provide the required information of the exporter/notifier and importer/consignee (give registration number only where applicable, full name, address including the name of the country and telephone and fax numbers including the country code; contact person and e-mail address). The phone and fax numbers and the e-mail address should facilitate contact of all relevant persons at any time regarding an incident during shipment.

In non-EU States and non-OECD countries, information relating to the competent authority of export/dispatch may be given instead of the exporter/notifier.

EU specific instructions:

In accordance with point 15 of Art. 2 of the EU-WSR the notifier/exporter may be a dealer or broker. In this case, provide a copy of the contract or evidence of the contract (or a declaration certifying its existence) between the producer/generator, new producer or collector and the broker or dealer in an annex (see Annex II, Part 1, point 23 of the EU-WSR).

OECD and EU specific instruction:

Normally, the importer/consignee would be the disposal or recovery facility given in block 10. In some cases, however, the importer may be another person, for example a recognised trader/broker, a dealer, a broker, or a corporate body, such as the headquarters or mailing address of the receiving disposal or recovery facility in block 10. In order to act as an importer, a recognised trader, dealer, broker or corporate body must be under the jurisdiction of the country of import and possess or have some other form of legal control over the waste at the moment the shipment arrives in the country of import. In such cases, information relating to the recognised trader, dealer, broker or corporate body should be completed in block 2.

Block 3[80]: When issuing a notification document, a competent authority will, according to its own system, provide an identification number which will be

---

[79] https://ec.europa.eu/environment/pdf/waste/shipments/correspondents_guidelines11_en.pdf

[80] In Germany, the form is either issued by the competent authority, created via software or obtained from a publisher. For clear identification, the publishers (including competent authorities or software) have a 4-digit "publisher number" which precedes the notification number.

printed in this block (see above).

The appropriate boxes should be ticked to indicate:

Under A, whether the notification covers one shipment (individual shipment/single notification) or multiple shipments (general notification);

Under B, give the type of operation the waste being shipped is destined for;

OECD and EU specific instructions:

Under C, if the waste is destined to a pre-consented recovery facility in an EU State (see Art. 14 of the EU-WSR) or OECD country (see chapter II, section D of the OECD Decision).

Blocks 4, 5 and 6: Give the number of shipments in block 4 and the intended date of a single shipment or, for multiple shipments, the dates of the first and last shipments, in block 6. In block 5, give the estimated minimum and maximum weight in tonnes (1 tonne equals 1 megagram (Mg) or 1 000 kg) of the waste. In some countries, giving the volume in cubic metres (1 cubic metre equals 1 000 litres) or other metric units, such as kilograms or litres, may also be acceptable. When other metric units are used, the unit of measure may be indicated and the unit in the document may be crossed out.

The total quantity shipped must not exceed the maximum quantity declared in block 5. The intended period of time for shipments in block 6 may not exceed one year, *with the exception of multiple shipments to pre-consented recovery facilities in OECD countries or EU States, for which the intended period of time may not exceed three years.*

EU specific:

*In the EU all shipments **must take place within the validity period** of the written or tacit consents of all competent authorities concerned issued by the competent authorities according to Art. 9(6) of the EU-WSR.*

In the case of multiple shipments, some non-EU States or non-OECD countries may, based on the Basel Convention, require the expected dates or the expected frequency and the estimated quantity of each shipment to be quoted in blocks 5 and 6 or attached in an annex. Where a competent authority issues a written consent to the shipment and the validity period of that consent in block 20 differs from the period indicated in block 6, the decision of the competent authority overrides the information in block 6.

Block 7: Types of packaging should be indicated using the codes provided in the list of abbreviations and codes attached to the notification document. If special handling precautions are required, such as those required by producers' handling instructions for employees, health and safety information, including information on dealing with spillage, and instructions in writing for the transport of dangerous goods, tick the appropriate box and attach the information in an annex.

Block 8: Provide the required information (give registration number only where applicable, full name, address including the name of the country and telephone and fax numbers including the country code; e-mail address and the name of a contact person responsible for the shipment). If more than one carrier is involved, append to the notification document a complete list giving the required information for each carrier. Where the transport is organised by a forwarding agent, the agent's details and the respective information on actual carriers should be provided in an annex.

Means of transport should be indicated using the abbreviations provided in the list of abbreviations and codes attached to the notification document.

EU specific instructions:

Provide evidence of registration of the carrier(s) regarding waste transports (e.g. a declaration certifying its existence) in an annex (see Annex II, Part 1, point 15 of the EU-WSR).

Block 9:   Provide the required information on the generator/producer of the waste. This information is required under the EU-WSR and the Basel Convention and many countries may require it under their national legislation.

Such information is not required, however, for movements of wastes destined for recovery under the OECD Decision.

The registration number of the generator/producer should be given where applicable. If the exporter/notifier is the generator/producer of the waste then write 'Same as block 1'. If the waste has been generated/produced by more than one generator/producer, write 'See attached list' and append a list providing the requested information for each generator/producer. Where the generator/producer is not known, give the name of the person in possession or control of such waste (holder). Also provide information on the process by which the waste was generated/produced and the site of generation/production.

Some non-EU States may accept that information on the generator be given in a separate annex which would only be available to the competent authorities.

Block 10:   Provide the required information on the destination of the shipment by ticking either disposal or recovery facility. Where applicable, the registration number and actual site of disposal or recovery should be given if it is different from the address of the facility.

If the disposer or recoverer is also the importer/consignee, state here 'Same as block 2'. If the disposal or recovery operation is a D13–D15 or R12 or R13 operation[81], the facility performing the operation should be mentioned in block 10, as well as the location where the operation will be performed. In such a case, corresponding information on the subsequent facility or facilities, where any subsequent R12/R13 or D13–D15 operation and the D1–D12 or R1–R11 operation or operations takes or take place or may take place should be provided in an annex.

EU specific instructions:

If the recovery or disposal facility is listed in Annex I, Category 5 of Di-

---

[81]   according to the definitions of operations set out in the list of abbreviations and codes attached to the notification document or in the EU according to Annexes IIA or IIB of the Waste Framework Directive

I. Explanatory Notes

> *rective 96/61/EC of 24 September 1996 on integrated pollution and prevention control, evidence (e.g. a declaration certifying its existence) of a valid permit issued in accordance with Articles 4 and 5 of that Directive must be provided in an annex in case a facility is located in the EU.*

Block 11: Indicate the type of recovery or disposal operation by the using R-codes or D-codes provided in the list of abbreviations and codes attached to the notification document[82] or *in the EU by using R-codes or D-codes of Annexes IIA or IIB of the EU-WFD.*

If the disposal or recovery operation is a D13–D15 or R12 or R13 operation, corresponding information on the subsequent operations (any R12/R13 or D13–D15 as well as D1–D12 or R1–R11) should be provided in an annex. Also indicate the technology to be employed.

EU specific instructions:

If the waste is destined for recovery, provide the planned method of disposal for the non-recoverable fraction after recovery, the amount of recovered material in relation to non-recoverable waste, the estimated value of the recovered material and the cost of recovery and the cost of disposal of the non-recoverable fraction in an annex. In addition, in cases of imports into the EU of wastes destined for disposal, indicate a prior duly motivated request from the country of export/dispatch according Art. 41(4) of the EU-WSR under 'reason for export' and attach this request in an annex.

Some countries outside the OECD may, based on the Basel Convention, also require that the reason for export is specified.

Block 12: Give the name or names by which the material is commonly known or the commercial name and the names of its major constituents (in terms of quantity and/or hazard) and their relative concentrations (expressed as a percentage), if known. In the case of a mixture of wastes, provide the same information for the different fractions and indicate which fractions are destined for recovery. A chemical analysis of the composition of the waste may be required in accordance with national legislation or in the case of EU Member States in accordance with Annex II Part 3 point 7 of the EU-WSR.

Attach further information in an annex if necessary.

Block 13: Indicate physical characteristics of the waste at normal temperatures and pressures by using the codes provided in the list of abbreviations and codes attached to the notification document.

Block 14: Give the code according to the system adopted under the Basel Convention (under subheading (i) in block 14) and, where applicable, the systems adopted in the OECD Decision (under subheading (ii)) and other accepted classification systems (under subheadings (iii) to (xii)).

According to the OECD Decision, only one waste code (from either the

---

Basel or OECD systems) should be given, except in the case of mixtures of wastes for which no individual entry exists. In such a case, the code of each fraction of the waste should be provided in order of importance (in an annex if necessary).

EU specific instructions:

State the code that identifies the waste according to Annexes III, IIIA, IIIB, IV or IVA of the EU-WSR. According to the second subparagraph, point 6 of Art. 4 of the EU-WSR, give only one waste code (from Annexes III, IIIA, IIIB, IV or IVA of the EU-WSR) with the following two exceptions: In the case of wastes not classified under one single entry in either Annex III, IIIB, IV or IVA, give only one type of waste. In the case of mixtures of wastes not classified under one single entry in either Annex III, IIIB, IV or IVA, unless listed in Annex IIIA, provide the code of each fraction of the waste in order of importance (in an annex if necessary).

Subheading (i):

cf. appendix 1

Basel Convention Annex VIII codes should be used for wastes that are subject to control under the Basel Convention or the OECD Decision (see Part I of Appendix 4 in the OECD Decision) or are subject to the procedure of prior written notification and consent (see Part I of Annex IV of the EU-WSR);

Basel Annex IX codes should be used for wastes that are not usually subject to control under the Basel Convention and the OECD Decision or subject to the procedure of prior written notification and consent under the EU-WSR. These codes should be also used, if, for a specific reason such as contamination by hazardous substances (in the EU see paragraph 1 to Annex III of the EU-WSR) or different classification according to national regulations, these wastes are subject to such control (see Part I of Appendix 3 in the OECD Decision) or according to Art. 63 of the EU-WSR are subject to the procedure of prior written notification and consent (see Part I of Annex III to the EU-WSR).

Basel Annexes VIII and IX can be found in the text of the Basel Convention as well as in the Instruction Manual available from the Secretariat of the Basel Convention and in Annex V to the EU-WSR. If a waste is not listed in Annexes VIII or IX to the Basel Convention, insert "not listed".

Subheading (ii):

cf. appendix 1

OECD Member Countries should use OECD codes for wastes listed in Part II of Appendices 3 and 4 of the OECD Decision and EU Member States should use OECD codes for wastes listed in in Part II of Annexes III and IV to the EU-WSR, i.e. wastes that have no equivalent listing in the Basel Convention or that have a different level of control under the OECD Decision or the EU-WSR from the one required by the Basel Convention.

If a waste is not listed in Part II of Appendices 3 and 4 of the OECD Decision or in in Part II of Annexes III and IV to the EU-WSR, insert "not listed".

I. Explanatory Notes

Fig. 4   Front Page of the Notification Document

Notification document for transboundary movements/shipments of waste

<table>
<tr>
<td colspan="2">

1. Exporter - notifier Registration No:
Name:
Address:

Contact person:
Tel: Fax:
E-mail:
</td>
<td colspan="2">

3. Notification No:
Notification concerning
A.(i) Individual shipment: ☐ (ii) Multiple shipments: ☐
B.(i) Disposal (1): ☐ (ii) Recovery : ☐
C. Pre-consented recovery facility (2;3) Yes ☐ No ☐
</td>
</tr>
<tr>
<td colspan="2" rowspan="2">

2. Importer - consignee Registration No:
Name:
Address:

Contact person:
Tel: Fax:
E-mail:
</td>
<td colspan="2">**4. Total intended number of shipments:**</td>
</tr>
<tr>
<td colspan="2">

5. Total intended quantity (4):
Tonnes (Mg):
m³:
</td>
</tr>
<tr>
<td colspan="2" rowspan="2">

8. Intended carrier(s) Registration No:
Name(7):
Address:

Contact person:
Tel: Fax:
E-mail:
Means of transport (5):
</td>
<td colspan="2">

6. Intended period of time for shipment(s) (4):
First departure: Last departure:
</td>
</tr>
<tr>
<td colspan="2">

7. Packaging type(s) (5):
Special handling requirements (6): Yes: ☐ No: ☐
</td>
</tr>
<tr>
<td colspan="2" rowspan="2">

9. Waste generator(s) - producer(s) (1;7;8) Registration No:
Name:
Address:

Contact person:
Tel: Fax:
E-mail:
Site and process of generation (6)
</td>
<td colspan="2">

11. Disposal / recovery operation(s) (2)
D-code / R-code (5) :
Technology employed (6):

Reason for export (1;6):
</td>
</tr>
<tr>
<td colspan="2">

12. Designation and composition of the waste (6):
</td>
</tr>
<tr>
<td colspan="2" rowspan="2">

10. Disposal facility (2): ☐ **or recovery facility (2):** ☐
Registration No:
Name:
Address:

Contact person:
Tel: Fax:
E-mail:
Actual site of disposal/recovery:
</td>
<td colspan="2">

13. Physical characteristics (5):
</td>
</tr>
<tr>
<td colspan="2">

14. Waste identification (fill in relevant codes)
(i) Basel Annex VIII (or IX if applicable):
(ii) OECD code (if different from (i)):
(iii) EC list of wastes:
(iv) National code in country of export:
(v) National code in country of import:
(vi) Other (specify):
(vii) Y-code:
(viii) H-code (5):
(ix) UN class (5):
(x) UN Number:
(xi) UN Shipping name:
(xii) Customs code(s) (HS):
</td>
</tr>
</table>

15. (a) Countries/states concerned, (b) Code no. of competent authorities where applicable, (c) Specific points of exit or entry (border crossing or port)

	State of export - dispatch	State(s) of transit (entry and exit)	State of import - destination
(a)			
(b)			
(c)			

16. Customs offices of entry and/or exit and/or export (European Community):

Entry:	Exit:	Export:

17. Exporter's - notifier's / generator's - producer's (1) declaration:

I certify that the information is complete and correct to my best knowledge. I also certify that legally enforceable written contractual obligations have been entered into and that any applicable insurance or other financial guarantee is or shall be in force covering the transboundary movement.

		18. Number of annexes attached
Exporter's - notifier's name:	Date: Signature:	
Generator's - producer's name:	Date: Signature:	

FOR USE BY COMPETENT AUTHORITIES

19. Acknowledgement from the relevant competent authority of countries of import - destination / transit (1) / export - dispatch (9):	**20. Written consent (1;8) to the movement provided by the competent authority of (country):**
Country: Notification received on: Acknowledgement sent on: Name of competent authority: Stamp and/or signature:	Consent given on: Consent valid from: until: Specific conditions: No: ☐ If Yes, see block 21 (6): ☐ Name of competent authority: Stamp and/or signature:

21. Specific conditions on consenting to the movement document or reasons for objecting

(1) Required by the Basel Convention
(2) In the case of an R12/R13 or D13-D15 operation, also attach corresponding information on any subsequent R12/R13 or D13-D15 facilities and on the subsequent R1-R11 or D1-D12 facilit(y)ies when required
(3) To be completed for movements within the OECD area and only if B(ii) applies
(4) Attach detailed list if multiple shipments

(5) See list of abbreviations and codes on the next page
(6) Attach details if necessary
(7) Attach list if more than one
(8) If required by national legislation
(9) If applicable under the OECD Decision

Subheading (iii):

cf.
appendix 2

European Union Member States should use the codes included in the European Waste List (see Commission Decision 2000/532/EC as amended)[83].

Subheadings (iv) and (v):

Where applicable, national identification codes other than the European Waste List used in the country of dispatch and, if known, in the country of destination should be used.

Subheading (vi):

If useful or required by the relevant competent authorities, add here any other code or additional information that would facilitate the identification of the waste.

EU specific instructions:

Such codes may be included in Annexes IIIA, IIIB, IV (EU48) or IVA to the EU-WSR. In that case, the Annex number should be stated in front of the codes. As regards Annex IIIA, the relevant code(s) as indicated in Annex IIIA should be used, as appropriate in sequence. Certain Basel entries such as B1100 and B3020 are restricted to particular waste streams only, as indicated in Annex IIIA.

Subheading (vii):

cf. appendix
1 and 2

State the appropriate Y-code or Y-codes according to the 'Categories of wastes to be controlled' (see Annex I to the Basel Convention and Appendix 1 to the OECD Decision), or according to the 'Categories of wastes requiring special consideration' given in Annex II to the Basel Convention (see Annex IV Part I to the EU-WSR or Appendix 2 to the Basel Instruction Manual), if it or they exist(s).

Y-codes are not required by the EU-WSR and the OECD Decision except where the waste shipment falls under one of the 'Categories requiring special consideration' under the Basel Convention (Annex II wastes), in which case the Basel Y-code should be indicated.

Nevertheless, indicate the Y-code or Y-codes for wastes defined as hazardous according Article 1(1) (a) of the Basel Convention in order to fulfil the reporting requirements under the Basel Convention.

Subheading (viii):

cf. table 10

If applicable, state here the appropriate H-code or HP-codes, i.e. the codes indicating the hazardous characteristics exhibited by the waste (see the list of abbreviations and codes attached to the notification document).

EU specific instructions:

If there is no hazardous characteristic covered by the Basel Convention, but the waste is hazardous according to Annex III to the Waste Framework Directive, state the HP-code or HP-codes according to Annex III to the EU-WSR and insert 'EU' after the HP code (e.g. HP14 EU).

---

[83] See http://europa.eu.int/eur-lex/en/consleg/main/2000/en_2000D0532_index.html

I. Explanatory Notes

The following table 10 can be used as an aid for the completion of block 14, sub-item viii. It lists the correlations between the criteria of the Basel Convention and the EU-WFD. For sub-items i and ii cf. Appendix 1 and for sub-item iii cf. Appendix 2 to these explanatory notes. There are also aids for sub-items vii and viii given.

Table 10: Correlation table of the hazard criteria from the Basel Convention and the hazard properties of Annex III to the EU-WFD

BC	EU	Hazard statement Code (EU)	Comments
H 1	HP 1	H 200, 201, 202, 203, 204	
H 3	HP 3	H 224, 225, 226	Other flash points apply in the BC
--	HP 3	H 220, 221, 222, 223	Flammable gases and aerosols
		H 242	Heating can cause fire
H 4.1	HP 3	H 228	
H 4.2	HP 3	H 252 (250, 251)	
H 4.3	HP 3	H 260, 261	
H 5.1	HP 2	H 270, 271, 272	
H 5.2	HP 1	H 240, 241	
H 6.1	HP 6	H 300, 301, 302, 310, 311, 312, 330, 331, 332	
H 6.2	HP 9	--	According to national regulations
--	HP 4	H 314, 318, 315, 319	Irritant is not implemented in the BC
H 8	HP 8	H 314	
H 10	HP 12	EUH 029, EUH 031, EUH 032	
H 11	HP 5	H 370, 371, 372, 373, 304	
	HP 7	H 350, 351	
	HP 10	H 360, 361	
	HP 11	H 340, 341	
H 12	HP 14	H 400, 410, 411, 412, 413	
--	HP 14	H 420	Ozone-depleting substances are to be assigned to HP14 in the EU
H 13	HP 15	H 205, EUH 001, EUH 019, EUH 044	
--	HP 13	H 317	May cause allergic skin reactions
		H 334	May cause allergies, asthma like symptoms or breathing difficulties if inhaled
--	HP 5	H 335	May irritate the respiratory tract

Fig. 5   Reverse Side of the Notification Document

List of abbreviations and codes used in the notification document

DISPOSAL OPERATIONS (block 11)

D1   Deposit into or onto land, (e.g., landfill, etc.)
D2   Land treatment, (e.g., biodegradation of liquid or sludgy discards in soils, etc.)
D3   Deep injection, (e.g., injection of pumpable discards into wells, salt domes or naturally occurring repositories, etc.)
D4   Surface impoundment, (e.g., placement of liquid or sludge discards into pits, ponds or lagoons, etc.)
D5   Specially engineered landfill, (e.g., placement into lined discrete cells which are capped and isolated from one another and the environment, etc.)
D6   Release into a water body except seas/oceans
D7   Release into seas/oceans including sea-bed insertion
D8   Biological treatment not specified elsewhere in this list which results in final compounds or mixtures which are discarded by means of any of the operations in this list
D9   Physico-chemical treatment not specified elsewhere in this list which results in final compounds or mixtures which are discarded by means of any of the operations in this list (e.g., evaporation, drying, calcination, etc.)
D10   Incineration on land
D11   Incineration at sea
D12   Permanent storage, (e.g., emplacement of containers in a mine, etc.)
D13   Blending or mixing prior to submission to any of the operations in this list
D14   Repackaging prior to submission to any of the operations in this list
D15   Storage pending any of the operations numbered in this list

RECOVERY OPERATIONS (block 11)

R1   Use as a fuel (other than in direct incineration) or other means to generate energy (Basel/OECD) - Use principally as a fuel or other means to generate energy (EU)
R2   Solvent reclamation/regeneration
R3   Recycling/reclamation of organic substances which are not used as solvents
R4   Recycling/reclamation of metals and metal compounds
R5   Recycling/reclamation of other inorganic materials
R6   Regeneration of acids or bases
R7   Recovery of components used for pollution abatement
R8   Recovery of components from catalysts
R9   Used oil re-refining or other reuses of previously used oil
R10   Land treatment resulting in benefit to agriculture or ecological improvement
R11   Uses of residual materials obtained from any of the operations numbered R1-R10
R12   Exchange of wastes for submission to any of the operations numbered R1-R11
R13   Accumulation of material intended for any operation in this list.

PACKAGING TYPES (block 7)

1.   Drum
2.   Wooden barrel
3.   Jerrican
4.   Box
5.   Bag
6.   Composite packaging
7.   Pressure receptacle
8.   Bulk
9.   Other (specify)

MEANS OF TRANSPORT (block 8)

R = Road
T = Train/rail
S = Sea
A = Air
W = Inland waterways

PHYSICAL CHARACTERISTICS (block 13)

1.   Powdery/powder
2.   Solid
3.   Viscous/paste
4.   Sludgy
5.   Liquid
6.   Gaseous
7.   Other (specify)

H-CODE AND UN CLASS (block 14)

UN Class	H-code	Characteristics
1	H1	Explosive
3	H3	Flammable liquids
4.1	H4.1	Flammable solids
4.2	H4.2	Substances or wastes liable to spontaneous combustion
4.3	H4.3	Substances or wastes which, in contact with water, emit flammable gases
5.1	H5.1	Oxidizing
5.2	H5.2	Organic peroxides
6.1	H6.1	Poisonous (acute)
6.2	H6.2	Infectious substances
8	H8	Corrosives
9	H10	Liberation of toxic gases in contact with air or water
9	H11	Toxic (delayed or chronic)
9	H12	Ecotoxic
9	H13	Capable, by any means, after disposal of yielding another material, e. g., leachate, which possesses any of the characteristics listed above

Further information, in particular related to waste identification (block 14), i.e. on Basel Annexes VIII and IX codes, OECD codes and Y-codes, can be found in a Guidance/Instruction Manual available from the OECD and the Secretariat of the Basel Convention.

Subheading (ix):

If applicable, state here the United Nations class or classes which indicate the hazardous characteristics of the waste according to the United Nations classification (see the list of abbreviations and codes attached to the notification document) and are required to comply with international rules for the transport of dangerous goods (see the United Nations Recommendations on the Transport of Dangerous Goods. Model Regulations (Orange Book), latest edition)[84].

Subheadings (x and xi):

If applicable, state here the appropriate United Nations number or numbers and United Nations shipping name or names. These are used to identify the waste according to the United Nations classification system and are required to comply with international rules for transport of dangerous goods (see the United Nations Recommendations on the Transport of Dangerous Goods. Model Regulations (Orange Book), latest edition).

Subheading (xii):

If applicable, state here customs code or codes, which allow identification of the waste by customs offices (see the list of codes and commodities in the 'Harmonised commodity description and coding system' produced by the World Customs Organisation).

Block 15: On line (a) of block 15, provide the name of the countries[85] of export/dispatch, transit and import/destination or the codes for each country by using the ISO standard 3166 abbreviations[86]. On line (b), provide, where applicable, the code number of the respective competent authority for each country and on line (c) insert the name of the border crossing or port and, where applicable, the customs office code number as the point of entry to or exit from a particular country. For transit countries give the information in line (c) for points of entry and exit. If more than three transit countries are involved in a particular shipment, attach the appropriate information in an annex.

Block 16[87]: *Provide the required information in case shipments enter, pass through or leave the European Union.*

Block 17: Each copy of the notification document is to be signed and dated by the exporter/notifier (or by dealer or broker/recognised trader if acting as an exporter/notifier) before being forwarded to the competent authorities of the countries concerned. In some non-EU Member States and non-OECD-countries, the competent authority of export/dispatch may sign and date.

---

84 See http://www.unece.org/trans/danger/danger.htm

85 In the Basel Convention, the term 'State' is used instead of 'country'.

86 The Basel Convention uses the term "States", whereas the OECD Decision uses "Member countries" and the European Community Regulation uses "Member States".

87 This block should be completed for movements involving entering, passing through or leaving Member States of the European Union.

When the exporter/notifier is not the same person as the original producer/generator[88], this producer/generator, the new producer/generator or the collector is, where practicable, also to sign and date; it is noted that this may not be practicable in cases where there are several producers/generators (definitions regarding practicability may be contained in national legislation). Further, where the producer/generator is not known, the person in possession or control of the waste (holder) should sign. This declaration should also certify the existence of insurance against liability for damage to third parties. Some countries may require proof of insurance or other financial guarantees and a contract to accompany the notification document.

Block 18: Indicate the number of annexes containing any additional information supplied with the notification document (see blocks 5, 6, 7, 8, 9, 10, 11, 12, 14, 15, 20 or 21). Each annex must include a reference to the notification number to which it relates, which is indicated in the corner of block 3.

EU specific instruction:

This includes also the case when additional information and documentation, which are not covered by any block, is requested by the competent authorities; see points in Annex II Part 3 to the EU-WSR.

Block 19: Under the Basel Convention, the competent authority or authorities of the country or countries of import/destination (where applicable) and transit issue such an acknowledgement. Under the OECD Decision, the competent authority of the country of import/destination issues the acknowledgement. Some non-OECD countries may, according to their national legislation, require that the competent authority of export/dispatch also issues an acknowledgement.

Blocks 20 and 21: Block 20 is for use by competent authorities of any country concerned when providing a written consent. The Basel Convention (except if a country has decided not to require written consent with regard to transit and has informed the other Parties thereof in accordance with Art. 6 (4) of the BC) and certain countries always require a written consent (according Art. 9 (1) of the EU-WSR, a competent authority of transit may provide a tacit consent) whereas the OECD Decision does not require a written consent. Indicate the name of the country (or its code by using the ISO standard 3166 abbreviations). If the shipment is subject to specific conditions, the competent authority in question should tick the appropriate box and specify the conditions in block 21 or in an annex to the notifycation document. If a competent authority wishes to object to the shipment it should do so by writing 'OBJECTION' in block 20. Block 21, or a separate letter, may then be used to explain the reasons for the objection.

---

[88] Under the Basel Convention, the waste generator is also required to sign the declaration; it is noted that this may not be practicable in cases where there are several generators.

4.9.3 Examination by the Competent Authorities

(Art. 6 of BC; Chapt. D.(2) OECD-Decision; Art. 4 to 6 of EU-WSR)

In accordance with the BC and the EU-WSR, the competent authority of export/dispatch first checks the completeness of the documentation (cf. sect. 4.7) before submitting it to other authorities. According to the OECD Decision the exporter submits the documentation directly to all authorities concerned. The time limits mentioned below originate from the OECD-Decision and have been implemented in the EU by the EU-WSR.

The EU-WSR specifies that the competent authority of export/dispatch checks if all "mandatory information" (in accordance with Annex II, Parts 1 and 2 to the EU-WSR) has been given. If the "mandatory information" is complete, the notification is considered to be "properly executed". If the transit authority raises objections within the meaning of Art. 11 or 12 of the EU-WSR within the period of three working days, it shall not proceed with the notification and shall immediately inform the notifier thereof.

Otherwise, the competent authority of export/dispatch shall forward copies of the notification documentation, to the competent authority of import/destination and to the transit authorities within three working days. This also applies if the "optional" information listed in Annex II, Part 3 to the EU-WSR is not yet available.

The competent authority of export/dispatch and also the transit state authorities can demand the submission of "optional data" from the notifier. If they request such documentation or information within three working days of receipt, they must inform the other authorities concerned, in particular the authority of import/destination. When the requesting authority has received all the requested "optional" information, the import/destination authority must be informed without delay.

Upon receipt of the notification by the import/destination authority, these authority checks the completeness of all "mandatory" and "optional" information and data (according to Annex II, Parts 1, 2 and 3 to the EU-WSR).

If the import/destination authority is satisfied that all the documentation and information has been received and has not been notified by a transit authority or an authority of transit requesting "optional" information from the notifier, the notification is considered to be "properly completed".

The authority of import/destination then acknowledges receipt by stamping block 19 of the notification document and sends it to the notifier with copies to the authority of export/dispatch and the authorities of transit. According to the OECD Decision and the EU-WSR, processing must take place within three working days.

Fig. 6: Flowchart of the Notification Procedure

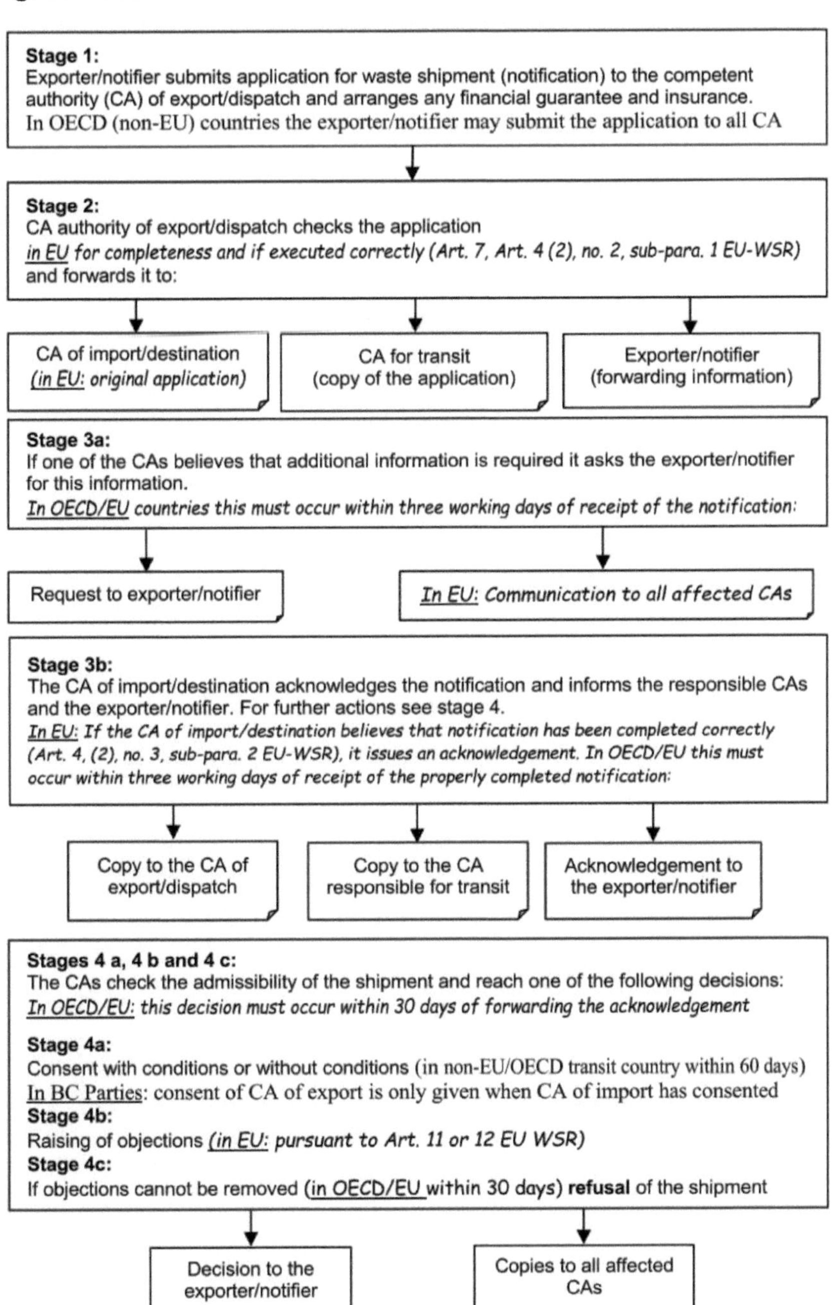

According to the OECD Decision and the EU-WSR, the 30-day period within which the authorities of export/dispatch, import/destination and transit must take a decision on the notification starts with the issue of the acknowledgement of receipt by the competent authority of import/destination.

The following three decision-making options are provided for:

- consent without conditions,

- consent with conditions (Art. 6 (2) of the BC, see also block 21 of notification document and in the EU according to Art. 10 of the EU-WSR) or

- in the EU raising an objection according to Art. 11 of the EU-WSR (for waste for disposal) or Art. 12 of the EU-WSR (for waste for recovery).

After expiry of the 30-day period without an official decision by the authority of transit, the unconditional tacit consent of the authority of transit is deemed to have been granted.

If the authorities give their written consent, conditions may be attached to the consent. In the EU these are listed exhaustively in Art. 10 of the EU-WSR. In particular, these are conditions relating to transport safety (e.g. handling of the waste, transport route).

4.9.3.1 Time Limits (OECD and EU-specific)

The time limits specified in the OECD Decision and the EU-WSR refer to working days, i.e. Monday to Friday, excluding public holidays that fall on Monday to Friday. In the EU the calculation shall be made in accordance with EC Regulation No. 1182/71[89]. Pursuant to Art. 3 (1) of the EU-WSR, the day on which the notification is received by the competent authority shall not be counted.

4.9.4 Financial Guarantee

(Art. 6 (11) BC; Chapt. II D.(1) b) OECD Decision; Articles 4 (5), 6, 10 (3), 22, 24, 35, 38, 42 and 44 EU-WSR)

The Basel Convention requires that "any transboundary movement of hazardous wastes or other wastes shall be covered by insurance, bond or other guarantee as may be required by the State of import or any State of transit which is a Party". These guarantees are intended to provide for immediate funds for alternative management of the waste in cases where shipment and

---

[89] Regulation (EEC, EURATOM) No 1182/71 of the Council of 3 June 1971 determining the rules applicable to periods, dates and time limits, OJ, L 124, p. 1 of 08.06.1971

disposal cannot be carried out as originally intended. These guarantees may take the form of an insurance policy, bank letters, bonds or other promise of compensation for damage, depending on the countries concerned. It is recommended that a notifier consult national legislation to determine the relevant requirements for the transfrontier shipment. The terms, "insurance", "bond", and "guarantee" are not defined in the BC. All are similar in that they denote financial instruments intended to assure the availability of funding for specified activities (e.g., transportation, storage, disposal of wastes) in cases where the shipment and subsequent disposal of wastes cannot be completed in accordance with the terms of the contract, is an illegal shipment or is otherwise not being undertaken in accordance with the terms of the respective legal instrument.

Also the OECD Decision requires that, where applicable, the exporter or the importer shall provide financial guarantees in accordance with national or international law requirements. The OECD guidance explains that "[a] financial guarantee may take the form of an insurance policy, bank letters, bonds or other means of compensation, depending on the countries concerned."[90]

Financial guarantees are known by many names, whose meanings and structure may vary from country to country. A meaningful definition of any particular instrument can only be provided with reference to national law. But in the end, the name attached to the guarantee is not important, as far as providing protection against mishaps in the shipment of wastes. What is important is the protection it provides, and especially how readily available it is, if needed, to finance a timely response to difficulties arising during the shipment of wastes. Financial guarantee requirements are imposed by countries of export primarily to help ensure that their responsibilities under the respective legal instrument can be discharged.

Within the EU according to Art. 6 of the EU-WSR, the notifier or another natural or legal person on its behalf has to establish the financial guarantee or equivalent insurance. This must cover the costs of a possible take-back of the waste. The provision of a financial guarantee (bank guarantee or insurance) is a pre-requisite for the granting of a permit or consent.

A financial guarantee complying with the requirements of the EU-WSR must contain at least the following information:

- name of the producer of the waste,

- name of the consignee of the waste,

- beneficiary (competent authority),

---

[90] Guidance Manual for the Implementation of Council Decision C(2001)107/FINAL as amended [2]

I. Explanatory Notes

- indication of the notification number,
- amount of the contractually agreed sum,
- period of validity.

A financial guarantee complying with the requirements of the EU-WSR must cover the following costs:

- transport costs
- costs of recovery and disposal, including any necessary interim procedures; and
- storage costs for 90 days

The determination of the financial guarantee is usually carried out by the competent authority of export/dispatch. The question of how to calculate the financial guarantee has been and continues to be the subject of discussion. The competent authorities use different calculation methods for this. Even in the EU, it was not possible to agree on a uniform basis for calculation. The following section presents the calculation method of LAGA Notice 25 [7], which meets the requirements of the EU-WSR.

4.9.4.1 Sample Calculation of the Financial Guarantee [7]

The competent authority shall determine the amount of the financial guarantee at its own discretion, with recourse to past values. For this purpose using the following calculation method is recommended, which uses so-called specific costs for transportation, recovery or disposal, and storage:

$$FG = (C_T * D * S_T + C_{RD} * S_{RD} + C_S * S_S) * M$$

FG = Amount of the bank guarantee or equivalent insurance [€]

C_T = Return transportation costs per km, per tonne [€ / (km * t]

C_{RD} = Recovery or disposal costs per tonne [€ / t]

C_S = Costs of storage for 90 days per tonne [€ / t]

D = Distance [km]

M = Quantity of waste [t]

S_T = Safety margin for return shipment (1.0 to 1.3)

S_{RD} = Safety margin for recovery/disposal (1.0 to 1.3)

S_S = Safety margin for storage (1.0 to 1.3)

The actual costs can only be ascertained on a case-by-case basis with due regard for the nature and hazardousness of the waste, particularly for the event of an illegal shipment.

For this reason, the waste which is the subject of a shipment application is generally used as the starting point for calculation purposes. Experience has shown that the aforementioned safety margins (S_T, S_{RD}, S_S) for the respective costs generally cover such unknown variables – cf. in particular nature and hazardousness. Average specific costs may also be used for calculation purposes, to avoid having to re-examine the costs for every subsequent or new notification concerning similar waste.

A quote – in relation to the recovery or disposal costs – in which the recovery or disposal operation is offered at a price significantly below the average costs can only be accepted as a basis subject to the submission of a binding declaration. In this declaration, the disposal company should undertake to guarantee the quoted prices and acceptance of the specified quantity until such time as the financial guarantee is released.

In the following, information is given on the calculation of the respective costs.

<u>Transport costs (C_T) and safety margin (S_T):</u>

The cost of return shipment per kilometre, per tonne can only be calculated with due regard for the waste type and the transport conditions required. For example, the calculation could be based on the transport costs for outward transportation, plus a safety margin of 10-30 %. Alternatively, the average costs per tonne for return transportation may be used; this is probably the most common case, so that the distance (D) may be omitted from the above calculation formula.

<u>Distance (D):</u>

If allowance has not already been made for distance in the aforementioned transport costs per tonne, and if no concrete information regarding the distance is available, then the approximate distance should be used, where applicable with a safety margin.

<u>Recovery or disposal costs (C_{RD}) and safety margin (S_{RD}):</u>

The cost of the non-interim recovery or disposal should be ascertained with due regard for the waste type and constituents as well as the required recovery or disposal procedure. These are subject to market-related variations, and often characterised by additional fees (e.g. low calorific value or special constituents).

The costs of "all required interim procedures" cited under the recovery or disposal costs should be confined to those costs which are necessary for proper recovery or disposal, including any required repackaging etc.

These costs may be covered by calculating a safety margin of e.g. 10-30 %. Further measures are impossible to calculate in practice, since these must be known in advance, i.e. when calculating the financial guarantee. As a rule, repatriation with direct recovery or disposal should be the aim.

<u>Storage costs for 90 days (C_s) and safety margin (S_s):</u>

In accordance with Art. 22 and 24 of the EU-WSR, the waste must be returned within 90 or 30 days of notification, or within another period of time to be unanimously specified by the authorities.

The storage costs can likewise only be ascertained with due regard for the nature and hazardousness of the waste. When calculating the financial guarantee, it is sufficient to use as a basis the average costs of interim storage in relation to the notified waste and the maximum period of 90 days, plus a safety surcharge up to 30 %.

The storage costs determined in this way, including the safety margin, should cover the storage costs for the cases pursuant to Art. 22 (9) and Art. 24 (7) of the EU-WSR from the date on which the competent authority of export/dispatch receives notification through to the date of return (cf. Art. 23 (1) and Art. 25 (1).

<u>Volume of waste (M):</u>

The total volume of waste is derived from the notification document.

In the case of general notifications, part-quantities may be used as a basis for calculation rather than the total volume. The competent authority shall determine the part-quantities in collaboration with the notifier. However, this method pre-supposes that the confirmation of recovery or disposal is available for every part-quantity shipped, so that the financial guarantee may be transferred to the next part-quantity.

The financial guarantee may also be made by a third party (another natural or legal person acting on its behalf), provided the competent authority has been granted access to this guarantee by the notifier by way of a power of attorney or contractual agreement.

If the financial guarantee or a copy thereof is to be submitted after consent has been granted, the collateral provisions should include a proviso whereby consent will be withdrawn in the event of non-submission of the financial guarantee pursuant to Art. 10 (3) of the EU-WSR. This is necessary because the competent authority which established the financial guarantee is unable to check it, particularly the amount of cover, validity and term (from the initial transportation until expiry of the notification plus maximum deadline for recovery or disposal) until after the notification has been issued.

The competent authority of destination is not responsible for establishing the financial guarantee pursuant to Art. 6 (4) sub-para. 1 of the EU-WSR, and in the aforementioned proviso within the meaning of Art. 10 (3) of the EU-WSR, it can therefore only stipulate the presentation of the financial guarantee to the competent authority of dispatch immediately prior to the first shipment.

Approval or establishment of the financial guarantee – including the form, wording and amount of the cover - should generally occur prior to or together with forwarding of the notification pursuant to Art. 7 (1) of the EU-WSR, but no later than the consent notice for notification by the competent authority of dispatch.

4.9.4.2 *Form and content of the financial guarantee or equivalent insurance:*

As a general rule, for every approved form of the required financial guarantee, it is important to ensure effective protection from bankruptcy and unconditional access for the recipient in accordance with the envisaged protection purpose.

Bank guarantees are issued on special forms by the respective banks. When checking the content, it is important to ensure that the bank gives a guarantee of payment as co-principal debtor to the competent authority for the set security amount, and also to ensure that the following points are taken into account (cf. sect. II.14: draft bank guarantee):

- As a rule, specification of the notification number

- Specification of the beneficiary (generally the competent authority of export/dispatch)

- Specification of the legal basis for the amount of the cover (Art. 6 (11) of the BC; Chapt. II D.(1) b of the OECD Decision; Art. 22 and 24 of the EU-WSR)

In general the financial guarantee shall be released once a certificate regarding the completion of the non-interim recovery or disposal has been presented for the shipped quantity of waste for which the financial guarantee was established.

EU specific

If, in accordance with Art. 15, letter f of the EU-WSR, one or more subsequent recovery or disposal operation(s) is/are implemented in another country and a new notification is required, the original financial guarantee may be released once consent has been granted by the affected competent authorities. Any take-back obligations arising from the new notifications will then no longer concern the original competent authority of dispatch.

In the event of the return of waste pursuant to Art. 22 or 24 of the EU-WSR, the competent authority which has access to the financial guarantee is obliged to reimburse the other affected authorities for the costs incurred in conjunction with the return (e.g. storage costs or if, for example, an alternative form of recovery or disposal in the receiving country has been specified by the affected authorities).

In order to reimburse the costs, the affected authorities should provide the competent authority which has access to the financial guarantee with corresponding receipts or evidence – in the form of invoices or quotes – so that these costs may be requested in writing from the bank or insurance company on the basis of the financial guarantee.

4.9.5 Contract

(Art. 6 (3) BC; Chapt. II D (1) OECD Decision; Art. 5 EU-WSR)

Prior to any transfrontier shipment of waste requiring prior written notification and consent a valid written contract between the exporter and the importer specifying environmentally sound management of the waste in question shall be in place. The contract shall clearly identify the waste generator/producer, each person who shall have legal control of the wastes and the disposal or recovery facility. All persons involved in the contract shall have appropriate legal status.

The parties to a contract must ensure that the contract complies with the requirements set out in the BC, the OECD Decision, and the EU-WSR or by relevant national legislation. The involved parties must be aware that in some countries, competent authorities may impose additional requirements concerning the contracts beyond those in the legal instruments mentioned above. For example, the involved parties may be required to submit the contracts (or portions thereof) to the competent authorities for review. It may therefore be useful to attach a copy of the contract to the notification.

In general, contracts should confirm that the carriers and disposal or recovery facilities operate under the legal jurisdiction of a Party to the Basel Convention, Member Country of the OECD or Member State of the EU and have appropriate legal status. All persons transporting or disposing of wastes must be – "authorized or allowed to perform such types of operations", – by the competent authorities of the country of export, countries of transit or country of import.

The assignment of legal responsibility and liability in contracts for any adverse consequences resulting from mishandling, accidents or other unforeseeable events, assists the competent authorities in identifying the responsible parties at any given moment during the movement of the waste(s), in accordance with national legislation. The contract should also specify which party shall assume responsibility for alternative arrangements in cases where the original terms of the contract cannot be fulfilled.

It should be noted that, according to Art. 8 of the BC, Chapt. II D.(4) of the OECD Decision or Art. 22 of the EU-WSR the country of export must ensure that the wastes are taken back into the country of export, by the exporter, when a transfrontier shipment of waste cannot be completed in accordance with the terms of the contract and if alternative arrangements cannot be made for the disposal of the waste in an environmentally sound manner.

(EU specific)

According to Art. 5 of the EU-WSR the contract to be concluded between the notifier/exporter and the consignee/importer shall include at least a

- a take-back obligation for the notifier pursuant to Art. 22 and Art. 24 (2), and

- the assurance of environmentally sound disposal by the consignee/importer.

The declaration of commitment contained in block 17 of notification document and block 15 of movement document indirectly indicates that the contract must be concluded in writing. A model can be found in sect. II.12. However, national legislation may require different or additional requirements with regard to the content of the contract, and should always be consulted.

4.9.6 Administrative Costs (Fees)

(Art. 29 EU-WSR)

The Administrative costs of processing the notification shall be set by the competent authority according to national legislation.

4.9.7 Disagreement on Classification Issues

(Art. 6 (5) BC; Chapt. B.(3) and B.(4) OECD Decision; Art. 28 EU-WSR)

All three instruments contain provisions for cases in which the countries or authorities concerned have different views on the definition or classification of waste.

The BC recognises situations where one or two countries concerned (export, import or transit) classify waste differently. This is the case when

- one country classifies additional wastes as hazardous according to Art. 1 (1) b) of the BC, or

I. Explanatory Notes

- there are differences in the definition of waste.

According to the BC, in the case where only the country of export classifies waste as hazardous, this leads to the country of export assuming the obligations of the country of import and transit (so-called unilateral notification).

In the case where only the country of import or transit classifies the waste as subject to notification, the notification obligation falls to these countries.

If only the country of transit considers waste to be hazardous, it is the country of transit that has the notification obligation.

Like the BC, the OECD Decision applies the mutatis mutantis principle in the case of different classifications of waste by the states concerned. This means that if only the country of export or the country of export and the country of transit consider waste to be subject to notification, this leads to a unilateral notification procedure by the country of export, in which the exporter must ensure through contractual arrangements that the recovery facility fulfils the obligations of the OECD Decision.

In the case where only the country of import classifies the waste as subject to notification, the mutatis mutandis principle applies and the duty to notify shall shift to the recovery facility or the importer. This means that either the recovery facility or the importer shall provide the notification to the competent authorities concerned. In practice, the recovery facility or the importer may arrange, within the contract for example, that the exporter provide the notification to competent authorities in accordance with the OECD Decision.

(EU specific)

According to Art. 28 of the EU-WSR, in case of disagreements on classification issues, only the exporting and importing countries are in charge to decide on the classification.

If the competent authorities of dispatch and of destination cannot agree

- on the classification as regards the distinction between waste and non-waste, the subject matter shall be treated as if it were waste.

- on the classification of the notified waste as being green or amber listed, the waste shall be regarded as amber.

- on the classification of the waste treatment operation notified as being recovery or disposal, the provisions regarding disposal shall apply.

4.9.8 Grounds for Objection

(Chapt. II D.(2) OECD Decision; Art. 11 and 12 EU-WSR)

While the Basel Convention and the OECD Decision do not contain specific grounds for objection, the EU-WSR sets out the grounds for objection in detail.

Objections to the planned shipment can be raised by the competent authorities if they have reason to believe that the proposed shipment violates national or EU standards or regulations. The grounds for objection and the objection procedure are regulated for waste destined for disposal in Art. 11 of the EU-WSR and waste destined for recovery in Art. 12 of the EU-WSR (cf. Table 11).

A special feature is the shipment of mixed municipal waste collected from households (EWL code 20 03 01), which is subject to the same conditions as waste for disposal, so that this waste in accordance with Art. 3 (5) of the EU-WSR is also subject to the principles of proximity and self-sufficiency when shipped for recovery.

If a competent authority raises objections, the 30-day period that begins with the issue of the confirmation of receipt is not interrupted or inhibited. The reasons that led to the objections must be removed by the notifier within the 30-day period, otherwise a new notification is required (Art. 11 (5) and Art. 12 (4) of the EU-WSR).

Further explanations of the grounds for objection can be found in the German Enforcement Guidance issued by LAGA [7] or by Oexle and Backes in [6], pp. 405 to 492.

4.9.9 Conditions

(Art. 6 BC; Art. 10 EU-WSR)

The competent authorities may attach conditions to their consent. Such conditions are possible with regard to both waste management requirements and the transport of waste. Conditions are all ancillary provisions that a competent authority imposes together with the consent to a notification. They may also be conditions subsequent, for example, if the authority makes a revocation in the event that, for example, the guarantee or insurance is not valid at the latest at the start of the shipment.

Table 11: Grounds for objection according to Art. 11 and 12 of the EU-WSR

Grounds for objection	Art. 11 of the EU-WSR Disposal	Art. 12 of the EU-WSR Recovery
Infringement of measures implementing the principles of proximity, priority of recovery and self-sufficiency at EU and national level.	Para 1 letter a Para 3	
Shipment or disposal is not in compliance with national legislation on the protection of the environment or public health, or on the maintenance of public order and safety.	Para 1 letter b	Para 1 letter b
Conviction of the notifier or consignee for illegal shipments or other unlawful acts in the field of environmental protection	Para 1 letter c	Para 1 letter d
Violation of Art. 15 and 16 of the EU-WSR by the notifier or the disposal facility	Para 1 letter d	Para 1 letter e
Violation of a Member State ban on the import of hazardous waste or of waste listed in Annex II to the Basel Convention	Para 1 letter e	
Violation of obligations under international conventions	Para 1 letter f	Para 1 letter f
Shipment or disposal does not comply with the EU-WFD, in particular with Art. 5 and 7	Para 1 letter g	Para 1 letter a
Disposal facility does not apply best available techniques	Para 1 letter h	Para 1 letter i
The waste is mixed municipal waste from private households (EWL waste code 20 03 01);	Para 1 letter i	
Waste is not treated in accordance with binding environmental protection standards for disposal under EU legislation.	Para 1 letter j	Para 1 letter j
Shipments of waste destined for recovery are not in compliance with national legislation.		Para 1 letter c
Recovery is not justified from an economic and environmental point of view.		Para 1 letter g
The waste is destined for disposal and not for recovery		Para 1 letter h
The waste in question is not treated in accordance with waste management plans to comply with mandatory recovery and recycling obligations.		Para 1 letter k

4.9.10 Subsequent Demand for Documents

(Art. 6 BC; Chapt. II D.(2) c) OECD Decision; Art. 4 No. 3 EU-WSR)

In order to avoid unnecessary additional demands for documents, it is advisable — as already mentioned above — to find out from the authority which documents are necessary from the authority's point of view before submitting the documents. For example, it is advisable to submit copies of the insurance, the financial guarantee, the carriers' declarations or authorisations, etc. together with the notification.

(Germany specific)

The possibilities of the authorities to request further documents in accordance with No. 14 of Annex II, Part 3 to the EU-WSR were limited by Art. 3 (3) of the German Waste Shipment Act to the information that is relevant and necessary for the assessment of the notification. This can be certificates of good conduct, EfB-certificates (certified specialised waste management facilities), and proof of a lack of domestic disposal capacities or similar documents.

4.10 Procedure after Receipt of Written Consents

The relevant competent authority sends a copy of its decision on written consent to the requested shipment to all other authorities concerned. In the case of exports of waste from the EU, it also sends a copy to the EU customs office of export and exit or, in the case of imports into the EU, to the EU customs office of entry. The exporter/notifier shall receive the consent for the transfrontier shipment in the following form:

(a) In the case of written consent:

- Notification document with the stamp of the competent authority in block 20,

- Consent to the shipment with specific conditions and ancillary provisions in block 21 and/or separately as an annex,

- Movement document containing the information provided prior to the application.

(b) In the case of so-called tacit consent by a transit authority:

- Notification document with acknowledgement of receipt by the competent authority in block 19,

- Tacit consent shall be deemed to have been given after a period of 30 days (or 60 days according to Art. 6 (4) of the BC) — calculated from the date of the acknowledgement of receipt — if no objections have been raised,

- Conditions and ancillary provisions in the form of a decision by other authorities (e.g. transit authorities),

- • Movement document with the information provided before the application was submitted.

The respective conditions and ancillary provisions must be complied with by the notifier and the other companies involved in the shipment.

The shipment is only permissible (according to Art. 6 of the BC; Chapt. II D.(2) of the OECD Decision or Art. 9 (6) of the EU-WSR) if the written consent of the competent authority of export/dispatch, the written consent of the competent authority of import/destination and, in the case of countries of transit, the written or tacit consent of the competent authorities of transit have been obtained. It is the responsibility of the exporter/notifier to verify that all necessary consents have been given.

4.10.1 Tracking Procedure

The transport and whereabouts of the waste must be documented using the movement document (cf. Fig. 7). At the time of notification, the exporter/notifier is to complete blocks 3, 4 and 9–14.

After receipt of the consents from the competent authorities of export/dispatch, import/destination and transit or, in relation to the competent authority of transit, tacit consent can be assumed, and before the actual start of the shipment, the exporter/notifier is to complete blocks 2, 5–8 (except the means of transport, the date of transfer and the signature), 15 and, if appropriate, 16.

Block 1:	The competent authority of export/dispatch is to enter the notification number (this is to be copied from block 3 in the notification document).
Blocks 3 and 4:	Reproduce the same information on the exporter/notifier[91] and importer/consignee as given in blocks 1 and 2 in the notification document.
Block 9:	Reproduce the information given in block 9 of the notification document.
Blocks 10 and 11:	Reproduce the information given in blocks 10 and 11 in the notification document. If the disposer or recoverer is also the importer/consignee, write in block 10: 'Same as block 4'. If the disposal or recovery operation is a D13–D15 or R12 or R13 operation (according to Annexes IIA or IIB to Directive 2006/12/EC on waste), the information on the facility performing the operation provided in block 10 is sufficient. No further information on any subsequent facilities performing R12/R13 or D13–D15 operations and the subsequent facility(ies) performing the D1–D12 or R1–R11 operation(s) needs to be included in the movement document.
Blocks 12, 13 and 14:	Reproduce the information given in blocks 12, 13 and 14 in the notification document.

---

[91] In some non-EU Member States and non-OECD countries, information relating to the competent authority of export/dispatch may be given instead.

Fig. 7: Front Page of the Movement Document

Movement document for transboundary movements/shipments of waste

<table>
<tr><td colspan="2">1. Corresponding to notification No:</td><td colspan="2">2. Serial/total number of shipments: /</td></tr>
<tr><td colspan="2">3. Exporter - notifier Registration No:
Name:

Address:

Contact person:
Tel: Fax:
E-mail:</td><td colspan="2">4. Importer - consignee Registration No:
Name:

Address:

Contact person:
Tel: Fax:
E-mail:</td></tr>
<tr><td colspan="2">5. Actual quantity: Tonnes (Mg):   m³:</td><td colspan="2">6. Actual date of shipment:</td></tr>
<tr><td colspan="4">7. Packaging Type(s) (1):   Number of packages:
Special handling requirements: (2)   Yes: ☐   No: ☐</td></tr>
<tr><td>8.(a) 1st Carrier (3):
Registration No:
Name:
Address:

Tel:
Fax:
E-mail:</td><td>8.(b) 2nd Carrier:
Registration No:
Name:
Address:

Tel:
Fax:
E-mail:</td><td colspan="2">8.(c) Last Carrier:
Registration No:
Name:
Address:

Tel:
Fax:
E-mail:</td></tr>
<tr><td colspan="2" align="center">- - - - - - - To be completed by carrier's representative - - - - - - -</td><td colspan="2" align="right">More than 3 carriers (2) ☐</td></tr>
<tr><td>Means of transport (1):
Date of transfer:
Signature:</td><td>Means of transport (1):
Date of transfer:
Signature:</td><td colspan="2">Means of transport (1):
Date of transfer:
Signature:</td></tr>
<tr><td colspan="2">9. Waste generator(s) - producer(s) (4;5;6):
Registration No:
Name:
Address:

Contact person:
Tel: Fax:
E-mail:
Site of generation (2):</td><td colspan="2">12. Designation and composition of the waste (2):</td></tr>
<tr><td colspan="2" rowspan="2">10. Disposal facility ☐   or recovery facility ☐
Registration No:
Name:
Address:

Contact person:
Tel: Fax:
E-mail:
Actual site of disposal/recovery (2)</td><td colspan="2">13. Physical characteristics (1):</td></tr>
<tr><td colspan="2">14. Waste identification (fill in relevant codes)
(i) Basel Annex VIII (or IX if applicable):
(ii) OECD code (if different from (i)):
(iii) EC list of wastes:
(iv) National code in country of export:
(v) National code in country of import:
(vi) Other (specify):
(vii) Y-code:
(viii) H-code (1):
(ix) UN class (1):
(x) UN Number:
(xi) UN Shipping name:
(xii) Customs code(s) (HS):</td></tr>
<tr><td colspan="2">11. Disposal/recovery operation(s)
D-code / R-code (1):</td><td colspan="2"></td></tr>
<tr><td colspan="4">15. Exporter's - notifier's / generator's - producer's (4) declaration:
I certify that the above information is complete and correct to my best knowledge. I also certify that legally enforceable written contractual obligations have been entered into, that any applicable insurance or other financial guarantee is in force covering the transboundary movement and that all necessary consents have been received from the competent authorities of the countries concerned.
Name: Date: Signature:</td></tr>
<tr><td colspan="4">16. For use by any person involved in the transboundary movement in case additional information is required</td></tr>
<tr><td colspan="4">17. Shipment received by importer - consignee (if not facility):   Date: Name: Signature:</td></tr>
<tr><td colspan="4" align="center">TO BE COMPLETED BY DISPOSAL / RECOVERY FACILITY</td></tr>
<tr><td colspan="3">18. Shipment received at disposal facility ☐   or recovery facility ☐
Date of reception: Accepted: ☐   Rejected*: ☐
Quantity received: Tonnes (Mg):   m³: *immediately contact competent authorities
Approximate date of disposal/recovery:
Disposal/recovery operation (1):
Name:
Date:
Signature:</td><td>19. I certify that the disposal/recovery of the waste described above has been completed.
Name:

Date:
Signature and stamp:</td></tr>
</table>

(1) See list of abbreviations and codes on the next page
(2) Attach details if necessary
(3) If more than 3 carriers, attach information as required in blocks 8 (a,b,c).
(4) Required by the Basel Convention
(5) Attach list if more than one
(6) If required by national legislation

In some non-EU Member States and non-OECD Countries, the competent authority of export/dispatch may complete these blocks instead of the exporter/notifier. At the time of taking possession of the consignment, the carrier or its representative is to complete the means of transport, the date of transfer and the signature, which appear in blocks 8(a) to 8(c) and, if appropriate, 16.

4.10.1.1   Announcement of Transport Prior to Shipment

On the movement document, which must be carried along with each shipment of waste, the companies involved in the shipment acknowledge the transfer, receipt and disposal of the waste. The individual steps required are described in detail below.

1. Completion of blocks 2, 5 to 8 (where possible) and 15 of the movement document by the exporter/notifier, including information on the quantity supplied.

Block 2:	For a general notification for multiple shipments, enter the serial number of the shipment and the total intended number of shipments indicated in block 4 in the notification document (for example, enter '4/11' for the fourth shipment out of eleven intended shipments under the general notification in question). In the case of a single notification, enter '1/1'.
Block 5:	Give the actual weight in tonnes (1 tonne equals 1 megagram (Mg) or 1 000 kg of the waste. In some third countries, giving the volume in cubic metres (1 cubic metre equals 1 000 litres) or other metric units, such as kilograms or litres, may be acceptable. When other metric units are used, the unit of measure may be indicated and the unit in the form may be crossed out. Attach, wherever possible, copies of weighbridge tickets.
Block 6:	Enter the date when the shipment actually starts. The **starting dates of all shipments should be within the validity period** issued by the competent authorities. Where the different competent authorities involved have granted different validity periods, the shipment or shipments may only take place in the time period during which the consents of all competent authorities are simultaneously valid.

EU specific instruction:

*All shipments must **take place within the validity period** of the written or tacit consents of all competent authorities concerned issued by the competent authorities according to Art. 9(6) of the EU-WSR.*

Block 7:	Types of packaging should be indicated using the codes provided in the list of abbreviations and codes attached to the movement document. If special handling precautions are required, such as those prescribed by producers' handling instructions for employees, health and safety information, including information on dealing with spillage, and transport emergency cards, tick the appropriate box and attach the information in an annex. Also enter the number of packages making up the consignment.

Fig. 8: Reverse Side of the Movement Document

FOR USE BY CUSTOMS OFFICES (if required by national legislation)	
20. Country of export - dispatch or customs office of exit The waste described in this movement document left the country on: Signature: Stamp:	**21. Country of import - destination or customs office of entry** The waste described in this movement document entered the country on: Signature: Stamp:

22. Stamps of customs offices of transit countries			
Name of country: Entry:	Exit:	Name of country: Entry:	Exit:
Name of country: Entry:	Exit:	Name of country: Entry:	Exit:

List of Abbreviations and Codes Used in the Movement Document

DISPOSAL OPERATIONS (block 11)

D1   Deposit into or onto land, (e.g., landfill, etc.)
D2   Land treatment, (e.g. biodegradation of liquid or sludgy discards in soils, etc.)
D3   Deep injection, (e.g., injection of pumpable discards into wells, salt domes or naturally occurring repositories, etc.)
D4   Surface impoundment, (e.g., placement of liquid or sludge discards into pits, ponds or lagoons, etc.)
D5   Specially engineered landfill, (e.g., placement into lined discrete cells which are capped and isolated from one another and the environment), etc.
D6   Release into a water body except seas/oceans
D7   Release into seas/oceans including sea-bed insertion
D8   Biological treatment not specified elsewhere in this list which results in final compounds or mixtures which are discarded by means of any of the operations in this list
D9   Physico-chemical treatment not specified elsewhere in this list which results in final compounds or mixtures which are discarded by means of any of the operations in this list (e.g., evaporation, drying, calcination, etc.)
D10   Incineration on land
D11   Incineration at sea
D12   Permanent storage, (e.g., emplacement of containers in a mine, etc.)
D13   Blending or mixing prior to submission to any of the operations in this list
D14   Repackaging prior to submission to any of the operations in this list
D15   Storage pending any of the operations in this list

RECOVERY OPERATIONS (block 11)

R1   Use as a fuel (other than in direct incineration) or other means to generate energy (Basel/OECD) - Use principally as a fuel or other means to generate energy (EU)
R2   Solvent reclamation/regeneration
R3   Recycling/reclamation of organic substances which are not used as solvents
R4   Recycling/reclamation of metals and metal compounds
R5   Recycling/reclamation of other inorganic materials
R6   Regeneration of acids or bases
R7   Recovery of components used for pollution abatement
R8   Recovery of components from catalysts
R9   Used oil re-refining or other reuses of previously used oil
R10   Land treatment resulting in benefit to agriculture or ecological improvement
R11   Uses of residual materials obtained from any of the operations numbered R1-R10
R12   Exchange of wastes for submission to any of the operations numbered R1-R11
R13   Accumulation of material intended for any operation in this list

PACKAGING TYPES (block 7)

1. Drum
2. Wooden barrel
3. Jerrican
4. Box
5. Bag
6. Composite packaging
7. Pressure receptacle
8. Bulk
9. Other (specify)

MEANS OF TRANSPORT (block 8)

R = Road   A = Air
T = Train/rail   W = Inland waterways
S = Sea

PHYSICAL CHARACTERISTICS (block 13)

1. Powdery / powder   5. Liquid
2. Solid   6. Gaseous
3. Viscous / paste   7. Other (specify)
4. Sludgy

H-CODE AND UN CLASS (block 14)

UN class	H-code	Characteristics
1	H1	Explosive
3	H3	Flammable liquids
4.1	H4.1	Flammable solids
4.2	H4.2	Substances or wastes liable to spontaneous combustion
4.3	H4.3	Substances or wastes which, in contact with water, emit flammable gases
5.1	H5.1	Oxidizing
5.2	H5.2	Organic peroxides
6.1	H6.1	Poisonous (acute)
6.2	H6.2	Infectious substances
8	H8	Corrosives
9	H10	Liberation of toxic gases in contact with air or water
9	H11	Toxic (delayed or chronic)
9	H12	Ecotoxic
9	H13	Capable, by any means, after disposal of yielding another material, e. g., leachate, which possesses any of the characteristics listed above

Further information, in particular related to waste identification (block 14), i.e. on Basel Annexes VIII and IX codes, OECD codes and Y-codes, can be found in a Guidance/Instruction Manual available from the OECD and the Secretariat of the Basel Convention.

Block 15:   Block 15 (See Annex II, Part 2, point 9):

At the time of shipment, the exporter/notifier (or the dealer or broker/recognised trader if acting as an exporter/notifier) shall sign and date the movement document. In some non-EU Member States and non-OECD Countries, the competent authority of export/dispatch, or the generator of the waste according to the Basel Convention, may sign and date the movement document.

Some countries may require copies or originals of the notification document containing the written consent, including any conditions, of the competent authorities concerned to be enclosed with the movement document.

According to Article 16(c) of the EU-WSR, enclose copies of the notification document containing the written consent, including any conditions, of the competent authorities concerned with the movement document.

2. Transmit a copy of the movement document to the competent authorities concerned at least three working days before the movement takes place (transport notification).

Note: If for organisational reasons it is not possible to fill in certain fields at the indicated time, the procedure must be agreed with the competent authorities. Block 16 is provided for unforeseeable cases.

Block 16:   This block can be used by any person involved in a shipment (notifier or the competent authority of dispatch, as appropriate, consignee, any competent authority, carrier) in specific cases where more detailed information is required by national legislation concerning a particular item (for example, information on the port where a transfer to another transport mode occurs, the number of containers and their identification number, or additional proof or stamps indicating that the shipment has been consented by the competent authorities).

Additional EU specific instruction:

Give the routing (point of exit from and entry into each country concerned, including customs offices of entry into and/or exit from and/or export from the Community) and route (route between points of exit and entry), including possible alternatives, also in case of unforeseen circumstances either in block 16 or attach it in an annex.

4.10.1.2   Transport of the Waste

1. Completion of the information in block 8 of the movement document by the relevant carrier.

**Block 8
(a), (b)
and (c)**   Provide the required information (give registration number only where applicable, address including the name of the country and telephone and fax numbers including the country code). When more than three carriers are involved, appropriate information on each carrier should be attached to the movement document. The means of transport, the date of transfer and a signature should be provided by the carrier or carrier's representative taking

possession of the consignment. A copy of the signed movement document is to be retained by the notifier. Upon each successive transfer of the consignment, the new carrier or carrier's representative taking possession of the consignment will have to comply with the same request and also sign the document. A copy of the signed document is to be retained by the previous carrier.

2. Carry the movement document and a copy of the notification document and copies of the conditions/approval decisions of the competent authorities.

3. In case of shipment out of or into the EU: hand over a certified copy of the notification document and the movement document to the customs office of exit or entry. In case of transit of waste through the EU: Presentation of the above mentioned documents to the customs office of exit by the carrier (not shown in Fig. 9).

4. If necessary, fields 20, 21 and 23 of the movement document must be completed and stamped by the respective customs office.

4.10.1.3   Arrival of the Waste

1. Completion of the movement document by the importer/consignee in box 17 (if the importer/consignee does not operate a disposal facility) and box 18 by the operator of the disposal facility (on arrival of the waste at the final disposal facility).

Block 17:   This block is to be completed by the importer/consignee in the event that it is not the disposer or recoverer (cf. paragraph 15 above) and in case the importer/consignee takes charge of the waste after the shipment arrives in the country of destination.

Not required under the OECD Decision.

Block 18:   This block is to be completed by the authorised representative of the disposal or recovery facility upon receipt of the waste consignment. Tick the box of the appropriate type of facility. With regard to the quantity received, please refer to the specific instructions on block 5. A signed copy of the movement document is given to the last carrier. If the shipment is rejected for any reason, the representative of the disposal or recovery facility must immediately contact his or her competent authority.

According to Article 16(d) or, if appropriate, 15(c) of the EU-WSR and the OECD Decision, signed copies of the movement document must be sent within three days to the notifier and the competent authorities in the countries concerned (with the exception of those OECD transit countries which have informed the OECD Secretariat that they do not wish to receive such copies of the movement document).

The original movement document shall be retained by the disposal or recovery facility.

> Receipt of the waste consignment must be certified by any facility performing any disposal or recovery operation, including any D13–D15 or R12 or R13 operation. A facility performing any D13–D15 or R12/R13 operation or a D1–D12 or R1–11 operation subsequent to a D13–D15 or R12 or R13 operation in the same country, is not, however, required to certify receipt of the consignment from the D13–D15 or R12 or R13 facility. Thus, block 18 does not need to be used for the final receipt of the consignment in such a case. Indicate also the type of disposal or recovery operation by using the list of abbreviations and codes attached to the movement document[92] and the approximate date by which the disposal or recovery of waste will be completed (this is not required by the OECD Decision).

2. Transmission of a copy of the movement document by the consignee or the operator of the facility to the exporter/notifier and to the competent authorities concerned no later than three days after receipt of the waste.

3. In the case of <u>transit of waste for disposal through the EU</u>, the arrival of the waste at its destination shall be confirmed by the exporter/notifier to the last transit authority as early as 42 days after leaving the EU.

4.10.1.4   Disposal/Recovery of the Waste

1. Confirmation of final disposal or recovery of the waste shall be provided by the disposer no later than one calendar year after receipt of the waste by making an appropriate entry in block 19 of the movement document.

Block 19:   This block is to be completed by the disposer or recoverer to certify the completion of the disposal or recovery of the waste.

Under the Basel Convention, signed copies of the document with block 19 completed should be sent to the exporter and competent authorities of the country of export.

According to Article 16(e) or, if appropriate, 15(d) of the EU-WSR and the OECD Decision, signed copies of the movement document with block 19 completed should be sent to the notifier and competent authorities of export/dispatch, transit (not required by the OECD Decision) and import/destination as soon as possible, but no later than 30 days after the completion of the recovery or disposal and no later than one calendar year following the receipt of the waste.

Some non-EU Member States and non-OECD Countries may require in accordance with the Basel Convention that signed copies of the document with block 19 completed must be sent to the exporter/notifier and the competent authority of export/dispatch.

For disposal or recovery operations D13–D15 or R12 or R13, the information on the facility performing such an operation provided in block 10 is sufficient, and no further information on any subsequent facilities performing R12/R13 or D13–D15 operations and the subsequent facility(ies)

---

[92]   In the EU by using R-codes or D codes of Annexes IIA or IIB of the EU-WFD

performing the D1–D12 or R1–R11 operation(s) need be included in the movement document.

The disposal or recovery of waste must be certified by any facility performing any disposal or recovery operation, including a D13–D15 or R12 or R13 operation. Therefore, a facility performing any D13–D15 or R12/R13 operation or a D1–D12 or R1–R11 operation, subsequent to a D13–D15 or R12 or R13 operation in the same country, should not use block 19 to certify the recovery or disposal of the waste, since this block will already have been completed by the D13–D15 or R12 or R13 facility. The means of certifying disposal or recovery in this particular case must be ascertained by each country.

Blocks 20, 21 and 22: *The blocks must be used for control by customs offices at the borders of the EU.* Not required by the Basel Convention or by the OECD Decision. The blocks may be used for control by customs offices at the borders of country of export, transit and import if so required by national legislation.

2. Transmit the certificate of disposal to the exporter/notifier and the competent authority of export (according to the EU-WSR it shall be submitted to all authorities concerned) no later than 30 days after the disposal has been carried out. This evidence is not required for shipments of waste through the EU.

   <u>Note:</u> If a shorter deadline is agreed in the contract between the exporter/notifier and the importer/consignee, this must be taken into account.

   The receipt of certification of disposal closes the loop on the shipment and assures the country of export that the shipment has been carried out in accordance with relevant approvals (cf. sect. I.4.9.5 on contracts and sect I.4.3 on illegal shipments).

4.11 Controls *(EU specific)*

(Art. 50 EU-WSR)

The authorities responsible for the enforcement of transfrontier waste shipments are obliged to carry out inspections of such shipments with regard to compliance with the requirements of the EU-WSR. An amendment to the EU-WSR[93] introduced the mandatory preparation, review and, if necessary, updating of control plans.

According to Art. 50(2) of the EU-WSR, inspections of establishments, undertakings, brokers and dealers and inspections of waste shipments are to be carried out. The controls may be carried out

---

[93] Regulation (EU) No 660/2014 of 15. May 2014, OJ, L 189 of 27.06.2015, p. 137

Fig. 9:   Flowchart of the Tracking Procedure

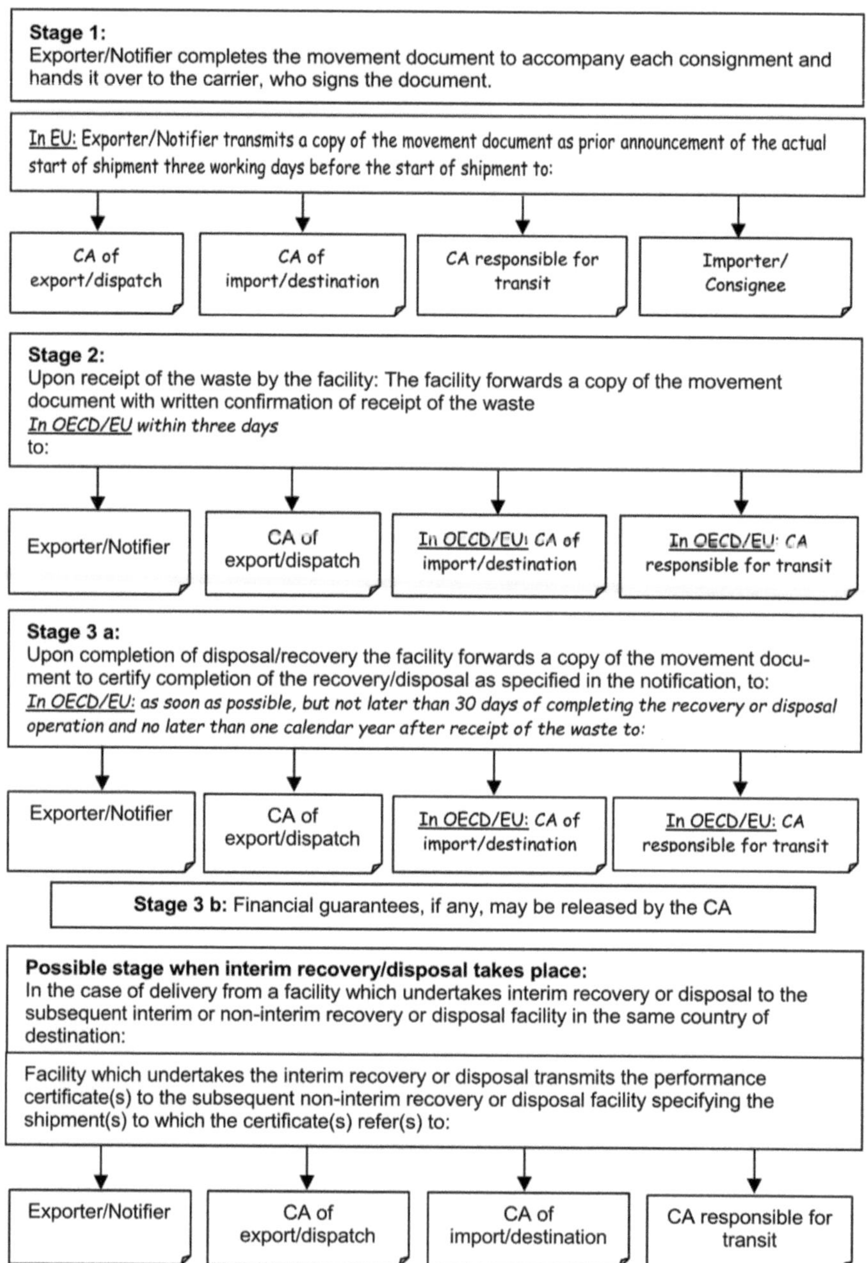

- at the place of origin at the producer's, holder's or notifier's premises,

- at the place of destination (including interim and non-interim recovery or disposal) at the importer/consignee or disposal facility

- at the external borders of the European Union, and/or

- during shipment within the European Union.

In the context of controls, natural or legal persons as holder or as person arranging for the transport of a substance or object bear the burden of proof to demonstrate that the substance or object being transported is not waste (cf. Art. 50 (4a) of the EU-WSR).

This means that the notifier, the person who arranges the shipment, the holder, the carrier, the consignee and/or the facility operator who receives the waste bears the burden of proof that the shipment of waste complies with the WSR (cf. Art. 50 Para. 4c of the EU-WSR).

In the event that the required evidence is not provided in due time or the evidence and information are insufficient, the shipment of the substance or object or the shipment of the waste shall be considered an illegal shipment and shall be dealt with in accordance with Articles 24 and 25 (cf. Art. 50 par. 4b and 4d of the EU-WSR).

For the transport of used electrical and electronic equipment, the provisions of Art. 23 and Annex VI to the EU Directive on Waste Electrical and Electronic Equipment apply in derogation of the provisions of Art. 50 Para. 4a and 4b of the EU-WSR.

The controls are determined independently by the competent authorities on the basis of control plans drawn up in accordance with Art. 50 (2a) of the EU-WSR taking into account current findings from practice and the tasks of the individual authorities involved in the controls. If necessary, the authorities shall determine controls jointly or together with authorities of other EU Member States.

4.11.1 Control Plans *(EU-specific)*

(Art. 50 Para. 2a EU-WSR)

The obligation to draw up control plans is directed at the individual EU Member States. The EU Member States regulates the responsibilities for drawing up, reviewing and updating these control plans.

The basis for a control plan is a risk assessment for specific waste streams and origins of illegal shipments, taking into account enforcement findings, in particular on detected illegal shipments of waste. Factors that may be considered in the risk assessment include:

- the origin of the waste and the method of waste generation,
- the composition and nature of the type of waste, with particular reference to the hazard properties,
- the parties involved in the shipment,
- the countries affected by the shipment (e.g. with regard to disposal capacities, facility standards, disposal costs), and
- the disposal measures envisaged.

The content of the control plan is limited to the elements listed in Art. 50 para. 2a of the EU-WSR. Thus, neither the risk assessment to be carried out prior to the preparation of the plan nor the minimum number of inspections to be determined from it need to be listed in the control plan.

4.11.2 Cooperation on Controls (IMPEL/TFS)

The European implementation network IMPEL also includes a team of experts consisting of representatives of enforcement authorities, police, customs and other authorities involved in the enforcement of transfrontier waste shipments and management of waste. The tasks of the Transfrontier Waste Shipment Working Group (TFS[94]) are the practical implementation and enforcement of international and European waste shipment and waste management legislation. The aim of the network is to promote compliance with the EU-WSR and the Waste Directives, to strengthen the exchange of knowledge, best practices and experience in the enforcement of the regulations and directives, to stimulate a uniform enforcement system and to strengthen it with joint projects. This is done through awareness raising and capacity building, facilitation of inter-agency and cross-border cooperation and operational enforcement activities.

The TFS Working Group carries out projects on waste shipments with a focus on the control of waste shipments and waste treatment facilities and develops guidance documents on different areas of shipments. One example is a guide on how to deal with illegal waste shipments. Further information and materials are available on the IMPEL webpage[95].

---

[94] TFS = Transfrontier Shipment of Waste
[95] IMPEL Webpage: https://www.impel.eu/topics/waste-and-tfs/

4.12 **Customs Offices of Entry into and Exit from the EU**

(Art. 55 EU-WSR)

According to the EU-WSR, EU Member States may designate certain customs offices of entry into and exit from the Community for shipments of waste entering or leaving the EU. If a Member State has designated such a customs office, shipments of waste entering or leaving the EU may not pass through any other border crossing point in that Member country. The customs offices that may be designated in EU Member States pursuant to Art. 55 of the EU-WSR are listed on the website of the Commission (DG Environment).[96]

The customs offices designated are customs offices of entry (Art. 2 No. 29 of the EU-WSR) and customs offices of exit (Art. 2 No. 28 of the EU-WSR). As a rule, these are customs offices at the external border of the EU (border customs offices in ports, on land roads, at airports and in railway stations). In maritime, rail, postal and air transport, however, the customs office of exit is the customs office responsible for the place where the waste is taken over under a through contract of carriage with destination in a third country.

The EU Commission has published guidelines[97] for customs controls on waste shipments.

5. Additional Requirements

5.1 Transport Rules and Regulations

(Art. 4 (7) b) BC; Chapt. II B.(1) c OECD Decision; EU Directive 2008/68/EC[98])

The Basel Convention, the OECD Decision and the EU law places obligations on countries to require that wastes subject to transfrontier shipment shall be packaged, labelled, and transported in conformity with generally accepted and recognized international rules and standards in the field of packaging, labelling, and transport, and that due account is taken of relevant internationally recognized practices. In particular, while national rules and regulations should be consistent to the extent possible with the "United Nations Recommendations on the Transport of Dangerous Goods" and the

---

[96] Customs Offices of the EU: https://ec.europa.eu/environment/waste/shipments/links.htm

[97] European Commission: "Guidelines for customs controls on transboundary shipments of waste", OJ, C 157, p. 1 of 12.05.2015

[98] Directive 2008/68/EC of the European Parliament and of the Council of 24 September 2008 on the inland transport of dangerous goods, OJ L 260, p13
This Directive applies to the transport of dangerous goods by road, by rail or by inland waterway within or between Member States of the EU

most recent revision of the "Model Regulations on the Transport of Dangerous Goods"[99] it is useful to verify the specifics in the national legislation of the countries where you are operating.

The transportation of hazardous goods is a complex and regulated process involved in all transport stages with widely applied regulatory schemes for all modes of transportation. The basis for most regional, national and international regulatory schemes has been published by the United Nations Economic and Social Council under the UN Recommendations on the Transport of Dangerous Goods.

For instance, the transportation of dangerous goods by sea is governed and regulated by the International Maritime Organization (IMO) under the International Maritime Dangerous Goods Code (IMDG Code) intended for protecting the ship's crew, prevent marine pollution and ensure safe transportation of dangerous goods by sea.

Similarly, the International Civil Aviation Organization has its own regulation for air shipment of hazardous materials that have been developed to accommodate special aspects of moving goods by air. Same applies for rail transport, with the Intergovernmental Organization for International Carriage by Rail that has developed its own regulatory system concerning the International Carriage of Dangerous Goods by Rail (RID).

Concerning road traffic in Europe the "European Agreement concerning the International Carriage of Dangerous Goods by Road" (ADR) was developed. It contains regulations for road traffic regarding packaging, load securing and labeling of goods.

5.1.1 The UN Orange Book

The UN Recommendations and Model Regulations are often referred to as the "Orange Book" due to the color of the cover. The book is updated biennially, in December of even numbered years, although it is not published until the following year. It is known by its revision number, e.g. 21st, and not by any year.

Although not directly applicable to shippers, the Orange Book is of considerable interest as changes to the Model Regulations are likely to be adopted by the international and national authorities in due course. The Model Regulations have, without doubt, reached the pre-eminent position in dangerous goods "legislation" and their use as a basis for worldwide modal regulations contributes both to safety and to the facilitation of trade.

---

[99]   https://unece.org/rev-21-2019

Fig. 10: Overview of Regulations on the Transport of Dangerous Goods

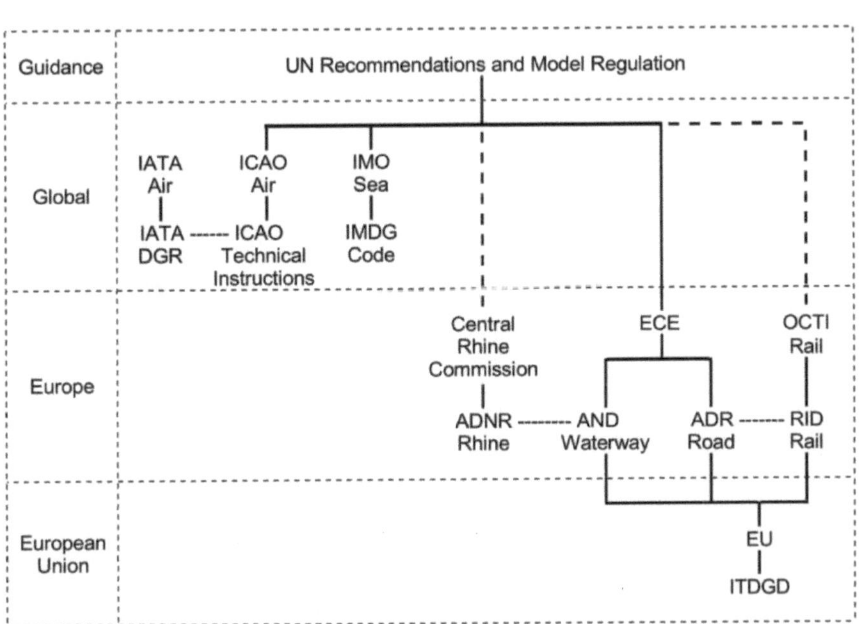

The basic concepts of the UN Model Regulations are listed below:

- Classification rules.

- Nine danger Classes (some split into Divisions).

- The allocation of proper shipping names and UN numbers which enable dangerous goods to be identified anywhere in the world.

- Packaging provisions.

- Intermediate bulk container provisions.

- Multi-modal tank provisions.

- Labelling, marking and placarding requirements.

- Rules for documentation.

5.1.2 ADR Agreement

The Agreement Concerning the International Carriage of Dangerous Goods by Road (ADR) is prepared by the Economic Commission for Europe (ECE), a UN agency which should not be confused with the European Community

(EC) nor with the Commission of the European Union. Details of these can be found on the UN ECE website.[100]

Unlike the international air and sea regulations, ADR generally had covered the European area only, but recently has begun to be acceded to by countries outside Europe.[101] Hence the reason "European" has been dropped from its title.

ADR was not originally based on the UN Recommendations but is now substantially harmonised with the UN. A special feature of ADR is the requirement for some shipments of dangerous goods to be accompanied by transport emergency Instructions in Writing (IiW).

Every truck driver who transports dangerous items needs to acquire an "ADR training certificate". The ADR certificate is valid for five years and then expires automatically if it is not extended before the end of its term. The driver must take a refresher course to do this.

Every company that regularly transports dangerous goods needs at least one dangerous goods officer. The officer is responsible for ensuring that the regulations and requirements relating to hazardous goods are complied with. Every truck carrying this type of goods needs special equipment. This includes the orange warning signs that can be opened and closed, helmet and safety goggles as well as two fire extinguishers.

5.1.3 RID Regulations

The Regulations Concerning the International Carriage of Dangerous Goods by Rail (RID)[102] are part of the international Convention Concerning International Carriage by Rail (COTIF) and covers several Countries.[103] They are developed by a UN Agency, the Central Office for International Rail Transport (OCTI).

Recognising that dangerous goods carried by rail are also likely to require carriage by road at the beginning or end (or both) of their rail journey, the

---

[100] https://unece.org/adr-2019-files

[101] Contracting states are: Albania, Andorra, Austria, Azerbaijan, Belarus, Belgium, Bosnia and Herzegovina, Bulgaria, Croatia, Cyprus, Czech, Denmark, Estonia, Finland, France, Georgia, Germany, Greece, Hungary, Iceland, Ireland, Italy, Kazakhstan, Latvia, Liechtenstein, Lithuania, Luxembourg, Malta, Montenegro, Morocco, the Netherlands, Nigeria, North Macedonia, Norway, Poland, Portugal, the Republic of Moldova, Romania, Russian Federation, San Marino, Serbia, Slovakia, Slovenia, Spain, Sweden, Switzerland, Tajikistan, Tunisia, Turkey, Ukraine, the United Kingdom and Uzbekistan.

[102] French title: Règlement Concernant le Transport International Ferroviaire des Marchandises Dangereuses

[103] Countries covered: Afghanistan, Albania, Algeria, Armenia, Austria, Azerbaijan, Belgium, Bosnia and Herzegovina, Bulgaria, Croatia, Czech Republic, Denmark, Estonia, Finland, France, Georgia, Germany, Greece, Hungary, Iran, Iraq, Ireland, Italy, Latvia, Liechtenstein, Lithuania, Luxembourg, Monaco, Montenegro, Morocco, the Netherlands, North Macedonia, Norway, Poland, Portugal, Romania, Russia, Serbia, Slovakia, Slovenia, Spain, Sweden, Switzerland, Syria, Tunisia, Turkey, Ukraine and the UK.

provisions of ADR, RID and ADN in the areas of classification, packaging, labelling and consignor's documentation have been harmonised. This is achieved through a process of joint meetings, the decisions of which are subsequently ratified in separate ADR and RID plenary sessions at which aspects unique to each mode are discussed and agreed.

5.1.4 ADN

The UN ECE Inland Transport Committee, which produces the ADR Agreement, is also responsible for ADN, the European Provisions Concerning the International Carriage of Dangerous Goods by Inland Waterway.

5.1.5 ICAO Technical Instructions

The International Civil Aviation Organisation (ICAO), a UN agency, is responsible for regulating the carriage of dangerous goods by air. The ICAO Technical Instructions for the Safe Transport of Dangerous Goods by Air (TIs) are published biennially. ICAO adopts the UN Recommendations very quickly and fully, adding additional measures where necessary for airfreight. For example, there are special considerations for receptacles containing liquids. Training provisions are mandatory and enforcement of TIs is strict.

5.1.6 IATA Dangerous Goods Regulations

The International Air Transport Association (IATA) had published its Restricted Articles Regulations for many years prior to the emergence of the ICAO TIs. These were essentially conditions of contract, but were widely enforced by the airlines. When the ICAO TIs were introduced, IATA decided to continue with its own publication, the Dangerous Goods Regulations. These regulations do not have the same official status as the TIs but are rules agreed by the airlines themselves.

The IATA Regulations adopted the provisions of the ICAO TIs, though in a format consistent with the previous IATA rules. IATA did, however, include some unique requirements of its own, including a special format (for the "shippers declaration for dangerous goods"), and some information not then in the TIs. The IATA Dangerous Goods Regulations are published annually.

5.1.7 IMDG Code

The International Maritime Dangerous Goods (IMDG) Code deals with the transport of dangerous goods by sea in packages, intermediate bulk containers (IBCs) and tanks. It is published by the International Maritime Organisation (IMO), a UN agency which was previously known as the Intergovernmental Maritime Consultative Organisation. Other IMO Codes deal with the shipment of solid bulk cargoes and dangerous chemicals in bulk.

I. Explanatory Notes

The IMDG Code is drawn up to assist compliance with the general requirements of the Convention for Safety of Life at Sea (SOLAS) and to assist with harmonisation of the international controls and the facilitation of trade by sea.

IMDG strongly follows the UN Recommendations but also applies its own specific requirements, such as the segregation of goods on board ship and the classification and marking of marine pollutants.

From January 2004 the Code became mandatory for use by all IMO member countries. Though certain aspects (e.g. training of on-shore personnel) of the Code are voluntary, for States to apply if they so wish.

The Code is published every two years and may be used from 1 January of the following odd numbered year and must be used from 1 January of the subsequent even numbered year.

5.1.7.1   IMDG Code Supplement

The IMDG Code Supplement volume contains various materials that are intended to be used in conjunction with the IMDG Code. These are listed below:

- Emergency Procedures (EMS)

- Medical First Aid Guide (MFAG).

- International Code for the Safe Carriage of Packaged Irradiated Nuclear Fuel.

- Reporting Procedures.

- Guidelines for Packing Cargo in Freight Containers or Vehicles.

- Use of Pesticides in Ships.

Some are also available as separate titles, so that they can be purchased and used where possession of the full set is not needed.

Many of these provisions (such as EMS and MFAG) are intended for use only on board ship although it is desirable that the shipper knows of their existence and application.

The IMO, UN ECE, and International Labour Organization joint Guidelines for Packing Cargo in Freight Containers or Vehicles are strongly recommended for use wherever vehicles and freight containers are loaded for subsequent sea transport. The IMDG Reporting Procedures are also helpful in understanding the type of information that is required to be transmitted when packaged dangerous goods are lost from a vessel.

6.   Literature

[1]   "Revised Guide to the Control System", 8 May 2014, UNEP-CHW-COMP-GUID-Control System, downloaded in March 2021:
http://www.basel.int/Implementation/LegalMatters/Compliance/GeneralIssuesActivities/Activities201415/Guidetothecontrolsystem/tabid/3561/Default.aspx

[2]   "Guidance Manual for the Control of Transboundary Movements of Recoverable Wastes", OECD 2009, downloaded in March 2021:
https://www.oecd.org/env/waste/guidance-manual-control-transboundary-movements-recoverable-wastes.pdf

[3]   Wuttke, J.: "Praxishandbuch zur grenzüberschreitenden Abfallverbringung", Verkehrs-Verlag J. Fischer GmbH & Co. KG, Düsseldorf 2021.

[4]   Kummer, Katharina: "International Management of Hazardous Waste – The Basel Convention and Related Legal Rules", Oxford Monographs in International Law, Oxford University Press, Oxford1995

[5]   Szelinski, B.-A., S. Schneider: "Grenzüberschreitende Abfallverbringung", Behr's Verlag, Hamburg 1995

[6]   Bundesministerium für Nachhaltigkeit und Tourismus (Austria): "Bundes-Abfallwirtschafsplan 2017 – Teil 2", Wien, Dezember 2017, downloaded in September 2020:
https://www.bmlrt.gv.at/umwelt/abfall-ressourcen/bundes-abfallwirtschaftsplan/BAWP2017-Final.html

[7]   LAGA Notice 25: "Enforcement Guidance for Regulation (EC) No. 1013/2006 of the European Parliament and of the Council of 14 June 2006 on shipments of waste and for the Waste Shipment Act (Ger.) of 19. July 2007 (WSA)", State May 2017, available in German only at:
http://www.laga-online.de

[8]   Oexle, Epiney, Breuer (Hrsg.): "EG-Abfallverbringungsverordnung – Kommentar", Verlag Carl Heymanns, Köln 2010

[11]   LAGA: "Technische Hinweise zur Einstufung von Abfällen nach ihrer Gefährlichkeit", Berlin März 2019

[12]   Commission Notice on Technical Guidance on the Classification of Waste of 9. April 2018, OJ, C 124, p.1

Appendix 1 – Consolidated Waste Lists (BC, OECD and EU-WSR)

Consolidated waste lists with suggested assignments to hazard properties and to customs codes

The wastes listed in Annexes II, VIII and IX to the Basel Convention, the wastes listed in appendixes 3 and 4 to the OECD-Decision and the wastes listed in Annexes III, IIIA, IIIB and IV to the EU-WSR were merged and consolidated on the following pages. The following is assigned to the respective waste codes and waste designations:

- In the third column, the code numbers of the customs tariff[1] (customs codes) are given where available or assignable. The first six digits of the customs codes are regulated worldwide by the World Customs Organisation (WCO) in the form of the HS Nomenclature.
  The "Combined Nomenclature" of the European Union (EU) integrates the HS Nomenclature and regulates the first eight digits of the customs tariff and applies both in the EU and in the associated states.
- In the fourth column, the respective waste streams and waste constituents (Y) of Annexes I and II to the Basel Convention (cf. Sect. III.1).
- In the fifth column, possible hazard criteria (H) of the Basel Convention (cf. Sect. III.1).
- Notes, comments and references can be found in the sixth column.

Waste codes highlighted in light grey are waste codes from the OECD Decision. The EU has adopted these waste codes. Waste codes highlighted in dark grey are EU-specific codes (see respective comment).

It should be noted that these assignments are suggested classifications. In individual cases, other allocations are possible for specific wastes, and the commentary column contains information on this in some cases.

The assignment of customs codes is not trivial, as customs codes are mostly very specific, whereas this is not always the case for waste codes. Therefore, it is not always possible to achieve a clear and complete correspondence between waste codes and customs codes. In addition, in a correspondence table[2] published by the EU Commission, several customs codes are often assigned to one waste code and vice versa.

---

[1] International Convention on the Harmonized Commodity Description and Coding System, done at Brussels on 14 June 1983

[2] Commission Implementing Regulation (EU) 2016/1245 of 28 July 2016 setting out a preliminary correlation table between codes of the Combined Nomenclature provided for in Council Regulation (EEC) No 2658/87 and entries of waste listed in Annexes III, IV and V to Regulation (EC) No 1013/2006 of the European Parliament and of the Council on shipments of waste, OJ, L 204, 29.7.2016, p. 11–69

("GREEN" LISTED WASTE)

This compilation is based on the legal instruments of the BC, the OECD-Decision and the EU-WSR. Additions and deletions to the basic Waste Lists of the Basel Convention by the OECD and the EU are highlighted by using different characters and a grey background
(light grey for the OECD and dark grey for the EU) throughout the completely consolidated list. The text below the legal instrument is the "chapeau" of the waste list - and the classification of wastes have to be made with respect to the chapeau.

Basel Convention Annex IX, List B

Wastes contained in the Annex will not be wastes covered by Article 1, paragraph 1 (a), of this Convention unless they contain Annex I material to an extent causing them to exhibit an Annex III characteristic.

OECD Decision C(2001)107/Final, Appendix III - List of Wastes Subject to the Green control procedure

Regardless of whether or not wastes are included on this list, they may not be subject to the Green control procedure if they are contaminated by other materials to an extent which

(a) increases the risks associated with the wastes sufficiently to render them appropriate for submission to the amber control procedure, when taking into account the criteria in Appendix 6[3], or

(b) prevents the recovery of the wastes in an environmentally sound manner.

EU-WSR, Annex III - List of Wastes Subject to the General Information Requirements laid down in Article 18[4].

Regardless of whether or not wastes are included on this list, they may not be subject to the general information requirements laid down in Article 18 if they are contaminated by other materials to an extent which

(a) increases the risks associated with the wastes sufficiently to render them appropriate for submission to the procedure of prior written notification and consent, when taking into account the hazardous characteristics listed in Annex III to Directive 91/689/EEC; or

(b) prevents the recovery of the wastes in an environmentally sound manner.

The following wastes shall be subject to the general information requirements laid down in Art. 18[5]:

Wastes listed in Annex IX to the Basel Convention

---

[3] refers to the OECD-Decision C(2001)107/Final
[4] refers to the EU-WSR
[5] refers to the EU-WSR

Consolidated Waste Lists

("GREEN" LISTED WASTE)[6]

List B (Annex IX of Basel Convention[7])

Code	Waste Designation	Customs Code	BC (Y)	BC (H)	Comments
B1	**Metal and metal-bearing wastes**				
B1010	Metal and metal-alloy wastes in metallic, non-dispersible form:				
	- Precious metals (gold, silver, the platinum group, but not mercury)	ex 711230			
	- Iron and steel scrap	ex 720429			
	- Copper scrap	ex 740400			
	- Nickel scrap	ex 750300			
	- Aluminium scrap	ex 60200			
	- Zinc scrap	ex 790200			
	- Tin scrap	ex 800200			
	- Tungsten scrap	ex 810197			
	- Molybdenum scrap	ex 810297			
	- Tantalum scrap	ex 810330			
	- Magnesium scrap	ex 810420			
	- Cobalt scrap	ex 810530			
	- Bismuth scrap	ex 810600			
	- Titanium scrap	ex 810830			
	- Zirconium scrap	ex 810930			
	- Manganese scrap	ex 81110019			
	- Germanium scrap	ex 811292			
	- Vanadium scrap	ex 811292			
	- Scrap of hafnium, indium, niobium, rhenium and gallium	ex 81129221			Custom code for other metals
	- Thorium scrap				
	- Rare earths scrap	ex 811292			
	- Chromium scrap	ex 811222			
B1020	Clean, uncontaminated metal scrap, including alloys, in bulk finished form *i.e. non-dispersible form[8]* (sheet, plate, beams, rods, etc.), of:	ex 811020			
	- Antimony scrap	ex 811213			
	- Beryllium scrap	ex 810730			
	- Cadmium scrap	ex 7802			

---

[6] OECD and EU waste listing system

[7] Any reference in Annex IX of the Basel Convention to list A shall be understood as a reference to Annex IV to the EU-WSR

[8] "Non-dispersible" does not include any wastes in the form of powder, sludge, dust or solid items containing encased hazardous waste liquids

Code	Waste Designation	Customs Code	BC (Y)	BC (H)	Comments
	- Lead scrap (but excluding lead-acid batteries)	ex 811292			
B1030	Refractory metals containing residues	ex 811292			
B1031	Molybdenum, tungsten, titanium, tantalum, niobium and rhenium metal and metal alloy wastes in metallic dispersible form (metal powder), excluding such wastes as specified in list A under entry A1050, Galvanic sludges	ex 810210 (Mo) ex 810110 (W) ex 810820 (Ti) ex 810330 (Ta) ex 811292 (Ni) ex 811292 (Re)			
B1040	Scrap assemblies from electrical power generation not contaminated with lubricating oil, PCB or PCT to an extent to render them hazardous	ex 8500 ex 7204			
B1050	Mixed non-ferrous metal, heavy fraction scrap, not containing Annex I materials in concentrations sufficient to exhibit Annex III characteristics[9]				Customs code according to metal shredder light fraction see A3120
B1060	Waste selenium and tellurium in metallic elemental form including powder	ex 280490(Se) ex 280450(Te)			
B1070	Waste of copper and copper alloys in dispersible form, unless they contain Annex I constituents to an extent that they exhibit Annex III characteristics	ex 7404 ex 262030			
B1080	Zinc ash and residues including zinc alloys residues in dispersible form unless containing Annex I constituents in concentration such as to exhibit Annex III characteristics or exhibiting hazard characteristic H4.3[10]	ex 262019			
B1090	Waste batteries conforming to a specification, excluding those made with lead, cadmium or mercury	ex 854810			Customs code includes accumulators
B1100	Metal-bearing wastes arising from melting, smelting and refining of metals:	ex 2620			
	- Hard zinc spelter	ex 262011			
	- Zinc-containing drosses:	ex 262019			
	- Galvanizing slab zinc top dross (>90% Zn)	ex 262019			

---

[9]   Note that even where low level contamination with Annex I materials initially exists, subsequent processes, including recycling processes, may result in separated fractions containing significantly enhanced concentrations of those Annex I materials.

[10]   The status of zinc ash is currently under review and there is a recommendation with the United Nations Conference on Trade and Development (UNCTAD) that zinc ashes should not be dangerous goods.

Consolidated Waste Lists

Code	Waste Designation	Customs Code	BC (Y)	BC (H)	Comments
	- Galvanizing slab zinc bottom dross (>92% Zn=	ex 262019			
	- Zinc die casting dross (>85% Zn)	ex 262019			
	- Hot dip galvanizers slab zinc dross (batch)(>92% Zn)	ex 262019			
	- Zinc skimmings	ex 262019			
	- Aluminium skimmings (or skims) excluding salt slag	ex 262040			Salt slag is not listed
	— DELETED[11] and replaced by the following entry GB040	ex 62030			*cf. A1090 to A1140*
	- Wastes of refractory linings, including crucibles, originating from copper smelting	ex 6903			
	- Slags from precious metals processing for further refining	ex 711230			
	- Tantalum-bearing tin slags with less than 0.5% tin	ex 810330			
GB040	*Slags[12] from precious metals and copper processing for further refining*	*ex 262030* *ex 7112*			*customs code 7112 for precious metal containing slag; compare A1090 to A1140*
B1110	*DELETED[13], and replaced by the two following entries GC010 and GC020*	*ex 7112*			*Customs code from chap. 85 according to component or device*
GC010	*Electrical assemblies consisting only of metals or alloys*	*ex 7112* *ex 8501* *ex 850*			*Customs code from chap. 85 according to component or device*

---

[11] Deleted entry (in OECD and EU) in the Basel list:
"- Slags from copper processing for further processing or refining not containing arsenic, lead or cadmium to an extent that they exhibit Annex III hazard characteristics"

[12] This OECD entry applies instead of the part of Basel entry B1100 that refers to "slags from copper processing" and is mentioned in the above footnote.

[13] Deleted entry in the Basel list:
"B1110 Electrical and electronic assemblies:
— Electronic assemblies consisting only of metals or alloys
— Waste electrical and electronic assemblies or scrap (including printed circuit boards) not containing components such as accumulators and other batteries included on list A, mercury-switches, glass from cathode-ray tubes and other activated glass and PCB-capacitors, or not contaminated with Annex I constituents (e.g., cadmium, mercury, lead, polychlorinated biphenyl) or from which these have been removed, to an extent that they do not possess any of the characteristics contained in Annex III (note the related entry on list A A1180)
— Electrical and electronic assemblies (including printed circuit boards, electronic components and wires) destined for direct reuse, and not for recycling or final disposal"

Code	Waste Designation	Customs Code	BC (Y)	BC (H)	Comments
GC020	*Electronic scrap (e.g. printed circuit boards, electronic components, wire, etc.) and reclaimed electronic components suitable for base and precious metal recovery*	*ex 7112*			*Customs code from chap. 85 according to component or device*
GC050	*Spent Fluid Catalytic Cracking (FCC) Catalysts (e.g.: aluminium oxide, zeolites)*	*ex 3815* *ex 3825*			
B1115	Waste metal cables coated or insulated with plastics, not included in list A1190, excluding those destined for Annex IVA operations or any other disposal operations involving, at any stage, uncontrolled thermal processes, such as open burning.	ex 7408			
B1120	Spent catalysts excluding liquids used as catalysts, containing any of:	ex 7204 (Fe) ex 7404 (Cu)			Custom code according to material
	— Transition metals, excluding waste catalysts (spent catalysts, liquid used catalysts or other catalysts) on list A: Scandium Titanium Vanadium Chromium Manganese Iron Cobalt Nickel Copper Zinc Yttrium Zirconium Niobium Molybdenum Hafnium Tantalum Tungsten Rhenium	ex 7503 (Ni) ex 7902 (Zn) ex 8101 (W) ex 8102 (Zn) ex 8103 (Ta) ex 8104 (Fe) ex 8105 (Co) ex 8108 (Ti) ex 8109 (Zr) ex 8111 (Mn) ex 8112 (V, Hf, Nb, Re, Cr)			cf. A2030
	— Lanthanides (rare earth metals): Lanthanum Cerium Praseodymium Neodym Samarium Europium Gadolinium Terbium Dysprosium Holmium Erbium Thulium Ytterbium Lutetium	ex 8112			
B1130	Cleaned spent precious-metal-bearing catalysts	ex 7112			Customs code for precious scrap
B1140	Precious-metal-bearing residues in solid form which contain traces of inorganic cyanides	ex 7112			Customs code for precious scrap cyanides $\leq 0.1\%$

Consolidated Waste Lists

Code	Waste Designation	Customs Code	BC (Y)	BC (H)	Comments
B1150	Precious metals and alloy wastes (gold, silver, the platinum group, but not mercury) in a dispersible, non-liquid form with appropriate packaging and labelling	ex 7112			
B1160	Precious-metal ash from the incineration of printed circuit boards (note the related entry on list A A1150)	ex 711230			
B1170	Precious-metal ash from the incineration of photographic film	ex 711230			cf. AD090
B1180	Waste photographic film containing silver halides and metallic silver	ex 711299			cf. AD090
B1190	Waste photographic paper containing silver halides and metallic silver	ex 711299			cf. AD090
B1200	Granulated slag arising from the manufacture of iron and steel	ex 261800			cf. AD090
B1210	Slag arising from the manufacture of iron and steel including slags as a source of TiO_2 and vanadium	ex 261900			cf. AD090
B1220	Slag from zinc production, chemically stabilized, having a high iron content (above 20%) and processed according to industrial specifications (e.g., DIN 4301) mainly for construction	ex 2620			
B1230	Mill scaling arising from the manufacture of iron and steel	ex 261900			
B1240	Copper oxide mill-scale	ex 262030			
B1250	Waste end-of-life motor vehicles, containing neither liquids nor other hazardous components	ex 9705			
GC030	*Vessels and other floating structures for breaking up, properly emptied of any cargo and other materials arising from the operation of the vessel which may have been classified as a dangerous substance or waste*	*ex 890800*			
B2	**Wastes containing principally inorganic constituents, which may contain metals and organic materials**				
B2010	Wastes from mining operations in *non-dispersible form[14]:*				
	- Natural graphite waste	ex 250490			

---

[14] "Non-dispersible" does not include any wastes in the form of powder, sludge, dust or solid items containing encased hazardous waste liquids

Code	Waste Designation	Customs Code	BC (Y)	BC (H)	Comments
	- Slate waste, whether or not roughly trimmed or merely cut, by sawing or otherwise	ex 2514			
	- Mica waste	ex 252530			
	- Leucite, nepheline and nepheline syenite waste	ex 2529.30			
	- Feldspar waste	ex 252910			
	- Fluorspar waste	ex 252921			
	- Silica wastes in solid form excluding those used in foundry operations	ex 250590			
B2020	Glass waste in non-dispersible form:				
	- Cullet and other waste and scrap of glass except for glass from cathode-ray tubes and other activated glasses	ex 700100			
GE020	*Glass Fibre wastes in non-dispersible form[15]*	*ex 7019*			
B2030	Ceramic wastes in non-dispersible form:	ex 811300			
	- Cermet wastes and scrap (metal ceramic composites)	ex 6806			
	- Ceramic based fibres not elsewhere specified or included				
GF010	*Ceramic wastes in non-dispersible form[16] which have been fired after shaping, including ceramic vessels (before and/or after use)*	*ex 69*			
B2040	Other wastes containing principally inorganic constituents:				
	- Partially refined calcium sulphate produced from flue-gas desulphurization (FGD)	ex 382569			
	- Waste gypsum wallboard or plasterboard arising from the demolition of buildings	ex 6809			
	- Slag from copper production, chemically stabilized, having a high iron content (above 20%) and processed according to industrial specifications (e.g., DIN 4301 and DIN 8201) mainly for	ex 262190			

---

[15] "Non-dispersible" does not include any wastes in the form of powder, sludge, dust or solid items containing encased hazardous waste liquids

[16] as previous footnote 8

Consolidated Waste Lists

Code	Waste Designation	Customs Code	BC (Y)	BC (H)	Comments
	construction and abrasive applications				
	- Sulphur in solid form	ex 250300			
	- Limestone from the production of calcium cyan amide (having a pH less than 9)	ex 252100			
	- Sodium, potassium, calcium chlorides	ex 2827			
	- Carborundum (silicon carbide)	ex 284920			
	- Broken concrete	ex 253090			Customs code for other mineral materials
	- Lithium-tantalum and lithium-niobium containing glass scraps	ex 700100			
GG030	*Bottom ash and slag tap from coal-fired power plants*	*ex 262190*			
B2050	*DELETED[17], and replaced by the following entry GG040*	*ex 262190*			
GG040	*Coal-fired power plants fly ash*	*ex 262190*			
B2060	Spent activated carbon not containing any Annex I constituents to an extent they exhibit Annex III characteristics, for example, carbon resulting from the treatment of potable water and processes of the food industry and vitamin production (note the related entry on list A, A4160)	ex 382569			
B2070	Calcium fluoride sludge	ex 382569			Customs code for other wastes from chemical Industry
B2080	Waste gypsum arising from chemical industry processes not included on list A (note the related entry on list A A2040)	ex 382569			Customs code for other wastes from chemical Industry
B2090	Waste anode butts from steel or aluminium production made of petroleum coke or bitumen and cleaned to normal industry specifications (excluding anode butts from chlor alkali electrolyses and from metallurgical industry)	ex 382569			

---

[17]   Deleted entry (OECD and EU) in the Basel list:
"B2050 Coal-fired power plant fly-ash, not included on list A (note the related entry on list A A2060)"

Code	Waste Designation	Customs Code	BC (Y)	BC (H)	Comments
B2100	Waste hydrates of aluminium and waste alumina and residues from alumina production excluding such materials used for gas cleaning, flocculation or filtration processes	ex 262040			
B2110	Bauxite residue ("red mud") (pH moderated to less than 11.5)	ex 62099			Customs code for other slags, ashes and residues
B2120	Waste acidic or basic solutions with a pH greater than 2 and less than 11.5, which are not corrosive or otherwise hazardous (note the related entry on list A A4090)	ex 382569			Since the BC does not apply the criterion "irritant", the pH-values were used for differentiation
B2130	Bituminous material (asphalt waste) from road construction and maintenance, not containing tar[18] (note the related entry on list A, A3200)	ex 2517			
B3	**Wastes containing principally organic constituents, which may contain metals and inorganic materials**				
B3011	Plastic waste (note the related entries Y48 in Annex II and on list A A3210):				For shipments within the EU, EU3011 applies. *OECD countries may apply the old Basel entry B3010*
	— Plastic waste listed below, provided it is destined for recycling[19] in an environmentally sound manner and almost free from contamination and other types of wastes:[20]				
	— Plastic waste almost exclusively[21] consisting of one non-halogenated polymer, including but not limited to the following polymers:				
	– Polyethylene (PE)	ex 391510			
	– Polypropylene (PP)	ex 391590			
	– Polystyrene (PS)	ex 391520			
	– Acrylonitrile butadiene styrene (ABS)	ex 391590			
	– Polyethylene terephthalate (PET)	ex 391590 ex 390760			

---

[18] The concentration level of Benzo [a] pyrene should not be 50 mg/kg or more.

[19] Recycling/reclamation of organic substances that are not used as solvents (R3 in Annex IV, sect. B) or, if needed, temporary storage limited to one instance, provided that it is followed by operation R3 and evidenced by contractual or relevant official documentation.

[20] In relation to "almost free from contamination and other types of wastes", international and national specifications may offer a point of reference.

[21] In relation to "almost exclusively", international and national specifications may offer a point of reference.

Consolidated Waste Lists

Code	Waste Designation	Customs Code	BC (Y)	BC (H)	Comments
	– Polycarbonates (PC)	391590 ex 390740			
	– Polyethers	ex 391590 ex 390720			
	— Plastic waste almost exclusively[22] consisting of one cured resin or condensation product, including but not limited to the following resins:				
	– Urea formaldehyde resins	ex 390910			
	– Phenol formaldehyde resins	ex 390940			
	– Melamine formaldehyde resins	ex 390920			
	– Epoxy resins	ex 390730			
	– Alkyd resins	ex 390750			
	— Plastic waste almost exclusively[23] consisting of one of the following fluorinated polymers:[24]				
	– Perfluoroethylene/propylene (FEP)	ex 390469			
	– Perfluoroalkoxy alkanes:	ex 390469			
	– Tetrafluoroethylene/perfluoroalkylvinyl ether (PFA)	ex 390469			
	– Tetrafluoroethylene/perfluoromethyl vinyl ether (MFA)	ex 390469			
	– Polyvinylfluoride (PVF)	ex 390469			
	– Polyvinylidenefluoride (PVDF)	ex 390469			
	— Mixtures of plastic waste, consisting of polyethylene (PE), polypropylene (PP) and/or polyethylene terephthalate (PET), provided they are destined for separate recycling[25] of each material and in an environmentally sound manner, and almost free from contamination and other types of wastes.[26]				
EU3011	Plastic waste (note the related entry AC300 in part II of Annex IV, and the related entry EU48 in part I of Annex IV):	ex 3915			Entry only applies to waste shipped within the EU
	Plastic waste listed below, provided it is almost free from contamination and other types of waste:[27]				Entry B3011 applies to

---

[22] In relation to "almost exclusively", international and national specifications may offer a point of reference.

[23] In relation to "almost exclusively", international and national specifications may offer a point of reference.

[24] Post-consumer wastes are excluded

[25] Recycling/reclamation of organic substances that are not used as solvents (R3 in Annex IV, sect. B), with prior sorting and, if needed, temporary storage limited to one instance, provided that it is followed by operation R3 and evidenced by contractual or relevant official documentation.

[26] In relation to "almost free from contamination and other types of wastes", international and national specifications may offer a point of reference.

[27] In relation to 'almost free from contamination and other types of wastes', international and national specifications may offer a point of reference

Code	Waste Designation	Customs Code	BC (Y)	BC (H)	Comments
	— Plastic waste almost exclusively[28] consisting of one non-halogenated polymer, including but not limited to the following polymers:				exports to and imports from non EU Member States. *OECD countries may apply the old Basel entry B3010*
	– Polyethylene (PE)	ex 391510			
	– Polypropylene (PP)	ex 391590			
	– Polystyrene (PS)	ex 391520			
	– Acrylonitrile butadiene styrene (ABS)	ex 391590			
	– Polyethylene terephthalate (PET)	ex 391590 ex 390760			
	– Polycarbonates (PC)	391590 ex 390740			
	– Polyethers	ex 391590 ex 390720			
	— Plastic waste almost exclusively[29] consisting of one cured resin or condensation product, including but not limited to the following resins:				
	– Urea formaldehyde resins	ex 390910			
	– Phenol formaldehyde resins	ex 390940			
	– Melamine formaldehyde resins	ex 390920			
	– Epoxy resins	ex 390730			
	– Alkyd resins	ex 390750			
	— Plastic waste almost exclusively[30] consisting of one of the following fluorinated polymers[31]:				
	– Perfluoroethylene/propylene (FEP)	ex 390469			
	– Perfluoroalkoxy alkanes:	ex 390469			
	– Tetrafluoroethylene/perfluo roalkyl vinyl ether (PFA)	ex 390469			
	– Tetrafluoroethylene/perfluo romethyl vinyl ether (MFA)	ex 390469			
	– Polyvinylfluoride (PVF)	ex 390469			
	– Polyvinylidenefluoride (PVDF)	ex 390469			
	– Polytetrafluoroethylene (PTFE)	ex 3904			
	– Polyvinyl chloride (PVC)	ex 391530			
GH013	DELETED in the EU[32], *as included in EU3011*				

---

[28] In relation to 'almost exclusively', international and national specifications may offer a point of reference.

[29] In relation to 'almost exclusively', international and national specifications may offer a point of reference.

[30] In relation to 'almost exclusively', international and national specifications may offer a point of reference

[31] Post consumer wastes are excluded

[32] Deleted OECD-Entry:

Consolidated Waste Lists

Code	Waste Designation	Customs Code	BC (Y)	BC (H)	Comments
B3020	Paper, paperboard and paper product wastes	ex 4707			
	The following materials, provided they are not mixed with hazardous wastes: Waste and scrap of paper or paperboard of:	ex 470790			
	— unbleached paper or paperboard or of corrugated paper or paper-board	ex 470710			
	— other paper or paperboard, made mainly of bleached chemical pulp, not coloured in the mass	ex 470720			
	— paper or paperboard made mainly of mechanical pulp (for example, newspapers, journals and similar printed matter)	ex 470730			
	— other, including but not limited to 1) laminated paperboard 2) unsorted scrap	ex 470790			
B3026	The following waste from the pre-treatment of composite packaging for liquids, not containing Annex I materials in concentrations sufficient to exhibit Annex III characteristics:				
	— Non-separable plastic fraction				
	— Non-separable plastic-aluminium fraction				
B3027	Self-adhesive label laminate waste containing raw materials used in label material				
B3030	Textile wastes				
	The following materials, provided they are not mixed with other wastes and are prepared to a specification:				
	— Silk waste (including cocoons unsuitable for reeling, yarn waste and garnetted stock)	ex 5003			
	– not carded or combed	ex 5003			
	– other	ex 5003			

"GH013   391530   ex 390410-40   Polymers of vinyl chloride"

Code	Waste Designation	Customs Code	BC (Y)	BC (H)	Comments
	— Waste of wool or of fine or coarse animal hair, including yarn waste but excluding garnetted stock	ex 5103			
	− noils of wool or of fine animal hair	ex 510310			
	− other waste of wool or of fine animal hair	ex 510320			
	− waste of coarse animal hair	ex 510330			
	— Cotton waste (including yarn waste and garnetted stock)	ex 5202			
	− yarn waste (including thread waste)	ex 520210			
	− garnetted stock	ex 520291			
	− other	ex 520299			
	— Flax tow and waste	ex 530130			
	— Tow and waste (including yarn waste and garnetted stock) of true hemp (Cannabis sativa L.)	ex 530290			
	— Tow and waste (including yarn waste and garnetted stock) of jute and other textile bast fibres (excluding flax, true hemp and ramie)	ex 530390			
	— Tow and waste (including yarn waste and garnetted stock) of sisal and other textile fibers of the genus Agave	ex 5311			
	— Tow, noils and waste (including yarn waste and garnetted stock) of coconut	ex 5305			
	— Tow, noils and waste (including yarn waste and garnetted stock) of abaca (Manila hemp or Musa textilis Nee)	ex 5305			
	— Tow, noils and waste (including yarn waste and garnetted stock) of ramie and other vegetable textile fibers, not elsewhere specified or included	ex 5305			
	— Waste (including noils, yarn waste and garnetted stock) of man-made fibers	ex 5505			
	− of synthetic fibers	ex 550510			
	− of artificial fibers	ex 550520			
	— Worn clothing and other worn textile articles	ex 6309			

Consolidated Waste Lists

Code	Waste Designation	Customs Code	BC (Y)	BC (H)	Comments
	— Used rags, scrap twine, cordage, rope and cables and worn out articles of twine, cordage, rope or cables of textile materials	ex 6310			
	– sorted	ex 631010			
	– other	ex 631090			
B3035	Waste textile floor coverings, carpets	ex 5701			
B3040	Rubber wastes	ex 3901			
	The following materials, provided they are not mixed with other wastes:				
	— Waste and scrap of hard rubber (e.g., ebonite)	ex 401700			
	— Other rubber wastes (excluding such wastes specified elsewhere)	ex 4004			cf. B3080 and B3140
B3050	Untreated cork and wood waste:	ex 4401			
	— Wood waste and scrap, whether or not agglomerated in logs, briquettes, pellets or similar forms	ex 440130			
	— Cork waste: crushed, granulated or ground cork	ex 450190			
B3060	Wastes arising from agro-food industries provided it is not infectious:				
	— Wine lees	ex 2307			
	— Dried and sterilized vegetable waste, residues and by-products, whether or not in the form of pellets, of a kind used in animal feeding, not elsewhere specified or included	ex 230890			
	— Degras: residues resulting from the treatment of fatty substances or animal or vegetable waxes	ex 1522			
	— Waste of bones and horn-cores, unworked, defatted, simply pre-pared (but not cut to shape), treated with acid or degelatinised	ex 0506			
	— Fish waste	ex 0511			
	— Cocoa shells, husks, skins and other cocoa waste	ex 1802			

Code	Waste Designation	Customs Code	BC (Y)	BC (H)	Comments
	— Other wastes from the agro-food industry excluding by-products which meet national and international requirements and standards for human or animal consumption	ex 210690			
B3065	Waste edible fats and oils of animal or vegetable origin (e.g. frying oils), provided they do not exhibit an Annex III characteristic	ex 1500			
B3070	The following wastes:				
	— Waste of human hair	ex 050100			
	— Waste straw	ex 1213			
	— Deactivated fungus mycelium from penicillin production to be used as animal feed	ex 382561			
B3080	Waste parings and scrap of rubber	ex 4004 ex 4017			
GN010	*Waste of pigs', hogs' or boars' bristles and hair or of badger hair and other brush making hair*	*ex 0502*			
GN020	*Horsehair waste, whether or not put up as a layer with or without supporting material*	*ex 050300*			
GN030	*Waste of skins and other parts of birds, with their feathers or down, of feathers and parts of feathers (whether or not with trimmed edges) and down, not further worked than cleaned, disinfected or treated for preservation*	*ex 050590*			
B3090	Paring and other wastes of leather or of composition leather not suitable for the manufacture of leather articles, excluding leather sludges, not containing hexavalent chromium compounds and biocides (note the related entry on list A A3100)	ex 411520			
B3100	Leather dust, ash, sludges or flours not containing hexavalent chromium compounds or biocides (note the related entry on list A A3090)	ex 411520			
B3110	Fellmongery wastes not containing hexavalent chromium compounds or biocides or infectious substances (note the related entry on list A A3110)	ex 0511			Customs code for other goods of animal origin n.o.s.
B3120	Wastes consisting of food dyes	ex 3203			

Consolidated Waste Lists

Code	Waste Designation	Customs Code	BC (Y)	BC (H)	Comments
B3130	Waste polymer ethers and waste non-hazardous monomer ethers incapable of forming peroxides	ex 3907			
B3140	Waste pneumatic tyres, excluding those destined for Annex IVA operations	ex 401220			
B4	**Wastes which may contain either inorganic or organic constituent**				
B4010	Wastes consisting mainly of water-based/latex paints, inks and hardened varnishes not containing organic solvents, heavy metals or biocides to an extent to render them hazardous (note the related entry on list A A4070)	ex 382561			Customs code for other waste from the chemical or related industries
B4020	Wastes from production, formulation and use of resins, latex, plasticizers, glues/adhesives, not listed on list A, free of solvents and other contaminants to an extent that they do not exhibit Annex III characteristics, e.g., water-based, or glues based on casein starch, dextrin, cellulose ethers, polyvinyl alcohols (note the related entry on list A A3050)	ex 382561			Customs code for other waste from the chemical or related industries
B4030	Used single-use cameras, with batteries not included on list A	ex 9006			

<u>EU-specific:</u>

ANNEX IIIA[33]

MIXTURES OF TWO OR MORE WASTES LISTED IN ANNEX III AND NOT CLASSIFIED UNDER ONE SINGLE ENTRY AS REFERRED TO IN ARTICLE 3(2)

1. Regardless of whether or not mixtures are included on this list, they may not be subject to the general information requirements laid down in Article 18 if they are contaminated by other materials to an extent which:

(a) increases the risks associated with the wastes sufficiently to render them appropriate for submission to the procedure of prior written notification and consent, when taking into account the hazardous characteristics listed in Annex III to Directive 91/689/EEC; or

(b) prevents the recovery of the wastes in an environmentally sound manner.

2. The following mixtures of waste are included in this Annex:

(a) mixtures of waste classified under Basel entries B1010 and B1050;

(b) mixtures of waste classified under Basel entries B1010 and B1070;

(c) mixtures of wastes classified under Basel entries B3040 and B3080;

(d) mixtures of waste classified under (OECD) entry GB040 and under Basel entry B1100 restricted to hard zinc spelter, zinc-containing drosses, aluminium skimmings (or skims) excluding salt slag and wastes of refractory linings, including crucibles, originating from copper smelting;

(e) mixtures of waste classified under (OECD) entry GB040, under Basel entry B1070 and under Basel entry B1100 restricted to wastes of refractory linings, including crucibles, originating from copper smelting.

The entries referred to in points (d) and (e) shall not apply for exports to countries to which the OECD Decision does not apply.

3. The following mixtures of wastes classified under separate indents or sub-indents of one single entry are included in this Annex:

(a) mixtures of wastes classified under Basel entry B1010;

(b) mixtures of wastes classified under Basel entry B2010;

(c) mixtures of wastes classified under Basel entry B2030;

---

[33] Green listed waste according to EU-WSR

d) to f) have been deleted

(g)   mixtures of wastes classified under Basel entry B3020 restricted to unbleached paper or paperboard or of corrugated paper or paperboard, other paper or paperboard, made mainly of bleached chemical pulp, not coloured in the mass, paper or paperboard made mainly of mechanical pulp (for example, newspapers, journals and similar printed matter);

(h)   mixtures of wastes classified under Basel entry B3030;

(i)   mixtures of wastes classified under Basel entry B3040;

(j)   mixtures of wastes classified under Basel entry B3050.

4.   The following mixtures of wastes classified under separate indents or sub-indents of one single entry are included in this Annex only for the purposes of shipments within the Union:

(a)   mixtures of wastes classified under entry EU3011 and listed under the indent referring to non-halogenated polymers;

(b)   mixtures of wastes classified under entry EU3011 and listed under the indent referring to cured resins or condensation products;

(c)   mixtures of wastes classified under entry EU3011 and listed under "perfluoroalkoxy alkanes";

<u>EU-specific:</u>

ANNEX IIIB[34]

ADDITIONAL GREEN LISTED WASTE AWAITING INCLUSION IN THE RELEVANT ANNEXES TO THE BASEL CONVENTION OR THE OECD DECISION AS REFERRED TO IN ARTICLE 58(1)(B)

1. Regardless of whether or not wastes are included on this list, they may not be subject to the general information requirements laid down in Article 18 if they are contaminated by other materials to an extent which:

(a) increases the risks associated with the wastes sufficiently to render them appropriate for submission to the procedure of prior written notification and consent, when taking into account the hazardous characteristics listed in Annex III to Directive 2008/98/EC of the European Parliament and of the Council[35]; or

(b) prevents the recovery of the wastes in an environmentally sound manner.

2. The following wastes are included in this Annex:

BEU04	Composite packaging consisting of mainly paper and some plastic, not containing residues and not covered by Basel entry B3020
BEU05	Clean biodegradable waste from agriculture, horticulture, forestry, gardens, parks and cemeteries

3. The shipments of waste listed in this Annex are without prejudice to the provisions of Council Directive 2000/29/EC[36], including measures adopted pursuant to Article 16(3) thereof.

---

[34] Green listed waste according to EU-WSR
[35] OJ L 312, 22.11.2008, p. 3.
[36] OJ L 169, 10.7.2000, p. 1.

("AMBER" LISTED WASTE)

This compilation is based on the legal instruments of the BC, the OECD-Decision and the EU-WSR. Additions and deletions to the basic Waste Lists of the Basel Convention by the OECD and the EU are highlighted by using different characters and a grey background (light grey for the OECD and dark grey for the EU) throughout the completely consolidated list. The text below the legal instrument is the "chapeau" of the waste list and the classification of wastes have to be made with respect to the chapeau.

Basel Convention Annex VIII, List A

Wastes contained in this Annex are characterized as hazardous under Article 1, para-graph 1 (a), of this Convention and their designation on this Annex does not preclude the use of Annex III to demonstrate that a waste is not hazardous.

OECD Decision C(2001)107/Final, Appendix III - List of Wastes Subject to the Amber control procedure

EU-WSR, Annex IV - List of Wastes Subject to the Procedure of Prior Written Notification and Consent

The following wastes will be subject to the procedure of prior written notification and consent:

Wastes listed in Annexes II and VIII to the Basel Convention[37]:

---

[37]   Any reference in Annex VIII of the Basel Convention to list B shall be understood as a reference to Annex III of the EU-WSR.

("AMBER" LISTED WASTE)[38]

Annex II Basel Convention

Code	Waste Designation	Customs Code	BC (Y)	BC (H)	Comments
Y46	Wastes collected from households	ex 382510	Y46	----	EWL-Code: 20 03 01
Y47	Residues arising from the incineration of household wastes	ex 262110	Y47	----	EWL-Group 19 01 with exception of 19 01 17* and 19 01 18
Y48[39]	Plastic waste, including mixtures of such waste, with the exception of the following:		Y48	---	This entry applies to shipments outside the EU and shipments of EU Member States with non-EU Member States
	— Plastic waste that is hazardous waste (see entry A3210 in list A)				
	— Plastic waste listed below, provided it is destined for recycling[40] in an environmentally sound manner and almost free from contamination and other types of wastes:[41]				
	— Plastic waste almost exclusively[42] consisting of one non-halogenated polymer, including but not limited to the following polymers:				
	– Polyethylene (PE)				
	– Polypropylene (PP)				
	– Polystyrene (PS)				
	– Acrylonitrile butadiene styrene (ABS)				
	– Polyethylene terephthalate (PET)				
	– Polycarbonates (PC)				
	– Polyethers				
	— Plastic waste almost exclusively[43] consisting of one cured resin or condensation product, including but not limited to the following resins:				

---

[38]   The EU-WSR integrates the wastes of Annex II of Basel Convention into the amber list

[39]   In relation to "almost free from contamination and other types of wastes", international and national specifications may offer a point of reference.

[40]   Recycling/reclamation of organic substances that are not used as solvents (R3 in Annex IV, sect. B) or, if needed, temporary storage limited to one instance, provided that it is followed by operation R3 and evidenced by contractual or relevant official documentation.

[41]   In relation to "almost free from contamination and other types of wastes", international and national specifications may offer a point of reference.

[42]   In relation to "almost exclusively", international and national specifications may offer a point of reference.

[43]   In relation to "almost exclusively", international and national specifications may offer a point of reference.

Consolidated Waste Lists

Code	Waste Designation	Customs Code	BC (Y)	BC (H)	Comments
	– Urea formaldehyde resins				
	– Phenol formaldehyde resins				
	– Melamine formaldehyde resins				
	– Epoxy resins				
	– Alkyd resins				
	— Plastic waste almost exclusively[44] consisting of one of the following fluorinated polymers:[45]				
	– Perfluoroethylene/propylene (FEP)				
	– Perfluoroalkoxy alkanes:				
	– Tetrafluoroethylene/perfluoroalkyl vinyl ether (PFA)				
	– Tetrafluoroethylene/perfluoromethyl vinyl ether (MFA)				
	– Polyvinylfluoride (PVF)				
	– Polyvinylidenefluoride (PVDF)				
	— Mixtures of plastic waste, consisting of polyethylene (PE), polypropylene (PP) and/or polyethylene terephthalate (PET), provided they are destined for separate recycling[46] of each material and in an environmentally sound manner, and almost free from contamination and other types of wastes[47].				
EU48	Plastic waste not covered by entry AC300 in part II or by entry EU3011 in part I of Annex III, as well as mixtures of plastic waste not covered by point 4 of Annex III A.		Y48		This entry only applies to waste shipped within the EU

[44] In relation to "almost exclusively", international and national specifications may offer a point of reference.

[45] Post-consumer wastes are excluded

[46] Recycling/reclamation of organic substances that are not used as solvents (R3 in Annex IV, sect. B) or, if needed, temporary storage limited to one instance, provided that it is followed by operation R3 and evidenced by contractual or relevant official documentation.

[47] In relation to "almost free from contamination and other types of wastes", international and national specifications may offer a point of reference.

List A (Annex VIII of Basel Convention

Code	Waste Designation	Customs Code	BC (Y)	BC (H)	Comments
A1	**Metal and metal-bearing wastes**				
A1010	Metal wastes and waste consisting of alloys of any of the following:				
	— Antimony	ex 811020	Y27	H8, H12	
	— Arsenic	ex 280480	Y24	H6.1, (H11)	
	— Beryllium	ex 811213	Y20	H6.1	
	— Cadmium	ex 810730	Y26	H6.1, H12	
	— Lead	ex 7802	Y31	H6.1	
	— Mercury	ex 2805	Y29	H6.1, (H11)	
	— Selenium	ex 280490	Y25	H6.1, H12 (Se-Verb.)	
	— Tellurium	ex 280450	Y28	H6.1	
	— Thallium but excluding such wastes specifically listed on list B *under entry B1020, and which are in non-dispersible form*[48]	ex 811252	Y30	H6.1	
A1020	Waste having as constituents or contaminants, excluding metal waste in massive form, any of the following:	ex 2620			
	— Antimony; antimony compounds	ex 262091	Y27	H8, H12, H6.1 (SbF$_3$)	
	— Beryllium; beryllium compounds	ex 262091	Y20	H6.1	
	— Cadmium; cadmium compounds	ex 262091	Y26	H6.1	
	— Lead; lead compounds	ex 262029	Y31	H11, (H6.1)	
	— Selenium; selenium compounds	ex 262099	Y25	H6.1	
	— Tellurium; tellurium compounds	ex 262099	Y28	H6.1	
A1030	Wastes having as constituents or contaminants any of the following:	ex 2620			
	— Arsenic; arsenic compounds	ex 262060	Y24	H6.1, (H11)	
	— Mercury; mercury compounds	ex 262060	Y29	H6.1, (H11)	
	— Thallium; thallium compounds	ex 262060	Y30	H6.1, (H11)	
A1040	Wastes having as constituents any of the following:	ex 3825			
	— Metal carbonyls		Y19	H6.1, H3	
	— Hexavalent chromium compounds		Y21	H6.1	

---

[48] "Non-dispersible" does not include any wastes in the form of powder, sludge, dust or solid items containing encased hazardous waste liquids

Consolidated Waste Lists

Code	Waste Designation	Customs Code	BC (Y)	BC (H)	Comments
AA010	*Dross, scalings and other wastes from the manufacture of iron and steel[49]*	*ex 26190090*	*--*	*H4.1, H11*	*classified according to Art. 1(1)b BC*
AA060	*Vanadium ashes and residues[50]*	*ex 26209995*	*--*	*H11*	*classified according to Art. 1(1)b BC*
AA190	*Magnesium waste and scrap that is flammable, pyrophoric or emits, upon contact with water, flammable gases in dangerous quantities*	*ex 810420*	*--*	*H4.3, H4.1*	*classified according to Art. 1(1)b BC*
A1050	Galvanic sludges	ex 2620	Y17	H12, (H6.1)	
A1060	Waste liquors from the pickling of metals	ex 382550	Y17,(Y21 Y23)	H8, (H6.1)	
A1070	Leaching residues from zinc processing, dust and sludges such as jarosite, hematite, etc.	ex 262019	Y23	H6.1	
A1080	Waste zinc residues not included on list B, containing lead and cadmium in concentrations sufficient to exhibit Annex III characteristics	ex 262019	Y31, Y26, Y23	H11, H12	cf. B1080 and B1100
A1090	Ashes from the incineration of insulated copper wire	ex 262030	Y22, Y44	H12, H11	cf. B1100 and GB040
A1100	Dusts and residues from gas cleaning systems of copper smelters	ex 262030	Y22	H12, H11	cf. B1100 and GB040
A1110	Spent electrolytic solutions from copper electrorefining and electrowinning operations	ex 26200	Y22	H11	cf. B1100 and GB040
A1120	Waste sludges, excluding anode slimes, from electrolyte purification systems in copper electrorefining and electrowinning operations	ex 262030	Y22	H12, H11	cf. B1100 and GB040
A1130	Spent etching solutions containing dissolved copper	ex 262030	Y22, Y34	H11, H8	cf. B1100 and GB040
A1140	Waste cupric chloride and copper cyanide catalysts	ex 262030	Y22	H12 (CuCl$_2$) H6.1 (CuCN)	cf. B1100 and GB040
A1150	Precious metal ash from incineration of printed circuit boards not included on list B[51]	ex 711230	Y22	H11	cf. B1160
A1160	Waste lead-acid batteries, whole or crushed	ex 854810	Y31, (Y34)	H6.1, H11, (H8)	customs code includes also primary batteries and – elements

[49] This listing includes wastes in the form of ash, residue, slag, dross, skimming, scaling, dust, powder, sludge and cake, unless a material is expressly listed elsewhere.

[50] as mentioned in the above footnote

[51] Note that mirror entry on list B (B1160) does not specify exceptions.

Code	Waste Designation	Customs Code	BC (Y)	BC (H)	Comments
A1170	Unsorted waste batteries excluding mixtures of only list B batteries. Waste batteries not specified on list B containing Annex I constituents to an extent to render them hazardous	ex 854810	Y26, Y29, Y31	H6.1, H11	cf. B1090 this includes lithium batteries (H4.2) according to Art. 1(1)b BC
A1180	*does not apply[52], and the OECD entries GC010 and GC020 apply instead when appropriate[53]*	*ex 7112*	*depending on the constituent*	*depending on the effect of the constituent*	*customs code from Chap. 85 depending on the component or device cf. B1110, GC010 and GC020*
A1190	Waste metal cables coated or insulated with plastics containing or contaminated with coal tar, PCB[54], lead, cadmium, other organohalogen compounds or other Annex I constituents to an extent that they exhibit Annex III characteristics.	ex 7408	Y11, Y10, Y31, Y26, Y43, Y44, Y45	H11	cf. B1115
A2	**Wastes containing principally inorganic constituents, which may contain metals and organic materials**				
A2010	Glass waste from cathode-ray tubes and other activated glasses	ex 7001	Y31	H6.1, H11	Customs code for cathode ray tube devices: 8528
A2020	Waste inorganic fluorine compounds in the form of liquids or sludges but excluding such wastes specified on list B	ex 382569	Y32	H6.1, (H8)	Customs code for other wastes from the chem. Industry or related industry n.o.s.

---

[52]　entry in the Basel list not applied (in OECD and EU):
　　"A1180　Waste electrical and electronic assemblies or scrap containing components such as accumulators and other batteries included on list A, mercury-switches, glass from cathode-ray tubes and other activated glass and PCB-capacitors, or contaminated with Annex I constituents (e.g., cadmium, mercury, lead, polychlorinated biphenyl) to an extent that they possess any of the characteristics contained in Annex III (note the related entry on list B B1110)"

[53]　OECD Member countries may control these wastes differently in accordance with Chapter II B 6 of Decision C(2001) 107/FINAL concerning wastes not listed in Appendices 3 or 4, and the chapeau of Appendix 3 to the OECD Decision. Concerning this entry the Correspondents' meeting of EU has approved Correspondents' Guideline No 6.

[54]　PCBs are at a concentration level of 50 mg/kg or more.

Consolidated Waste Lists

Code	Waste Designation	Customs Code	BC (Y)	BC (H)	Comments
A2030	Waste catalysts but excluding such wastes specified on list B	ex 2620 ex 3825	Y2, Y4, Y5, Y6, Y31, etc.	H11, H6.1, H4.2	Customs code depending on the constituent or origin; cf. B1120 and B1130
A2040	Waste gypsum arising from chemical industry processes, when containing Annex I constituents to the extent that it exhibits an Annex III hazardous characteristic (note the related entry on list B B2080)	ex 382569	Y18	H8, H11 and according to the effect of the respective constituent	Customs code for other wastes from the chem. Industry or related industry n.o.s.
A2050	Waste asbestos (dusts and fibres)	ex 2524	Y36	H11	
RB020	*Ceramic-based fibres of physico-chemical characteristics similar to those of asbestos*	*ex 6815*		*H11*	*classified according to Art. 1(1)b BC*
A2060	*does not apply[55] and the OECD entry GG040 applies instead when appropriate[56]*	*ex 262190*			
AB030	*Wastes from non-cyanide based systems which arise from surface treatment of metals*	*ex 3825*	*(Y17)*	*H8, H6.1*	*classified according to Art. 1(1)b BC*
AB070	*Sands used in foundry operations*	*ex 3824*	*--*	*H6.1, H8*	*classified according to Art. 1(1)b BC*
AB120[57]	*Inorganic halide compounds, not elsewhere specified or included*	*ex 3825* *ex 3824*	*(Y45)*	*H11*	*classified according to Art. 1(1)b BC*
AB130	*Used blasting grit*				
AB150	*Unrefined calcium sulphite and calcium sulphate from flue gas desulphurization (FGD)*				

---

[55] entry in the Basel list not applied:
"A2060 Coal-fired power plant fly-ash containing Annex I substances in concentrations sufficient to exhibit Annex III characteristics (note the related entry on list B B2050)"

[56] OECD Member countries may control these wastes differently in accordance with Chapter II B 6 of Decision C(2001) 107/FINAL concerning wastes not listed in Appendices 3 or 4, and the chapeau of Appendix 3 to the OECD Decision. Concerning this entry the Correspondence meeting of EU has approved Correspondence Guideline No 4.

[57] This entry includes spent potlinings from aluminium smelting without inorganic cyanides but containing Y32, inorganic fluorine compounds excluding calcium fluoride.

Code	Waste Designation	Customs Code	BC (Y)	BC (H)	Comments
A3	**Wastes containing principally organic constituents, which may contain metals and inorganic materials**				
A3010	Waste from the production or processing of petroleum coke and bitumen	ex 32710	Y11	H11, H6.1	Customs code for other waste from the chemical or related industries cf. B2090
A3020	Waste mineral oils unfit for their originally intended use	ex271099	Y8	H6.1, (H3)	
A3030	Wastes that contain, consist of or are contaminated with leaded anti-knock compound sludges	ex 262021	Y31	H6.1, H4.1	
A3040	Waste thermal (heat transfer) fluids	ex 3825	Y8, Y45, Y41	H11, H12	Customs code depending on constituent
AC060	*Hydraulic fluids*	*ex 3819*	*(Y8, Y9, Y10, Y42)*	*H3, H6.1, H10, H11*	*classified according to Art. 1(1)b BC*
AC070	*Brake fluids*	*ex 3819*	*(Y9, Y42)*	*H6.1, H10*	*classified according to Art. 1(1)b BC*
AC080	*Antifreeze fluids*	*ex 3820*	*(Y42)*	*H3*	*classified according to Art. 1(1)b BC*
A3050	Wastes from production, formulation and use of resins, latex, plasticizers, glues/adhesives excluding such wastes specified on list B (note the related entry on list B B4020)	ex 382561	Y13	H3, H4.1	Customs code for other waste from the chemical or related industries cf. B3010
A3060	Waste nitrocellulose	ex 3912	Y15	H1	Customs code for other waste from the chemical or related industries
A3070	Waste phenols, phenol compounds including chlorophenol in the form of liquids or sludges	ex 382561	Y39	H6.1, H11	Customs code for other waste from the chemical or related industries cf. B3010

Consolidated Waste Lists

Code	Waste Designation	Customs Code	BC (Y)	BC (H)	Comments
A3080	Waste ethers not including those specified on list B	ex 382561	Y40	H3	Customs code for other waste from the chemical or related industries cf. B3110
AC150	*Chlorofluorocarbons*	*ex 290339*	*--*	*(H12)*	*classified according to Art. 1(1)b BC: HP14*
AC160	*Halons*	*ex 3903*	*--*	*(H12)*	*classified according to Art. 1(1)b BC: HP14*
AC250	*Surface active agents (surfactants)*	*ex 3401*	*--*	*(H12)*	*classified according to Art. 1(1)b BC: HP14*
A3090	Waste leather dust, ash, sludges and flours when containing hexavalent chromium compounds or biocides (note the related entry on list B B3100)	ex 4115.20	Y21, Y4	H6.1, (H11)	
A3100	Waste paring and other waste of leather or of composition leather not suitable for the manufacture of leather articles containing hexavalent chromium compounds or biocides (note the related entry on list B B3090)	ex 411520	Y21, Y4	H6.1, (H11)	
A3110	Fellmongery wastes containing hexavalent chromium compounds or biocides or infectious substances (note the related entry on list B B3110)	ex 0511	Y21, Y4	H6.2, H11	Customs code for other waste from the chemical or related industries
A3120	FLUFF — light fraction from shredding	ex 3825	Y18	H4.1	Customs code depending on constituent
AC170	*Treated cork and wood wastes*	*ex 4401 ex 4501 ex 3825*	*--*	*(H11)*	*Parts of the material stream[58] are to be classified as hazardous waste according to Art. 1 (1) b BC*

---

[58] In Germany the hazardous waste wood groups according to the German Waste Wood Ordinance

Code	Waste Designation	Customs Code	BC (Y)	BC (H)	Comments
AC260	*Liquid pig manure; faeces*	*ex 3101*	--	--	*Subject to control in accordance with the OECD Decision*
AC270	*Liquid pig manure; faeces*	*ex 3401*	--	--	*Subject to control in accordance with the OECD Decision*
A3130	Waste organic phosphorous compounds	ex 382561	Y37	H6.1	
A3140	Waste non-halogenated organic solvents but excluding such wastes specified on list B	ex 382549	Y42, Y6, Y12	H12, (H3)	
A3150	Waste halogenated organic solvents	ex 382541	Y41, Y6	H6.1, (H3)	
A3160	Waste halogenated or unhalogenated non-aqueous distillation residues arising from organic solvent recovery operations	ex 382561	Y18, Y6	H3, H6.1, H12	Customs code for other waste from the chemical or related industries
A3170	Wastes arising from the production of aliphatic halogenated hydrocarbons (such as chloromethane, dichloroethane, vinyl chloride, vinylidene chloride, allyl chloride and epichlorhydrin)	ex 382541	Y45	H6.1, H3	
A3180	Wastes, substances and articles containing, consisting of or contaminated with polychlorinated biphenyl (PCB), polychlorinated terphenyl (PCT), polychlorinated naphthalene (PCN) or polybrominated biphenyl (PBB), or any other polybrominated analogues of these compounds, at a concentration level of 50 mg/kg or more[59]	ex 271091	Y10, Y45	H11, H6.1	
A3190	Waste tarry residues (excluding asphalt cements) arising from refining, distillation and any pyrolytic treatment of organic materials	ex 382561	Y11	H11, H6.1	Customs code for other waste from the chemical or related industries cf. B2130
A3200	Bituminous material (asphalt waste) from road construction and maintenance, containing tar (note the related entry on list B, B2130)	ex 2715	Y11	H11	

---

[59] The 50 mg/kg level is considered to be an internationally practical level for all wastes. However, many individual countries have established lower regulatory levels (e.g., 20 mg/kg) for specific wastes.

Consolidated Waste Lists

Code	Waste Designation	Customs Code	BC (Y)	BC (H)	Comments
A3210	Plastic waste, including mixtures of such waste, containing or contaminated with Annex I constituents, to an extent that it exhibits an Annex III characteristic (note the related entry B3011, in list B and entry Y48 in list A)		according to the respective constituent	according to the effect of the respective constituent	*Entry applies to import from non-OECD and non-EU Member States In the EU and OECD the entry AC300 applies instead*
AC300	*Plastic waste, including mixtures of such wastes, containing or contaminated with Annex I constituents, to an extent that it exhibits an Annex III characteristic (note the related entry EU3011 in part I of Annex III, and the related entry EU48 in part I)*				*Entry applies to shipments within the EU/OECD*
A4	**Wastes which may contain either inorganic or organic constituents**				
A4010	Wastes from the production, preparation and use of pharmaceutical products but excluding such wastes specified on list B	ex 300692	Y2	H3, H8, H6.1, H11	Customs code for pharmaceutical waste
A4020	Clinical and related wastes; that is wastes arising from medical, nursing, dental, veterinary, or similar practices, and wastes generated in hospitals or other facilities during the investigation or treatment of patients, or research projects	ex 382530			
A4030	Wastes from the production, formulation and use of biocides and phytopharmaceuticals, including waste pesticides and herbicides which are off-specification, outdated,[60] or unfit for their originally intended use	ex 382561	Y4	H6.1, H11	Customs code for other waste from the chemical or related industries
A4040	Wastes from the manufacture, formulation and use of wood-preserving chemicals[61]	ex 382561	Y5 (Y22, Y24)	H6.1, H11	Customs code for other waste from the chemical or related industries

---

[60] "Outdated" means unused within the period recommended by the manufacturer

[61] This entry does not include wood treated with wood preserving chemicals.

Code	Waste Designation	Customs Code	BC (Y)	BC (H)	Comments
A4050[62]	Wastes that contain, consist of or are contaminated with any of the following:				cf. B1140
	— Inorganic cyanides, excepting precious-metal-bearing residues in solid form containing traces of inorganic cyanides	ex 382569	Y33, Y7, Y17	H6.1	Customs code for other waste from the chemical or related industries
	— Organic cyanides	ex 382569	Y38, Y7	H6.1	
A4060	Waste oils/water, hydrocarbons/water mixtures, emulsions	ex 271099	Y9	H12	
A4070	Wastes from the production, formulation and use of Inks, dyes, pigments, paints, lacquers, varnish excluding any such waste specified on list B (note the related entry on list B B4010)	ex 382561	Y12	H3, H4.1, H6.1	Customs code for other waste from the chemical or related industries
AD090	*Wastes from production, formulation and use of reprographic and photographic chemicals and materials not elsewhere specified or included*	*ex 3825*	*(Y16)*	*H8, H6.1, H11*	*classified according to Art. 1(1)b BC, cf. B1180, B1190 and B4030*
AD100	*Wastes from non-cyanide based systems which arise from surface treatment of plastics*	*ex 3825*	*(Y17)*	*H11*	*classified according to Art. 1(1)b BC*
AD120	*Ion exchange resins*	*ex 3914*	*(Y34, Y35)*	*H8*	*classified according to Art. 1(1)b BC*
A4080	Wastes of an explosive nature (but excluding such wastes specified on list B)	ex 3601 ex 3602 ex 3604	Y15	H1	
A4090	Waste acidic or basic solutions, other than those specified in the corresponding entry on list B (note the related entry on list B B2120)	ex 382569	Y34 Y35	H8	Customs code for other waste from the chemical or related industries
A4100	Wastes from industrial pollution control devices for cleaning of industrial off-gases but excluding such wastes specified on list B	ex 382569	Y18	H6.1, H11, H12	Customs code for other waste from the chemical or related industries cf. GG040

---

62   This entry includes spent potlinings from aluminium smelting containing Y33, inorganic cyanides. If the cyanides have been destroyed, spent potlinings are assigned to Part II entry AB120 because they contain Y32, inorganic fluorine compounds excluding calcium fluoride.

Consolidated Waste Lists

Code	Waste Designation	Customs Code	BC (Y)	BC (H)	Comments
AD150	*Naturally occurring organic material used as a filter medium (such as bio-filters)*	*ex 3825*			*Subject to control in accordance with the OECD Decision*
A4110	Wastes that contain, consist of or are contaminated with any of the following:	ex 3825			
	— Any congenor of polychlorinated dibenzo-furan		Y43	H11, (H6.1)	
	— Any congenor of polychlorinated dibenzo-dioxin		Y44	H11, (H6.1)	
A4120	Wastes that contain, consist of or are contaminated with peroxides	ex 382561	Y2, Y4, Y5, Y6	H5.1	Customs code for other waste from the chemical or related industries
A4130	Waste packages and containers containing Annex I substances in concentrations sufficient to exhibit Annex III hazard characteristics	ex 3923	depending on contamination	according to the effect of the respective constituent	
A4140	Waste consisting of or containing off specification or outdated[63] chemicals corresponding to Annex I categories and exhibiting Annex III hazard characteristics	ex 28 ex 29	according to the respective constituent	according to the effect of the respective constituent	
A4150	Waste chemical substances arising from research and development or teaching activities which are not identified and/or are new and whose effects on human health and/or the environment are not known	ex 3824 ex 3825	according to the respective constituent	according to the effect of the respective constituent	
A4160	Spent activated carbon not included on list B (note the related entry on list B B2060)	ex 3825	Y2, Y6, Y12, Y18	depending on the substances accumulated	Customs code for other waste from the chemical or related industries

---

[63] "Outdated" means unused within the period recommended by the manufacturer

Appendix 2 - European Waste List with Allocation of Possible Constituents and Hazard Property

Preliminary remarks

The European Waste List (EWL) is a harmonised, non-exhaustive list of wastes, i.e. a list that is regularly reviewed and amended as necessary. However, the inclusion of a substance in the EWL does not mean that the substance is a waste in all circumstances. The entry is only relevant if the definition of waste applies.

Hazard criteria and substance streams/ingredients

Possible constituents/substances are assigned to each EWL-code in the respective column. In further columns, proposals for allocation to waste streams/waste constituents (Y) and typical Basel Convention hazard criteria (H), as well as to possible EU-WSR waste codes and typical EU hazardous properties (HP) are included.

If the term "Art. 1 (1) b" is found in the 7th column (Y, Basel), the waste is classified as hazardous according to the EWL, but not according to the criteria of the Basel Convention.

If there is a "partly" in the 8th column (H, Basel), this means that only part of the waste stream covered by the EWL is also to be classified as hazardous waste according to the Basel Convention. Other partial streams are only classified as hazardous according to Annex III of the Waste Framework Directive (EU-WFD).

Abbreviations and explanations of the column headings

Abbreviation	Explanation
*	Hazardous waste according to European Waste List (EWL)
Ag	Silver
Al	Aluminium
AOX	Absorbable organic halogen compounds
Art. 1 (1) b	Hazardous waste according to Article 1 (1) b of Basel Convention (national definition)
As	Arsenic
B	Boron
Ba	Barium
B[a]P	Benz[a]pyrene

European Waste List

Abbreviation	Explanation
BeO	Beryllium oxide
CH_4	Methane
CaO	Caustic lime
Cd	Cadmium
CdSe	Cadmium selenide
Co	Cobalt
Code	Waste code according to European Waste List (EWL)
Cr	Chromium
CN	Cyanide
EOX	Extractable organic halogens
psbl.	possibly
EWL-Code	see code
Fe_2O_3	Iron oxide
$Fe(OH)_2$	Iron hydroxide
H	The column "typical H-criteria" contain the most likely applicable H-criteria of Annex III to the Basel Convention. In specific cases also other H-criteria may apply. Please note that these criteria correspond to the classification system of UN-RTDG
H_2	Hydrogen
HC	Hydrocarbons
HCB	Hexachlorobenzene (Perchlorobenzene)
HF	Hydrofluoric acid
Hg	Mercury
HM	Heavy Metals
HP	The column "typical HP-criteria" contain the most likely applicable HP-criteria. In specific cases also other HP-criteria may apply. Please note that these criteria are based on the classification system of the EU-CLP Regulation
KOH	Potassium hydroxide
Mg	Magnesium
MnO	Manganesoxide
Mo	Molybdenum
NaF/AlF_3	Natriumfluoride/aluminiumfloride
NaOH	Sodium hydroxide
Na_2S	Sodium sulphide
NH_3	Ammonia
Ni	Nickel
NL	Not listed; this means that no appropriate code of Green or Amber List

Abbreviation	Explanation
	corresponds to the EWL-code of
Pb	Lead
PbS	Lead sulphide
PAH	Polycyclic Hydrocarbons
partly	only a part of the waste streams covered by the EWL is also classified as hazardous under the Basel Convention
PCDD/PCDF	Polychlorinated dibenzo-p-dioxins and dibenzofurans
PCN	Polychlorinated Naphthalene
PCP	Pentachlorophenol
POP	Persistant organic pollutant
Possible con-stituents / substances	The column "Possible constituents/substances" contains information on typical constituents and on possible/typical hazardous components of the wastes classified under the waste code
R-Cl	Chlorinated Hydrocarbons
R-X	Halogenated Hydrocarbons
Sb	Antimony
$SiCl_4$	Siliciumtetrachloride
Sn	Tin
TBT	Tributyltin
Tl	Thallium
V	Vanadium
WSR	Waste Shipment Regulation (EU-WSR)
WSR-Codes	The WSR codes in this column give only an indication where suitable codes may be found (no differentiation of different indents). In several cases also OECD Codes (green list codes GX, amber list codes AX) are listed too.
Y-code	The column "possible Y-code" gives reference to a typical Y-code (Annex I or II to the Basel Convention). In specific cases other Y-codes may apply
Zn	Zinc
$ZnCl_2$	Zinc cloride

Code	H	Designation	possible constituents/substances	EU		Basel	
				typical HP-Criteria	possible WSR-Codes	possible Y-Codes	typical H- Criteria
01 01 00		**WASTES RESULTING FROM EXPLORATION, MI-NING, QUARRYING, AND PHYSICAL AND CHEMI-CAL TREATMENT OF MINERALS**					
01 01 00		**wastes from mineral excavation**					
01 01 01		wastes from mineral metalliferous excavation			NL, B2010		
01 01 02		wastes from mineral non-metalliferous excavation			NL, B2010		
01 03 00		**wastes from physical and chemical processing of metalliferous minerals**					
01 03 04	*	acid-generating tailings from processing of sulphide ore	HM (As, Cu, Tl, etc.)	HP8, HP7, HP14, HP10	NL	Y24, Y22, Y30	H8, H11, H10, H12
01 03 05	*	other tailings containing hazardous substances	inorganic cyanides, HM, Hg	HP6, HP7, HP10	A1010, A1030, A4050	Y33, Y29, Y30	H6.1, H11
01 03 06		tailings other than those mentioned in 01 03 04 and 01 03 05			NL, B2010		
01 03 07	*	other wastes containing hazardous substances from physical and chemical processing of metal-liferous minerals	cyanidic sludges, inorgan c cyanide, alkaline, HM, As Cd, Hg,	HP6, HP8	A4050	Y35, Y33, Y29, Y24, Y26	**partly** (H8)
01 03 08		dusty and powdery wastes other than those mentioned in 01 03 07			NL, B2010		
01 03 09		red mud from alumina production other than the wastes mentioned in 01 03 10			B2110		
01 03 10	*	red mud from alumina production containing hazardous substances other than the wastes mentioned in 01 03 07		HP4, HP8, HP14	NL	Y35	H8, H12
01 03 99		wastes not otherwise specified			NL		
01 04 00		**wastes from physical and chemical processing of non-metalliferous minerals**					

Code	H	Designation	possible constituents/substances	EU		Basel	
				typical HP-Criteria	possible WSR-Codes	possible Y-Codes	typical H- Criteria
01 04 07	*	wastes containing hazardous substances from physical and chemical processing of non-metalliferous minerals	HM, As, Pb, Ni, Cr, …	HP15	NL	**Art. 1 (1) b**	(H13)
01 04 08		waste gravel and crushed rocks other than those mentioned in 01 04 07			NL, B2010		
01 04 09		waste sand and clays			NL, B2010		
01 04 10		dusty and powdery wastes other than those mentioned in 01 04 07			NL		
01 04 11		wastes from potash and rock salt processing other than those mentioned in 01 04 07			NL		
01 04 12		tailings and other wastes from washing and cleaning of minerals other than those mentioned in 01 04 07 and 01 04 11			NL, B2010		
01 04 13		wastes from stone cutting and sawing other than those mentioned in 01 04 07			NL, B2010		
01 04 99		wastes not otherwise specified			NL		
01 05 00		**drilling muds and other drilling wastes**					
01 05 04		freshwater drilling muds and wastes			NL		
01 05 05	*	oil-containing drilling muds and wastes	HC, PAH, oil	HP3, HP14	A3020	Y9	H3, H12, H4.1
01 05 06	*	drilling muds and other drilling wastes containing hazardous substances	HC, PAH, Ba, solvents, detergents	HP14, HP13	A3020	Y9	H12, H4.1
01 05 07		barite-containing drilling muds and wastes other than those mentioned in 01 05 05 and 01 05 06			NL		
01 05 08		chloride-containing drilling muds and wastes other than those mentioned in 01 05 05 and 01 05 06			NL		
01 05 99		wastes not otherwise specified			NL		

Code	H	Designation	possible constituents/substances	EU		Basel	
				typical HP-Criteria	possible WSR-Codes	possible Y-Codes	typical H- Criteria
02 00 00		**WASTES FROM AGRICULTURE, HORTICULTURE, AQUACULTURE, FORESTRY, HUNTING AND FISHING, FOOD PREPARATION AND PROCESSING**					
02 01 00		**wastes from agriculture, horticulture, aquaculture, forestry, hunting and fishing**					
02 01 01		sludges from washing and cleaning			NL, B3060		
02 01 02		animal-tissue waste			B3060, NL		
02 01 03		plant-tissue waste			B3060, NL		
02 01 04		waste plastics (except packaging)			B3011/EU3011, (B3010), B3130 (GH013)		
02 01 06		animal faeces, urine and manure (including spoiled straw), effluent, collected separately and treated off-site			AC260		
02 01 07		wastes from forestry			B3050		
02 01 08	*	agrochemical waste containing hazardous substances	pesticides, rodenticides, disinfectant agents, psbl, fertilisers	HP4, HP5, HP6, HP3, H7, HP14	A4140, A4030, A4040, A4130	Y4	H6.1, H3, H8, H11, H12
2		agrochemical waste other than those mentioned in 02 01 08			NL		
02 01 10		waste metal			B1010, B1020		
02 01 99		wastes not otherwise specified			NL		
02 02 00		**wastes from the preparation and processing of meat, fish and other foods of animal origin**					
02 02 01		sludges from washing and cleaning			NL, B3060		
02 02 02		animal-tissue waste			B3060, GN010, GN20, GN30, NL		
02 02 03		materials unsuitable for consumption or processing			B3060, GN010, GN20, GN30, NL		

Code	H	Designation	possible constituents/substances	EU		Basel	
				typical HP-Criteria	possible WSR-Codes	possible Y-Codes	typical H- Criteria
02 02 04		sludges from on-site effluent treatment			AC270		
02 02 99		wastes not otherwise specified			NL		
02 03 00		**wastes from fruit, vegetables, cereals, edible oils, cocoa, coffee, tea and tobacco preparation and processing; conserve production; yeast and yeast extract production, molasses preparation and fermentation**					
02 03 01		sludges from washing, cleaning, peeling, centrifuging and separation			NL, B3060		
02 03 02		wastes from preserving agents			NL		
02 03 03		wastes from solvent extraction			NL, B3060, B3120		
02 03 04		materials unsuitable for consumption or processing			NL, B3060		
02 03 05		sludges from on-site effluent treatment			AC270		
02 03 99		wastes not otherwise specified			NL		
02 04 00		**wastes from sugar processing**					
02 04 01		soil from cleaning and washing beet			NL		
02 04 02		off-specification calcium carbonate			NL		
02 04 03		sludges from on-site effluent treatment			AC270		
02 04 99		wastes not otherwise specified			NL		
02 05 00		**wastes from the dairy products industry**					
02 05 01		materials unsuitable for consumption or processing			B3060, NL		
02 05 02		sludges from on-site effluent treatment			AC270		
02 05 99		wastes not otherwise specified			NL		
02 06 00		**wastes from the baking and confectionery industry**					
02 06 01		materials unsuitable for consumption or processing			B3060, NL		
02 06 02		wastes from preserving agents			NL		
02 06 03		sludges from on-site effluent treatment			AC270		

Code	H	Designation	possible constituents/substances	EU		Basel	
				typical HP-Criteria	possible WSR-Codes	possible Y-Codes	typical H- Criteria
02 06 99		wastes not otherwise specified			NL		
02 07 00		**wastes from the production of alcoholic and non-alcoholic beverages (except coffee, tea and cocoa)**					
02 07 01		wastes from washing, cleaning and mechanical reduction of raw materials			B3060, NL		
02 07 02		wastes from spirits distillation			NL, B3060		
02 07 03		wastes from chemical treatment			NL		
02 07 04		materials unsuitable for consumption or processing			B3060, NL		
02 07 05		sludges from on-site effluent treatment			AC270		
02 07 99		wastes not otherwise specified			NL		
03 00 00		**WASTES FROM WOOD PROCESSING AND THE PRODUCTION OF PANELS AND FURNITURE, PULP, PAPER AND CARDBOARD**					
03 01 00		**wastes from wood processing and the production of panels and furniture**					
03 01 01		waste bark and cork			B3050		
03 01 04	*	sawdust, shavings, cuttings, wood, particle board and veneer containing hazardous substances	wood preservatives, carba-mates, tar oil (creosote), PAH, PCP, B[a]P, HM, Cr, F, B, As, Cu, Hg	HP3, HP6, HP7	AC170	Y5, Y11, Y21, Y24, Y31	H4.1, H3, H6.1, H11
03 01 05		sawdust, shavings, cuttings, wood, particle board and veneer other than those mentioned in 03 01 04			B3050		
03 01 99		wastes not otherwise specified			NL		

172

Code	H	Designation	possible constituents/substances	EU		Basel	
				typical HP-Criteria	possible WSR-Codes	possible Y-Codes	typical H- Criteria
03 02 00		**wastes from wood preservation**					
03 02 01	*	non-halogenated organic wood preservatives	carbamates, phosphoric ester, pyrethrum, pyrethro-ide, etc.	HP7, HP6, HP5, HP4, HP14	A4040, A4030	Y5, Y4	H6.1, H11, H12
03 02 02	*	organochlorinated wood preservatives	lindan, fluazinam, etc., psbl, chlorophenoles (PCP)	HP7, HP6, HP5, HP4, HP14, HP11, HP10	A4040, A4030	Y5, Y4	H6.1, H11, H12
03 02 03	*	organometallic wood preservatives	Zn-carbamate, Cu-carba-mate, dithiocarbamate, etc.	HP7, HP6, HP5, HP4, HP14	A4040, A4030	Y5, Y4, Y23	H6.1, H11, H12
03 02 04	*	inorganic wood preservatives	chromium salts, Cr-, Cu-ar-senic compounds, fluoride compounds, borate, etc.	HP6, HP5, H7, HP14, HP4	A4040, A4030	Y5, Y4, Y22, Y23, Y24	H6.1, H11, H12
03 02 05	*	other wood preservatives containing hazardous substances	tar oil (creosote), PAH	HP6, HP7, HP14, HP4, HP11, HP10	A4040, A4030	Y5, Y4, Y11,	H6.1, H11, H12
03 02 99		wood preservatives not otherwise specified			NL		
03 03 00		**wastes from pulp, paper and cardboard production and processing**					
03 03 01		waste bark and wood			B3050		
03 03 02		green liquor sludge (from recovery of cooking liquor)			NL		
03 03 05		de-inking sludges from paper recycling			NL, AC270		
03 03 07		mechanically separated rejects from pulping of waste paper and cardboard			NL		
03 03 08		wastes from sorting of paper and cardboard destined for recycling			B3020		
03 03 09		lime mud waste			NL		
03 03 10		fibre rejects, fibre-, filler- and coating-sludges from mechanical separation			NL, AC270		

Code	H	Designation	possible constituents/substances	EU		Basel	
				typical HP-Criteria	possible WSR-Codes	possible Y-Codes	typical H- Criteria
03 03 11		sludges from on-site effluent treatment other than those mentioned in 03 03 10			AC270		
03 03 99		wastes not otherwise specified			NL		
04 00 00		**WASTES FROM THE LEATHER, FUR AND TEXTILE INDUSTRIES**					
04 01 00		**wastes from the leather and fur industry**					
04 01 01		fleshings and lime split wastes			NL, B3090		
04 01 02		liming waste			NL, B3090		
04 01 03	*	degreasing wastes containing solvents without a liquid phase	HC, solvent containing residues	HP3, HP6, HP5	A4060, A3140	Y6, Y9, Y42	H4.1, H6.1, H11
04 01 04		tanning liquor containing chromium			NL		
04 01 05		tanning liquor free of chromium			NL		
04 01 06		sludges, in particular from on-site effluent treatment containing chromium			B3100, A3090, AC270		
04 01 07		sludges, in particular from on-site effluent treatment free of chromium			B3100, AC270		
04 01 08		waste tanned leather (blue sheetings, shavings, cuttings, buffing dust) containing chromium			B3090, A3100		
04 01 09		wastes from dressing and finishing			B3060, B3100. B3110, GN010, GN020, GN030		
04 01 99		wastes not otherwise specified			NL		
04 02 00		**wastes from the textile industry**					
04 02 09		wastes from composite materials (impregnated textile, elastomer, plastomer)			B3030, B3035, B3090		
04 02 10		organic matter from natural products (for example grease, wax)			NL		
04 02 14	*	wastes from finishing containing organic solvents	HC, solvents, fluorine compounds	HP3, HP14	A3140	Y42	H3, H12

174

Code	H	Designation	possible constituents/substances	EU		Basel	
				typical HP-Criteria	possible WSR-Codes	possible Y-Codes	typical H- Criteria
04 02 15		wastes from finishing other than those mentioned in 04 02 14			NL, B3030		
04 02 16	*	dyestuffs and pigments containing hazardous substances	HM. azo dyes, solvents, aniline, diazonium com-pounds	HP7, HP3, HP4, HP14	A4140, A4070	Y12	H3, H11, H4.1, H12
04 02 17		dyestuffs and pigments other than those mentioned in 04 02 16			B4010		
04 02 19	*	sludges from on-site effluent treatment containing hazardous substances	HM, HC, CaO, diazonium compounds, solvents	HP4, HP14, HP6	AC270	Art. 1 (1) b	(H12, H6.1)
04 02 20		sludges from on-site effluent treatment other than those mentioned in 04 02 19			AC270		
04 02 21		wastes from unprocessed textile fibres			B3060		
04 02 22		wastes from processed textile fibres			B3060		
04 02 99		wastes not otherwise specified			NL		
05 00 00		**WASTES FROM PETROLEUM REFINING, NATURAL GAS PURIFICATION AND PYROLYTIC TREATMENT OF COAL**					
05 01 00		**wastes from petroleum refining**					
05 01 02	*	desalter sludges	HM, HC	HP4, HP5, HP14	A3020	Y9	H11, H12
05 01 03	*	tank bottom sludges	HC, PAH, Pb, Ni	HP6, HP10, HP14	NL, A4060	Y9	H6.1, H11, H12
05 01 04	*	acid alkyl sludges	HC, sulphuric acid	HP8, HP14	NL, A4090	Y9, Y34	H8, H12
05 01 05	*	oil spills	HC	HP3, HP14	A3020	Y8, Y9	H3, H12
05 01 06	*	oily sludges from maintenance operations of the plant or equipment	solvents, HC, PAH, Pb, metal particles (Cr, Ni), psbl, sulphides	HP3, HP7, HP10, HP6, HP14	A4060	Y9	H3, H6.1, H11, H12
05 01 07	*	acid tars	sulphuric acid, HC, PAH, tar	HP8, HP4, HP6, HP7, HP14	A3190, A4090	Y11	H8, H6.1, H11, H12

Code	H	Designation	possible constituents/substances	EU		Basel	
				typical HP-Criteria	possible WSR-Codes	possible Y-Codes	typical H- Criteria
05 01 08	*	other tars	PAHs, tar, HC, phenols	HP7, HP4, HP6, HP14	A3190	Y11	H6.1, H12
05 01 09	*	sludges from on-site effluent treatment containing hazardous substances	HC, PAHs, Cr, Ni, CaO	HP5, HP7, HP4, HP14, HP10	AC270	**Art. 1 (1) b**	(H11, H12)
05 01 10		sludges from on-site effluent treatment other than those mentioned in 05 01 09			AC270		
05 01 11	*	wastes from cleaning of fuels with bases	HC, free alkaline	HP4, HP8, HP14, (HP3)	A4060, A4090	Y9, Y35	H8, H12, (H3)
05 01 12	*	oil containing acids	HC, free acid, HM, Ni, Pb, V	HP7, HP8, HP3, HP4, HP10	A4090	Y9, Y34	H11, H8, H3
05 01 13		boiler feedwater sludges			NL		
05 01 14		wastes from cooling columns			NL		
05 01 15	*	spent filter clays	HC, PAHs	HP3, HP14,	A4060, A3020	Y9	H4.1, H12
05 01 16		sulphur-containing wastes from petroleum desulphurisation			B2040		
05 01 17		Bitumen			B2130		
05 01 99		wastes not otherwise specified			NL		
05 06 00		**wastes from the pyrolytic treatment of coal**					
05 06 01	*	acid tars	free acid, PAHs, HC	HP4, HP7, HP8, HP14	A4090	Y11	H8, H11, H12
05 06 03	*	other tars	phenols, tar oil, HC, PAH,	HP4, HP7, HP8, HP3, HP14, HP6	A3190, A3010	Y11	H8, H4.1, H3, H6.1, H11, H12
05 06 04		waste from cooling columns			NL		
05 06 99		wastes not otherwise specified			NL		
05 07 00		**wastes from natural gas purification and trans-portation**					

Code	H	Designation	possible constituents/substances	EU		Basel	
				typical HP-Criteria	possible WSR-Codes	possible Y-Codes	typical H- Criteria
05 07 01	*	wastes containing mercury	Hg compounds, activated carbon	HP6, HP12, HP14	A1030	Y29	H6.1, H11, H10, H12
05 07 02		wastes containing sulphur			B2040		
05 07 99		wastes not otherwise specified			NL		
06 00 00		**WASTES FROM INORGANIC CHEMICAL PROCES-SES**					
06 01 00		**wastes from the manufacture, formulation, supply and use (MFSU) of acids**					
06 01 01	*	sulphuric acid and sulphurous acid	acid	HP8	A4090	Y34	H8
06 01 02	*	hydrochloric acid	acid	HP8	A4090	Y34	H8
06 01 03	*	hydrofluoric acid	acid	HP8	A4090	Y34	H8
06 01 04	*	phosphoric and phosphorous acid	acid	HP8	A4090	Y34	H8
06 01 05	*	nitric acid and nitrous acid	acid	HP8	A4090	Y34	H8
06 01 06	*	other acids	acid	HP8	A4090	Y34	H8
06 01 99		wastes not otherwise specified			B2120		
06 02 00		**wastes from the MFSU of bases**					
06 02 01	*	calcium hydroxide	alkaline	HP8, HP4	A4090	Y35	H8
06 02 03	*	ammonium hydroxide	alkaline	HP8, HP4	A4090	Y35	H8
06 02 04	*	sodium and potassium hydroxide	alkaline	HP8, HP4	A4090	Y35	H8
06 02 05	*	other bases	alkaline	HP8, HP4	A4090	Y35	H8
06 02 99		wastes not otherwise specified			B2120		
06 03 00		**wastes from the MFSU of salts and their solutions and metallic oxides**					
06 03 11	*	solid salts and solutions containing cyanides	cyanides, HM	HP6, HP14	A4050	Y7, Y37, Y33	H6.1, H12
06 03 13	*	solid salts and solutions containing heavy metals	HM, Cr(VI), Cu, Zn, As, Cd, Sb, Pb, Ni	HP7, HP10, HP6, HP14	A1020, A1030	Y21, Y22, Y23, Y24, Y26, Y27, Y29, Y31	H11, H6.1, H12

Code	H	Designation	possible constituents/substances	EU		Basel	
				typical HP-Criteria	possible WSR-Codes	possible Y-Codes	typical H- Criteria
06 03 14		solid salts and solutions other than those mentioned in 06 03 11 and 06 03 13			NL, B2070, B2040		
06 03 15	*	metallic oxides containing heavy metals	oxides of Pb, Ni, As, Cr, Cd, (BeO)	HP6, HP7, HP10	A1020, A1030, AB030, AB100	Y21, Y22, Y23, Y24, Y26, Y27, Y29, Y31	**partly** (H11, H6.1)
06 03 16		metallic oxides other than those mentioned in 06 03 15			NL, B2100, B2040, B1080, B1240		
06 03 99		wastes not otherwise specified			NL		
06 04 00		**metal-containing wastes other than those mentioned in 06 03**					
06 04 03	*	wastes containing arsenic	As (V), As (III)	HP6, HP7, HP12	A1030	Y24	H6.1, H10, H11
06 04 04	*	wastes containing mercury	Hg	HP6, HP7	A1030	Y29	H6.1, H11
06 04 05	*	wastes containing other heavy metals	Ni, Cr, Cd, Pb, etc.	HP7, HP10, HP6,	A1020, A1030	Y20, Y21, Y22, Y23, Y24, Y26, Y27, Y29, Y31	**partly** (H11, H6.1)
06 04 99		wastes not otherwise specified			NL		
06 05 00		**sludges from on-site effluent treatment**					
06 05 02	*	sludges from on-site effluent treatment containing hazardous substances	HM, fluorides, borates	HP4, HP8	AC270	**Art. 1 (1) b**	(H8)
06 05 03		sludges from on-site effluent treatment other than those mentioned in 06 05 02			AC270		

178

Code	H	Designation	possible constituents/substances	EU		Basel	
				typical HP-Criteria	possible WSR-Codes	possible Y-Codes	typical H- Criteria
06 06 00		**wastes from the MFSU of sulphur chemicals, sulphur chemical processes and desulphurisation processes**					
06 06 02	*	wastes containing hazardous sulphides	alkali metals sulphides, NiS, CdS, PbS, Na$_2$S, etc.	HP12, HP6, HP7	A1020, A4140	Y 34	**partly** (H10, H11, H6.1)
06 06 03		wastes containing sulphides other than those mentioned in 06 06 02			NL, AB150		
06 06 99		wastes not otherwise specified			NL		
06 07 00		**wastes from the MFSU of halogens and halogen chemical processes**					
06 07 01	*	wastes containing asbestos from electrolysis	Asbestos, alkaline	HP7, HP4	A2050	Y36	H11
06 07 02	*	activated carbon from chlorine production	Hg, (AOX), psbl, PCDD/PCDF	HP6, HP11, HP3	A4160, A1030	Y29, Y43, Y44	H6.1, H11, H4.1, H4.2
06 07 03	*	barium sulphate sludge containing mercury	Hg	HP6, HP14	A1030	Y29	H6.1, H12
06 07 04	*	solutions and acids, for example contact acid	acid, HM	HP8, HP7	A4090	Y34	H8, H11
06 07 99		wastes not otherwise specified			NL		
06 08 00		**wastes from the MFSU of silicon and silicon derivatives**					
06 08 02	*	waste containing hazardous chlorosilanes	HCl, AOX, SiCl$_4$	HP3, HP4, HP6, HP8	A4140	Y13	H3, H8, H4.3, H6.1
06 08 99		wastes not otherwise specified			NL		
06 09 00		**wastes from the MSFU of phosphorous chemicals and phosphorous chemical processes**					
06 09 02		phosphorous slag			B1200		
06 09 03	*	calcium-based reaction wastes containing or contaminated with hazardous substances	caustic lime, soluble fluoride compounds, psbl, As, Ni, Cr, Cd	HP4, HP6, HP7	A1020, A1030, A2020, A2040	Y35	H6.1, H11
06 09 04		calcium-based reaction wastes other than those mentioned in 06 09 03			B2040, B2070, B2080		

Code	H	Designation	possible constituents/substances	EU		Basel	
				typical HP-Criteria	possible WSR-Codes	possible Y-Codes	typical H- Criteria
06 09 99		wastes not otherwise specified			NL		
06 10 00		**wastes from the MFSU of nitrogen chemicals, nitrogen chemical processes and fertiliser manufacture**					
06 10 02	*	wastes containing hazardous substances	halogenated solvents,	(HP2, HP5, HP3)	NL	**Art. 1 (1) b**	(H4.3, H5.1, H11)
06 10 99		wastes not otherwise specified			NL		
06 11 00		**wastes from the manufacture of inorganic pigments and opacificiers**					
06 11 01		calcium-based reaction wastes from titanium dioxide production			NL		
06 11 99		wastes not otherwise specified			NL		
06 13 00		**wastes from inorganic chemical processes not otherwise specified**					
06 13 01	*	inorganic plant protection products, wood-preserving agents and other biocides.	wood preservatives, bio-cides, Cr, As, B, F, Cu	HP6, HP7, HP3, HP14	A4030, A1020	Y4, Y45, Y5	H6.1, H11, H3, H12
06 13 02	*	spent activated carbon (except 06 07 02)	HM, active carbon, sol-vents	HP3	NL, A4160	**Art. 1 (1) b**	(H4.2, H4.1)
06 13 03		carbon black			NL		
06 13 04	*	wastes from asbestos processing	free asbestos fibres	HP7	A2050	Y36	H11
06 13 05	*	Soot	PAH, psbl, Cr, Ni, V, tar	HP7	A3190	Y11	H11
06 13 99		wastes not otherwise specified			NL, A3190	Y11	H11
07 00 00		**WASTES FROM ORGANIC CHEMICAL PROCESSES**					
07 01 00		**wastes from the manufacture, formulation, sup-ply and use (MFSU) of basic organic chemicals**					
07 01 01	*	aqueous washing liquids and mother liquors	HC, AOX, acids, alkaline, phenols	HP4, HP8, HP6, HP7, HP14	A4090	Y34, Y35	H11, H8, H6.1, H12

| Code | H | Designation | possible constituents/substances | EU | | Basel | |
				typical HP-Criteria	possible WSR-Codes	possible Y-Codes	typical H- Criteria
07 01 03	*	organic halogenated solvents, washing liquids and mother liquors	AOX, chlorophenols, aromatic R-Cl	HP2, HP3, HP5, HP6, HP7, H14	A3150	Y42, Y41	H11, H6.1, H5.1, H11, H12
07 01 04	*	other organic solvents, washing liquids and mother liquors	HC, AOX,	HP2, HP3, HP4, HP6, HP7, HP14	A3140	Y41, Y42	H11, H3, H6.1, H5.1, H12
07 01 07	*	halogenated still bottoms and reaction residues	PAH, R-Cl, HM	HP6, HP7, HP5, HP14	A3160	Y11, Y41	H6.1, H11, H12
07 01 08	*	other still bottoms and reaction residues	PAH, R-Cl, HM	HP3, HP7, HP6, HP5, HP14	A3160	Y11, Y42	H6.1, H3, H11, H12
07 01 09	*	halogenated filter cakes and spent absorbents	HC, AOX, acids, alkaline, etc.	HP2, HP3, HP6, HP7, HP14	A4160, A3170	Y41	H6.1, H3, H11, H5.1, H12
07 01 10	*	other filter cakes and spent absorbents	HC, acids, alkaline, etc.	HP3, HP6, HP7, HP5, HP14	A4160	Y42	H6.1, H3, H11
07 01 11	*	sludges from on-site effluent treatment containing hazardous substances	HC, AOX, CaO, HM	HP4, HP7, HP5. HP14	AC270	(Y18)	(H11, H12)
07 01 12		sludges from on-site effluent treatment other than those mentioned in 07 01 11			AC270		
07 01 99		wastes not otherwise specified			NL		
07 02 00		**wastes from the MFSU of plastics, synthetic rubber and man-made fibres**					
07 02 01	*	aqueous washing liquids and mother liquors	acids, alkaline, sulphides, metal compounds	HP4, HP8, HP7	A4090	Y34, Y35, Y13, Y41, Y42	H8, H11
07 02 03	*	organic halogenated solvents, washing liquids and mother liquors	AOX, chlorophenols, aromatic R-Cl	HP6, HP7, HP3, HP5, HP14	A3150	Y41, Y13	H6.1, H11, H3, H12

Code	H	Designation	possible constituents/substances	EU		Basel	
				typical HP-Criteria	possible WSR-Codes	possible Y-Codes	typical H- Criteria
07 02 04	*	other organic solvents, washing liquids and mother liquors	HC, AOX	HP3, HP4, HP6, HP7, HP5, HP14	A3140	Y42, Y13	H6.1, H11, H3, H12
07 02 07	*	halogenated still bottoms and reaction residues	HC, PAH, R-Cl, HM	HP6, HP7, HP5, HP3, HP14	A3160	Y11, Y41 Y13	H6.1, H3, H11, H12
07 02 08	*	other still bottoms and reaction residues	HC, PAH, HM	HP3, HP6, HP7, HP5, HP14	A3160	Y42, Y13	H6.1, H11, H3, H12
07 02 09	*	halogenated filter cakes and spent absorbents	HC, AOX, (acids and alka-line)	HP3, HP6, HP7, HP5, HP14	A4160	Y41, Y13	H6.1, H11, H3, H12
07 02 10	*	other filter cakes and spent absorbents	HC, AOX, acids, alkaline	HP3, HP6, HP7, HP5, HP14	A4160	Y42, Y13	H6.1, H3, H11, H12
07 02 11	*	sludges from on-site effluent treatment containing hazardous substances	HC, CaO, AOX, HM	HP4, HP7, HP6, HP14	AC270	Art. 1 (1) b	(H11, H6.1, H12)
07 02 12		sludges from on-site effluent treatment other than those mentioned in 07 02 11			AC270		
07 02 13		waste plastic			B3011/EU3011, (B3010), B3130, (GH013)		
07 02 14	*	wastes from additives containing hazardous substances	chlorinated paraffin's, plas-ticizers (phthalates), perox-ides, flame retardants, per-oxides, AOX, HM	HP5, HP6, HP7, HP14	A3050	Y13	H11, H6.1, H12
07 02 15		wastes from additives other than those mentioned in 07 02 14			NL		
07 02 16	*	waste containing hazardous silicones	silicon oil; not fully reacted chlorosilanes, tetraethyl silicate	HP3, HP6, HP14	A4140	Y13	H3, H4.1, H6.1, H12

Code	H	Designation	possible constituents/substances	EU		Basel	
				typical HP-Criteria	possible WSR-Codes	possible Y-Codes	typical H- Criteria
07 02 17		waste containing silicones other than those mentioned in 07 02 16			B3011/EU3011, B4020		
07 02 99		wastes not otherwise specified			NL		
07 03 00		**wastes from the MFSU of organic dyes and pigments (except 06 11)**					
07 03 01	*	aqueous washing liquids and mother liquors	inorg. fluorocompounds, alkaline, sulphides, metal compounds	HP4, HP8, HP6	A4070	Y12 Y34, Y35	H6.1, H8
07 03 03	*	organic halogenated solvents, washing liquids and mother liquors	AOX, chlorophenols, R-Cl	HP6, HP7, HP3, HP5	A3150, A4070	Y12, Y41	H3, H6.1, H11
07 03 04	*	other organic solvents, washing liquids and mother liquors	HC, solvents, Sn	HP3, HP6, HP7	A3140, A4070	Y12, Y42,	H3, H6.1, H11
07 03 07	*	halogenated still bottoms and reaction residues	HC, PAH, R-Cl, HM, Sn	HP6, HP7, HP3, HP5	A3150, A4070	Y11, Y12, Y41	H6.1, H11, H3
07 03 08	*	other still bottoms and reaction residues	HC, PAH, HM	HP3, HP7, HP6, HP5	A3160, A4070	Y11, Y12, Y42	H6.1, H3, H11
07 03 09	*	halogenated filter cakes and spent absorbents	HC, AOX, psbl, acids and alkaline	HP3, HP6, HP7, HP5	A4070	Y12, Y41	H6.1, H11, H3
07 03 10	*	other filter cakes and spent absorbents	HC, acids, alkaline etc.	HP3, HP6, HP7, HP5	A4070	Y12, Y42	H6.1, H11, H3
07 03 11	*	sludges from on-site effluent treatment containing hazardous substances	HC, CaO, AOX, HM	HP4, HP7, HP8, HP6, HP5	AC270	Y12, (Y18)	H11, H8, H6.1
07 03 12		sludges from on-site effluent treatment other than those mentioned in 07 03 11			AC270		
07 03 99		wastes not otherwise specified			NL, B4010		
07 04 00		**wastes from the MFSU of organic plant protection products (except 02 01 08 and 02 01 09), wood preserving agents (except 03 02) and other biocides**					
07 04 01	*	aqueous washing liquids and mother liquors	acids, alkaline, sulphides, metal compounds	HP4, HP8, HP6, HP7	A4040	Y34, Y35, Y4, Y5	H6.1, H11, H8

Code	H	Designation	possible constituents/substances	EU		Basel	
				typical HP-Criteria	possible WSR-Codes	possible Y-Codes	typical H- Criteria
07 04 03	*	organic halogenated solvents, washing liquids and mother liquors	AOX, chlorophenols, R-Cl	HP6, HP7, HP5	A3150, A4040	Y41, Y4	H6.1, H11
07 04 04	*	other organic solvents, washing liquids and mother liquors	HC, Solvents	HP3, HP6, HP7, HP5	A3140, A4040	Y42, Y4, Y5	H6.1, H11, H3
07 04 07	*	halogenated still bottoms and reaction residues	HC, PAH, R-Cl, HM	HP6, HP7, HP10, HP11	A3160, A4040	Y3, Y41, Y4, Y5	H6.1, H11
07 04 08	*	other still bottoms and reaction residues	HC, PAH, HM	HP3, HP6, HP7, HP5	A3160, A4040	Y6, Y42, Y4, Y5	H6.1, H11, H3
07 04 09	*	halogenated filter cakes and spent absorbents	HC, AOX, psbl, acids and alkaline	HP3, HP6, HP7, HP5	A4040	Y6, Y41, Y4, Y5	H6.1, H11, H3
07 04 10	*	other filter cakes and spent absorbents	HC, acids, alkaline, etc.	HP3, HP6, HP7, HP5	A4040	Y3, Y6, Y42, Y4, Y5	H6.1, H3, H11, H4.1
07 04 11	*	sludges from on-site effluent treatment containing hazardous substances	HC, CaO, AOX, HM	HP6, HP7, HP4, HP5	AC270	Y4, (Y18)	H6.1, H11
07 04 12		sludges from on-site effluent treatment other than those mentioned in 07 04 11			AC270		
07 04 13	*	solid wastes containing hazardous substances	HM, aromatic HC	HP6, HP7, HP5, HP3	A4040	Y4, Y5	H6.1, H11, H3
07 04 99		wastes not otherwise specified			NL		
07 05 00		**wastes from the MFSU of pharmaceuticals**					
07 05 01	*	aqueous washing liquids and mother liquors	acids, alkaline, sulphides, metal compounds	HP4, HP6, HP7, (HP8)	A4010	Y2, Y34, Y35,	H6.1, H11, (H8)
07 05 03	*	organic halogenated solvents, washing liquids and mother liquors	AOX, chlorophenols, aromatic R-Cl	HP6, HP7, HP5, HP3	A3150, A4010	Y2, Y41	H3, H6.1, H11
07 05 04	*	other organic solvents, washing liquids and mother liquors	HC, Solvents	HP3, HP6, HP7, HP5	A3140, A4010	Y2, Y42	H3, H6.1, H11
07 05 07	*	halogenated still bottoms and reaction residues	HC, PAH, aromatic R-Cl. HM	HP6, HP7, HP5, HP3	A3160, A4010	Y2, Y41	H6.1, H11, H3
07 05 08	*	other still bottoms and reaction residues	HC, PAH, HM	HP3, HP6, HP7, HP5	A3160, A4010	Y2, Y42	H6,1, H11, H3

Code	H	Designation	possible constituents/substances	EU		Basel	
				typical HP-Criteria	possible WSR-Codes	possible Y-Codes	typical H- Criteria
07 05 09	*	halogenated filter cakes and spent absorbents	HC, AOX, psbl, acids and alkaline	HP3, HP6, HP7, HP5	A4010	Y2, Y41	H6.1, H3, H11, H4.1
07 05 10	*	other filter cakes and spent absorbents	HC, acids, alkaline, etc.	HP3, HP6, HP7, HP5	A4010	Y42, Y2	H6.1, H3, H11, H4.1
07 05 11	*	sludges from on-site effluent treatment containing hazardous substances	HC, caustic lime, AOX, HM	HP4, HP7, H10, HP6	AC270	Art. 1 (1) b	(H11, H6.1)
07 05 12		sludges from on-site effluent treatment other than those mentioned in 07 05 11			AC270		
07 05 13	*	solid wastes containing hazardous substances	biocides	HP6, HP7, HP11	A4010	Y2, Y4	H11, H6.1
07 05 14		solid wastes other than those mentioned in 07 05 13			NL		
07 05 99		wastes not otherwise specified			NL		
07 06 00		**wastes from the MFSU of fats, grease, soaps, detergents, disinfectants and cosmetics**					
07 06 01	*	aqueous washing liquids and mother liquors	acids, alkaline, sulphides, metal compounds	HP4, HP8, HP6, HP7	A4090	Y34, Y35, Y4	H6.1, H8, H11
07 06 03	*	organic halogenated solvents, washing liquids and mother liquors	AOX, chlorophenols, aromatic R-Cl,	HP6, HP7, HP3, HP5	A3150	Y41, Y4	H3, H6.1, H11
07 06 04	*	other organic solvents, washing liquids and mother liquors	HC, Solvents	HP3, HP6, HP7, HP5	A3140	Y42, Y4	H3, H6.1, H11
07 06 07	*	halogenated still bottoms and reaction residues	HC, PAH, HM, aromatic R-Cl	HP6, HP7, HP14, HP5	A3160	Y11, Y41, Y4	H6.1, H11, H12
07 06 08	*	other still bottoms and reaction residues	HC, PAHs, HM	HP3, HP6, HP7, HP5	A3160	Y11 Y42, Y4	H6.1, H11, H3
07 06 09	*	halogenated filter cakes and spent absorbents	HC, AOX, psbl, acids and alkaline	HP3, HP6, HP7, HP5	A4160	Y41, Y4	H6.1, H11, H3
07 06 10	*	other filter cakes and spent absorbents	HC, acids, alkaline, etc.	HP3, HP6, HP7, HP5	A4160	Y42, Y4	H6.1, H11, H3
07 06 11	*	sludges from on-site effluent treatment containing hazardous substances	HC, CaO, AOX, HM	HP4, HP7, HP5, HP6	AC270	(Y4, Y18)	(H11, H6.1)

185

Code	H	Designation	possible constituents/substances	EU		Basel	
				typical HP-Criteria	possible WSR-Codes	possible Y-Codes	typical H- Criteria
07 06 12		sludges from on-site effluent treatment other than those mentioned in 07 06 11			AC270		
07 06 99		wastes not otherwise specified			NL		
07 07 00		**wastes from the MFSU of fine chemicals and chemical products not otherwise specified**					
07 07 01	*	aqueous washing liquids and mother liquors	acids, alkaline, sulphides, metal compounds	HP4, HP8, HP6, HP7	A4090	Y14, Y16, Y34, Y35	H8, H6.1, H11
07 07 03	*	organic halogenated solvents, washing liquids and mother liquors	AOX, chlorophenols, aromatic R-Cl	HP6, HP7, HP8, HP5	A3150	Y14, Y41	H6.1, H11, H8
07 07 04	*	other organic solvents, washing liquids and mother liquors	HC, Solvents	HP3, HP6, HP4, HP7, HP5	A3140	Y42	H11, H3, H6.1
07 07 07	*	halogenated still bottoms and reaction residues	HC, PAH, arom. R-Cl, HM	HP6, HP7, HP5	A3160	Y11, Y41	H6.1, H11
07 07 08	*	other still bottoms and reaction residues	HC, PAH, HM	HP3, HP5, HP6, HP7	A3160	Y11, Y42	H6.1, H3, H11
07 07 09	*	halogenated filter cakes and spent absorbents	HC, AOX, psbl, acids and alkaline	HP3, HP6, HP7, HP5	A3170, A4160	Y41	H6.1, H3, H11
07 07 10	*	other filter cakes and spent absorbents	HC, acids, alkaline, etc.	HP3, HP6, HP7, HP5	A4160	Y42	H6.1, H3, H11
07 07 11	*	sludges from on-site effluent treatment containing hazardous substances	HC, CaO, AOX, HM	HP4, HP7, HP6, HP5	AC270	**Art. 1 (1) b**	(H11, H6.1)
07 07 12		sludges from on-site effluent treatment other than those mentioned in 07 07 11			AC270		
07 07 99		wastes not otherwise specified			NL		
08 00 00		**WASTES FROM THE MANUFACTURE, FORMULA-TION, SUPPLY AND USE (MFSU) OF COATINGS (PAINTS, VARNISHES AND VITREOUS ENAMELS), ADHESIVES, SEALANTS AND PRINTING INKS**					
08 01 00		**wastes from MFSU and removal of paint and varnish**					
08 01 11	*	waste paint and varnish containing organic solvents or other hazardous substances	solvents, heavy metal pigments, biocides (Hg)	HP6 HP3, HP5, HP10	A4070, A3140, A3150	Y12, Y41, Y42	H6.1, H3, H11, H4.1

Code	H	Designation	possible constituents/substances	EU		Basel	
				typical HP-Criteria	possible WSR-Codes	possible Y-Codes	typical H-Criteria
08 01 12		waste paint and varnish other than those mentioned in 08 01 11			B4010		
08 01 13	*	sludges from paint or varnish containing organic solvents or other hazardous substances	HM-pigments, solvents, biocides (Hg), alkaline	HP6, HP8 HP3, HP5	A4070, A3140, A3150	Y12, Y41, Y42	H6.1, H8, H3, H4.1, H11
08 01 14		sludges from paint or varnish other than those mentioned in 08 01 13			B4010		
08 01 15	*	aqueous sludges containing paint or varnish containing organic solvents or other hazardous substances	HM-pigments, solvents, biocides (Hg), alkaline	HP3, HP4, HP6, HP5	A4070, A3140, A3150	Y12, Y41, Y42, (Y29)	H3, H6.1, H11
08 01 16		aqueous sludges containing paint or varnish other than those mentioned in 08 01 15			B4010		
08 01 17	*	wastes from paint or varnish removal containing organic solvents or other hazardous substances	HM-pigments, solvents, biocides (Hg), alkaline	HP3, HP4, HP5, HP6, HP8, HP10	A4070, A3140, A3150	Y12, Y41, Y42, Y35	H6.1, H8, H11, H3, H4.1
08 01 18		wastes from paint or varnish removal other than those mentioned in 08 01 17			B4010		
08 01 19	*	aqueous suspensions containing paint or varnish containing organic solvents or other hazardous substances	metal compounds	HP5, HP8	A4070, B3140, B3150	Y12, Y41, Y42	**partly** (H8, H11)
08 01 20		aqueous suspensions containing paint or varnish other than those mentioned in 08 01 19			B4010		
08 01 21	*	waste paint or varnish remover	solvents, alkaline	HP3, HP5, HP4, HP8	A3140, A4090	Y41, Y42, Y35	H3, H8, H4.1, H11
08 01 99		wastes not otherwise specified			NL		
08 02 00		**wastes from MFSU of other coatings (including ceramic materials)**					
08 02 01		waste coating powders			B4010		
08 02 02		aqueous sludges containing ceramic materials			NL, B2100		
08 02 03		aqueous suspensions containing ceramic materials			NL, B2100		
08 02 99		wastes not otherwise specified			NL		

Code	H	Designation	possible constituents/substances	EU		Basel	
				typical HP-Criteria	possible WSR-Codes	possible Y-Codes	typical H- Criteria
08 03 00		**wastes from MFSU of printing inks**					
08 03 07		aqueous sludges containing ink			B4010		
08 03 08		aqueous liquid waste containing ink			B4010		
08 03 12	*	waste ink containing hazardous substances	HM, phenols, solvents, HC	HP5, HP3, HP4, HP14	A4070	Y41, Y12, Y13	H3, H4.1, H11, H12
08 03 13		waste ink other than those mentioned in 08 03 12			B4010		
08 03 14	*	ink sludges containing hazardous substances	HM, phenols, solvents, HC	HP5, HP3, HP4, HP14	A4070	Y12, Y41	H3, H4.1, H11, H12
08 03 15		ink sludges other than those mentioned in 08 03 14			B4010		
08 03 16	*	waste etching solutions	metals (Ni, Cr, Cu, Al, Zn); acids	HP8, HP4	A4090, A3140	Y34, Y35, Y17	H8
08 03 17	*	waste printing toner containing hazardous substances	psbl, PAH, CdSe, etc.	HP3	A4070	Y12	**partly** (H3)
08 03 18		waste printing toner other than those mentioned in 08 03 17			B4010		
08 03 19	*	disperse oil	HC	HP5, HP7, HP3	A4070, A4060	Y12, Y9	H11, H3
08 03 99		wastes not otherwise specified			NL		
08 04 00		**wastes from MFSU of adhesives and sealants (including waterproofing products)**					
08 04 09	*	waste adhesives and sealants containing organic solvents or other hazardous substances	solvents, biocides, epoxies, amines	HP4, HP5, HP6, HP3	A3050	Y13, Y41, Y42	H3, H6.1, H11, H4.1
08 04 10		waste adhesives and sealants other than those mentioned in 08 04 09			B4020, B3130		
08 04 11	*	adhesive and sealant sludges containing organic solvents or other hazardous substances	solvents, biocides	HP4, HP5, HP3	A3050, A3140, A3150	Y13, Y41, Y42	H4.1, H11
08 04 12		adhesive and sealant sludges other than those mentioned in 08 04 11			B4020		

Code	H	Designation	possible constituents/substances	EU		Basel	
				typical HP-Criteria	possible WSR-Codes	possible Y-Codes	typical H- Criteria
08 04 13	*	aqueous sludges containing adhesives or sealants containing organic solvents or other hazardous substances	solvents, biocides	HP4, HP5	A3050, A3140, A3150	Y13, Y41, Y42	H11
08 04 14		aqueous sludges containing adhesives or sealants other than those mentioned in 08 04 13			B4020		
08 04 15	*	aqueous liquid waste containing adhesives or sealants containing organic solvents or other hazardous substances	solvents, biocides	HP4, HP5	A3050, A3140, A3150	Y13, Y41, Y42	H11
08 04 16		aqueous liquid waste containing adhesives or sealants other than those mentioned in 08 04 15			B4020		
08 04 17	*	rosin oil	HC, PAH	HP3, HP5, HP4	A3050, A3020, A4140	Y13, Y9	H3, H11
08 04 99		wastes not otherwise specified			NL		
08 05 00		**wastes not otherwise specified in 08**					
08 05 01	*	waste isocyanates	isocyanates	HP3, HP6, HP7	NL, A4130, A4140	**Art. 1 (1) b**	(H6.1, H3, H4.1, H11)
09 00 00		**WASTES FROM THE PHOTOGRAPHIC INDUSTRY**					
09 01 00		**wastes from the photographic industry**					
09 01 01	*	water-based developer and activator solutions	alkaline, phenols, amino-phenols, hydrazine, bora-tes, metal ions	HP4, HP5	AD090	**Art. 1 (1) b**	(H11)
09 01 02	*	water-based offset plate developer solutions	alkaline, phenols, amino-phenols, hydrazine, bora-tes, metal ions	HP4, HP5	AD090	**Art. 1 (1) b**	(H11)
09 01 03	*	solvent-based developer solutions	solvents, resin residues, monomers	HP3, HP4, HP6, H8	A3140	Y42	H6.1, H8, H3
09 01 04	*	fixer solutions	thiosulfates, metals (Ag, Cr), aromatics, biocides	HP8, HP14, (HP7)	AD090	(Y21)	**partly** (H8, H12)
09 01 05	*	bleach solutions and bleach fixer solutions	thiosulfates, metals (Ag, Cr), aromatics, biocides	HP4, HP8, HP14, (HP7)	A1040, AD90	(Y21)	**partly** (H8, H12)

Code	H	Designation	possible constituents/substances	EU		Basel	
				typical HP-Criteria	possible WSR-Codes	possible Y-Codes	typical H- Criteria
09 01 06	*	wastes containing silver from on-site treatment of photographic wastes	Ag-containing filter cake, Cr(VI)	HP14	AD090	(Y18)	**partly** (H12)
09 01 07		photographic film and paper containing silver or silver compounds			B1180, B1190		
09 01 08		photographic film and paper free of silver or silver compounds			B3010, B3020		
09 01 10		single-use cameras without batteries			B4030		
09 01 11	*	single-use cameras containing batteries included in 16 06 01, 16 06 02 or 16 06 03	alkaline, $ZnCl_2$	HP7, HP4	(A1030), A1170	Y29, Y35, Y26	H11
09 01 12		single-use cameras containing batteries other than those mentioned in 09 01 11			B4030		
09 01 13	*	aqueous liquid waste from on-site reclamation of silver other than those mentioned in 09 01 06	phenols, aminophenols	HP4, HP14, HP8	AD090	(Y18)	**partly** (H8, H12)
09 01 99		wastes not otherwise specified			NL		
10 00 00		**WASTES FROM THERMAL PROCESSES**					
10 01 00		**wastes from power stations and other combustion plants (except 19)**					
10 01 01		bottom ash, slag and boiler dust (excluding boiler dust mentioned in 10 01 04)			GG030		
10 01 02		coal fly ash			GG040		
10 01 03		fly ash from peat and untreated wood			NL		
10 01 04	*	oil fly ash and boiler dust	Ni, V, PAH, HC	HP5, HP6, HP7	AA060	**Art. 1 (1) b**	(H6.1, H11)
10 01 05		calcium-based reaction wastes from flue-gas desulphurisation in solid form			B2040, AB150		
10 01 07		calcium-based reaction wastes from flue-gas desulphurisation in sludge form			B2040, AB150		
10 01 09	*	sulphuric acid		HP8	A4090	Y34	H8
10 01 13	*	fly ash from emulsified hydrocarbons used as fuel	Ni, V, PAH, HC	HP5, HP7, HP6, HP4	AA060	**Art. 1 (1) b**	(H11, H6.1)

Code	H	Designation	possible constituents/substances	EU		Basel	
				typical HP-Criteria	possible WSR-Codes	possible Y-Codes	typical H- Criteria
10 01 14	*	bottom ash, slag and boiler dust from co-incineration containing hazardous substances	CaO, HM (especially Pb, Cr VI, Ni, Cd, psbl, As), psbl, PAH	HP7, HP10, HP8, HP6	A2060	Y18	H11, H8, H6.1
10 01 15		bottom ash, slag and boiler dust from co-incineration other than those mentioned in 10 01 14			GG030		
10 01 16	*	fly ash from co-incineration containing hazardous substances	HM (especially Pb, Cr VI, Ni, Cd, psbl, As), psbl, PAH, PCDD/PCDF	HP7, HP10, HP6, HP4	A4100, A2060	Y18, (Y43/44)	H11, H6.1
10 01 17		fly ash from co-incineration other than those mentioned in 10 01 16			GG040		
10 01 18	*	wastes from gas cleaning containing hazardous substances	CaO, psbl, Hg, PCDD/PCDF, etc.	HP4, HP6, HP15	NL, A2060	**Art. 1 (1) b**	(H6.1, H13)
10 01 19		wastes from gas cleaning other than those mentioned in 10 01 05, 10 01 07 and 10 01 18			NL		
10 01 20	*	sludges from on-site effluent treatment containing hazardous substances	heavy metal hydroxides, Hg, psbl, PCDD/PCDF, PAH, HC	HP10, HP7, HP6	AC270	Y18, (Y43, Y44)	H11, H6.1
10 01 21		sludges from on-site effluent treatment other than those mentioned in 10 01 20			AC270		
10 01 22	*	aqueous sludges from boiler cleansing containing hazardous substances	PAH, HC, HM (Pb), amido-sulfonic acid	HP4, HP8, HP6	AC270	**Art. 1 (1) b**	(H8, H6.1)
10 01 23		aqueous sludges from boiler cleansing other than those mentioned in 10 01 22			AC270		
10 01 24		sands from fluidised beds			NL		
10 01 25		wastes from fuel storage and preparation of coal-fired power plants			NL		
10 01 26		wastes from cooling-water treatment			NL		
10 01 99		wastes not otherwise specified			NL		

Code	H	Designation	possible constituents/substances	EU		Basel	
				typical HP-Criteria	possible WSR-Codes	possible Y-Codes	typical H- Criteria
10 02 00		**wastes from the iron and steel industry**					
10 02 01		wastes from the processing of slag			B1200, B1210, NL		
10 02 02		unprocessed slag			B1200, B1210		
10 02 07	*	solid wastes from gas treatment containing hazardous substances	Pb, Cd, Cr(VI), Hg, PCDD/PCDF	HP10, HP14, HP5, HP7, HP6	A4100, A1020, AA010	Y31, Y23, Y26, Y21	H11, H6.1, H12
10 02 08		solid wastes from gas treatment other than those mentioned in 10 02 07			NL		
10 02 10		mill scales			B1230		
10 02 11	*	wastes from cooling-water treatment containing oil	HC, psbl, Pb-, Cr-, Ni-Compounds	HP10, HP7, HP3	A4060	Y9	H11, H3
10 02 12		wastes from cooling-water treatment other than those mentioned in 10 02 11			NL		
10 02 13	*	sludges and filter cakes from gas treatment containing hazardous substances	Pb, Cd, Cr(VI), psbl, Ni, PCDD/PCDF	HP10, HP7, HP6	A4100, AA010	Y31, Y23, Y26, Y21	H11, H6.1
10 02 14		sludges and filter cakes from gas treatment other than those mentioned in 10 02 13			NL		
10 02 15		other sludges and filter cakes			NL		
10 02 99		wastes not otherwise specified			NL, B1210		
10 03 00		**wastes from aluminium thermal metallurgy**					
10 03 02		anode scraps			B2090, NL		
10 03 04	*	primary production slags	salts, metallic aluminium (fine particles), soluble fluoride compounds, free NaF/AlF$_3$	HP3, HP6	NL		(H3, H6.1)
10 03 05		waste alumina			B2100, B2040		

192

Code	H	Designation	possible constituents/substances	EU		Basel	
				typical HP-Criteria	possible WSR-Codes	possible Y-Codes	typical H- Criteria
10 03 08	*	salt slags from secondary production	salts, metallic aluminium (fine particles), aluminium nitrides/carbides, Pb- und Cd-compounds, F-compounds	HP3, HP6, HP7, HP10, HP12	NL	**Art. 1 (1) b**	(H10, H4.3, H6.1, H11)
10 03 09	*	black drosses from secondary production	salts, metallic aluminium (fine particles), aluminium nitrides/carbides, Pb- und Cd-compounds, F-compounds	HP3, HP6, HP7, HP10	NL	**Art. 1 (1) b**	(H6.1, H3, H11)
10 03 15	*	skimmings that are flammable or emit, upon contact with water, flammable gases in hazardous quantities	aluminium nitrides/carbides (NH_3- und CH_4-emission), fine particles of Al (H_2-emission)	HP3, HP6	NL	**Art. 1 (1) b**	(H4.3, H4.2, H6.1)
10 03 16		skimmings other than those mentioned in 10 03 15	aluminium nitrides (NH_3-emission)		B1100		
10 03 17	*	tar-containing wastes from anode manufacture	PAH, HC	HP5, HP7	A3190	Y11	H11
10 03 18		carbon-containing wastes from anode manufacture other than those mentioned in 10 03 17			B2090		
10 03 19	*	flue-gas dust containing hazardous substances	F-compounds, PCDD/PCDF, HCB	HP6	A4100	Y32, (Y43, Y44)	**partly** (H6.1)
10 03 20		flue-gas dust other than those mentioned in 10 03 19			NL		
10 03 21	*	other particulates and dust (including ball-mill dust) containing hazardous substances	may contain PCDD/PCDF, Al, HM-compounds	HP3, HP6, HP7	NL	**Art. 1 (1) b**	(H4.3, H6.1, H11)
10 03 22		other particulates and dust (including ball-mill dust) other than those mentioned in 10 03 21			B2100		
10 03 23	*	solid wastes from gas treatment containing hazardous substances	PAHs, F-compounds, psbl, PCDD/PCDF	HP5, HP6, HP14,	A4100	Y44, Y43, Y32, (Y11)	H6.1, H11, H12
10 03 24		solid wastes from gas treatment other than those mentioned in 10 03 23			NL		

Code	H	Designation	possible constituents/substances	EU		Basel	
				typical HP-Criteria	possible WSR-Codes	possible Y-Codes	typical H- Criteria
10 03 25	*	sludges and filter cakes from gas treatment containing hazardous substances	HM-hydroxides, psbl, PAH	HP6, HP7, HP10	A4110	Y32, (Y11)	H6.1, H11
10 03 26		sludges and filter cakes from gas treatment other than those mentioned in 10 03 25			NL		
10 03 27	*	wastes from cooling-water treatment containing oil	HC	HP15, HP3, HP14	A4060	Y9	H13, H3, H12
10 03 28		wastes from cooling-water treatment other than those mentioned in 10 03 27			NL		
10 03 29	*	wastes from treatment of salt slags and black drosses containing hazardous substances	may contain PCDD/PCDF, Al, HM-compounds	HP3, HP7, HP6	A4110	Y18	H4.3, H4.2, H6.1, H11
10 03 30		wastes from treatment of salt slags and black drosses other than those mentioned in 10 03 29			NL		
10 03 99		wastes not otherwise specified			NL		
10 04 00		**wastes from lead thermal metallurgy**					
10 04 01	*	slags from primary and secondary production	soluble lead compounds	HP10, HP6	A1020	Y31	H11, H6.1
10 04 02	*	dross and skimmings from primary and secondary production	insoluble lead compounds, Sb-, As- compounds, etc.	HP10, HP6	A1020	Y31	H11, H6.1
10 04 03	*	calcium arsenate	As- compounds	HP7, HP6, HP10	A1030, A4100	Y24, Y31	H6.1, H11
10 04 04	*	flue-gas dust	Pb-, Sb-, As- compounds	HP10, HP7, HP6	A1020, A4100	Y31, Y27	H6.1, H11
10 04 05	*	other particulates and dust	Pb-, Sb-, As-compounds	HP10, HP7, HP6	A1020	Y31, Y27	H6.1, H11
10 04 06	*	solid wastes from gas treatment	Pb-, Sb-, As-compounds	HP10, HP7, HP6	A1020, A4100	Y31, Y27	H11, H6.1
10 04 07	*	sludges and filter cakes from gas treatment	Pb-, Sb-, As-compounds	HP10, HP7, HP6	A1020, A4100	Y31, Y27	H11, H6.1
10 04 09	*	wastes from cooling-water treatment containing oil	HC	HP15, HP3	A4060	Y9	H13, H3
10 04 10		wastes from cooling-water treatment other than those mentioned in 10 04 09			NL		

Code	H	Designation	possible constituents/substances	EU		Basel	
				typical HP-Criteria	possible WSR-Codes	possible Y-Codes	typical H- Criteria
10 04 99		wastes not otherwise specified			NL		
10 05 00		**wastes from zinc thermal metallurgy**					
10 05 01		slags from primary and secondary production			B1220, B1100		
10 05 03	*	flue-gas dust	Cd, Pb, psbl, PCDD/PCDF	HP10, HP6	A4100, A1020, A1080	Y23, Y26, Y43, Y44	H11, H6.1
10 05 04		other particulates and dust			B1080		
10 05 05	*	solid waste from gas treatment	Cd, Pb, psbl, PCDD/PCDF	HP10, HP6	A4100, A1020, A1080	Y23, Y26, Y43, Y44	H11, H6.1
10 05 06	*	sludges and filter cakes from gas treatment	Cd, Pb	HP10, HP6	A4100, A1020	Y23, Y26	H11, H6.1
10 05 08	*	wastes from cooling-water treatment containing oil	HC	HP15, HP3	A4060	Y9	H13, H3
10 05 09		wastes from cooling-water treatment other than those mentioned in 10 05 08			NL		
10 05 10	*	dross and skimmings that are flammable or emit, upon contact with water, flammable gases in hazardous quantities	Zn	HP3	A1080	Y23	H4.3
10 05 11		dross and skimmings other than those mentioned in 10 05 10			B1100, B1080		
10 05 99		wastes not otherwise specified			NL		
10 06 00		**wastes from copper thermal metallurgy**					
10 06 01		slags from primary and secondary production			B1100, B2040, GB040		
10 06 02		dross and skimmings from primary and secondary production			B1100, B2040, B1070		
10 06 03	*	flue-gas dust	PCDD/PCDF, Pb, As, psbl, Ni, Zn	HP10, HP7	A1100	Y23, Y31, Y43	H11
10 06 04		other particulates and dust			B1070, B1240		
10 06 06	*	solid wastes from gas treatment	PCDD/PCDF, Pb, As, psbl, Ni, Zn	HP10, HP7, HP6	A1100, A4100, A1020	Y22, Y23, Y31, Y43, Y44	H6.1, H11

Code	H	Designation	possible constituents/substances	EU		Basel	
				typical HP-Criteria	possible WSR-Codes	possible Y-Codes	typical H- Criteria
10 06 07	*	sludges and filter cakes from gas treatment	PCDD/PCDF, Pb, As, psbl, Ni, Zn	HP10, HP7, HP6	A4100, A1020	Y22, Y23, Y31, Y43, Y44	H6.1, H11
10 06 09	*	wastes from cooling-water treatment containing oil	HC	HP15, HP3	A4060	Y9	H13, H3
10 06 10		wastes from cooling-water treatment other than those mentioned in 10 06 09			NL		
10 06 99		wastes not otherwise specified			NL, B1240, B2040		
10 07 00		**wastes from silver, gold and platinum thermal metallurgy**					
10 07 01		slags from primary and secondary production			B1100, GB040, B1150		
10 07 02		dross and skimmings from primary and secondary production			B1100, B1150		
10 07 03		solid wastes from gas treatment			NL, B1150		
10 07 04		other particulates and dust			B1150, B1160		
10 07 05		sludges and filter cakes from gas treatment			NL, B1150		
10 07 07	*	wastes from cooling-water treatment containing oil	HC	HP15, HP3	A4060	Y9	H13, H3
10 07 08		wastes from cooling-water treatment other than those mentioned in 10 07 07			NL		
10 07 99		wastes not otherwise specified			B1100		
10 08 00		**wastes from other non-ferrous thermal metal-lurgy**					
10 08 04		particulates and dust			NL, B1031		
10 08 08	*	salt slag from primary and secondary production	metal oxides, fine metal e.g. Mg	HP3, HP6	NL	**Art. 1 (1) b**	(H4.3, H6.1)
10 08 09		other slags			GB040, B1100		
10 08 10	*	dross and skimmings that are flammable or emit, upon contact with water, flammable gases in hazardous quantities	fine metal e.g. Mg,	HP3, HP6	AA190, NL	**Art. 1 (1) b**	(H4.3, H4.2, H6.1)

Code	H	Designation	possible constituents/substances	EU		Basel	
				typical HP-Criteria	possible WSR-Codes	possible Y-Codes	typical H- Criteria
10 08 11		dross and skimmings other than those mentioned in 10 08 10			NL		
10 08 12	*	tar-containing wastes from anode manufacture	PAH	HP7	A3010, A3190	Y11	H11
10 08 13		carbon-containing wastes from anode manufacture other than those mentioned in 10 08 12			NL		
10 08 14		anode scrap			NL		
10 08 15	*	flue-gas dust containing hazardous substances	metal compounds, psbl, PCDD/PCDF	HP7, HP10, HP6	A4100, A1020, A4110	(Y43, Y44)	**partly** (H11, H6.1)
10 08 16		flue-gas dust other than those mentioned in 10 08 15			NL		
10 08 17	*	sludges and filter cakes from flue-gas treatment containing hazardous substances	metal compounds, psbl, PCDD/PCDF	HP7, HP10	A4100, A1020	(Y43)	**partly** (H11)
10 08 18		sludges and filter cakes from flue-gas treatment other than those mentioned in 10 08 17			NL		
10 08 19	*	wastes from cooling-water treatment containing oil	HC	HP15, HP14	A4060	Y9	H13, H12
10 08 20		wastes from cooling-water treatment other than those mentioned in 10 08 19			NL		
10 08 99		wastes not otherwise specified			NL		
10 09 00		**wastes from casting of ferrous pieces**					
10 09 03		furnace slag			B1200, B1210		
10 09 05	*	casting cores and moulds which have not undergone pouring containing hazardous substances	binders (resins, oil, free phenols)	HP4, HP6, HP8, HP5	AB070	Y39	H6.1, H8, H11
10 09 06		casting cores and moulds which have not undergone pouring other than those mentioned in 10 09 05			AB070		
10 09 07	*	casting cores and moulds which have undergone pouring containing hazardous substances	PAH, metal compounds	HP4, HP6, HP8, HP7	AB070	Y39, Y11	H11, H6.1, H8
10 09 08		casting cores and moulds which have undergone pouring other than those mentioned in 10 09 07			AB070		

Code	H	Designation	possible constituents/substances	EU		Basel	
				typical HP-Criteria	possible WSR-Codes	possible Y-Codes	typical H- Criteria
10 09 09	*	flue-gas dust containing hazardous substances	Pb, Cd, etc. (psbl, Cr(VI), Zn	HP4, HP5, HP10, HP7, HP6	A4100, A1020	Y31, Y21, Y23	H11, H6.1
10 09 10		flue-gas dust other than those mentioned in 10 09 09			NL		
10 09 11	*	other particulates containing hazardous substances	Pb, Cd, etc. (psbl, Cr(VI); Zn	HP10, HP7, HP3, HP6	A4100, A1020	Y31, Y21, Y23	H11, H3, H6.1
10 09 12		other particulates other than those mentioned in 10 09 11			NL		
10 09 13	*	waste binders containing hazardous substances	resins, phenols, eventually acids	HP4, HP8, HP5, HP3	A3070	Y13, Y9, Y39	H8, H4.1, H11
10 09 14		waste binders other than those mentioned in 10 09 13			NL		
10 09 15	*	waste crack-indicating agent containing hazardous substances		HP3	NL	Art. 1 (1) b	(H3)
10 09 16		waste crack-indicating agent other than those mentioned in 10 09 15			NL		
10 09 99		wastes not otherwise specified			NL		
10 10 00		**wastes from casting of non-ferrous pieces**					
10 10 03		furnace slag			B1100		
10 10 05	*	casting cores and moulds which have not undergone pouring, containing hazardous substances	binders (resins, oil, free phenols), quartz sand	HP4, HP5	AB070	Art. 1 (1) b	(H11)
10 10 06		casting cores and moulds which have not undergone pouring, other than those mentioned in 10 10 05			AB070		
10 10 07	*	casting cores and moulds which have undergone pouring, containing hazardous substances	PAH, metal compounds, quartz sand	HP7, HP6, HP5	AB070	Art. 1 (1) b	(H11, H6.1)
10 10 08		casting cores and moulds which have undergone pouring, other than those mentioned in 10 10 07			AB070		

Code	H	Designation	possible constituents/substances	EU		Basel	
				typical HP-Criteria	possible WSR-Codes	possible Y-Codes	typical H- Criteria
10 10 09	*	flue-gas dust containing hazardous substances	Pb, Cd, etc. (psbl, Cr(VI), Ni, As)	HP10, HP7, HP6	A4100, A1020	Y31, Y23, Y26	H11, H6.1
10 10 10		flue-gas dust other than those mentioned in 10 10 09			NL		
10 10 11	*	other particulates containing hazardous substances	Pb, Cd, etc. (psbl, Cr(VI), Ni, As)	HP10, HP7, HP6, (HP3)	A4100, A1020, A1100	Y31, Y23, Y26	H11, H6.1, (H3)
10 10 12		other particulates other than those mentioned in 10 10 11			NL		
10 10 13	*	waste binders containing hazardous substances	resins, phenols, psbl, acids	HP4, HP8, HP5	A3070	Y39, Y13, Y9	H8, H11
10 10 14		waste binders other than those mentioned in 10 10 13			NL		
10 10 15	*	waste crack-indicating agent containing hazardous substances		HP3	NL	**Art. 1 (1) b**	(H3)
10 10 16		waste crack-indicating agent other than those mentioned in 10 10 15			NL		
10 10 99		wastes not otherwise specified			NL		
10 11 00		**wastes from manufacture of glass and glass products**					
10 11 03		waste glass-based fibrous materials			GE020		
10 11 05		particulates and dust			NL		
10 11 09	*	waste preparation mixture before thermal processing, containing hazardous substances	As-, Cr-, Co-, Pb-compounds	HP7, HP10	NL	(Y24, Y27, Y31)	**partly** (H11)
10 11 10		waste preparation mixture before thermal processing, other than those mentioned in 10 11 09			B2100		
10 11 11	*	waste glass in small particles and glass powder containning heavy metals (for example from cathode ray tubes)	lead compounds, psbl, mercury (glass from fluorescent tubes)	HP10	A1020, A2010	Y31, Y24, Y27	H11
10 11 12		waste glass other than those mentioned in 10 11 11			B2020, B2040		

Code	H	Designation	possible constituents/substances	EU		Basel	
				typical HP-Criteria	possible WSR-Codes	possible Y-Codes	typical H- Criteria
10 11 13	*	glass-polishing and -grinding sludge containing hazardous substances	Pb-compounds, fluorides	HP10, HP6, HP5	A1020	Y31, Y32, Y27	H11, H6.1
10 11 14		glass-polishing and -grinding sludge other than those mentioned in 10 11 13			NL		
10 11 15	*	solid wastes from flue-gas treatment containing hazardous substances	Hg, As, Pb, Hg, fluorides	HP6, HP7, HP10	A4100	Y31, Y29	H6.1, H11
10 11 16		solid wastes from flue-gas treatment other than those mentioned in 10 11 15			NL		
10 11 17	*	sludges and filter cakes from flue-gas treatment containing hazardous substances	Hg, As, Pb, Hg, fluorides	HP7, HP10	A4100	Y31, Y29	H11
10 11 18		sludges and filter cakes from flue-gas treatment other than those mentioned in 10 11 17			NL		
10 11 19	*	solid wastes from on-site effluent treatment containing hazardous substances	Pb-compounds, fluorides	HP10, HP6	AC270	Y31, Y32, (Y18)	H11, H6.1
10 11 20		solid wastes from on-site effluent treatment other than those mentioned in 10 11 19			NL		
10 11 99		wastes not otherwise specified			NL		
10 12 00		**wastes from manufacture of ceramic goods, bricks, tiles and construction products**					
10 12 01		waste preparation mixture before thermal processing			NL		
10 12 03		particulates and dust			NL		
10 12 05		sludges and filter cakes from gas treatment			NL		
10 12 06		discarded moulds			GF010		
10 12 08		waste ceramics, bricks, tiles and construction products (after thermal processing)			GF010		
10 12 09	*	solid wastes from gas treatment containing hazardous substances		HP6	NL, A4100	**Art. 1 (1) b**	(H6.1)
10 12 10		solid wastes from gas treatment other than those mentioned in 10 12 09			NL		

Code	H	Designation	possible constituents/substances	EU		Basel	
				typical HP-Criteria	possible WSR-Codes	possible Y-Codes	typical H- Criteria
10 12 11	*	wastes from glazing containing heavy metals	Pb, Sb, Cd, Co, Cr, Ni	HP7, HP10, HP6	A1020	Y27, Y26, (Y31)	H11, H6.1
10 12 12		wastes from glazing other than those mentioned in 10 12 11			NL		
10 12 13		sludge from on-site effluent treatment			AC270		
10 12 99		wastes not otherwise specified			NL		
10 13 00		**wastes from manufacture of cement, lime and plaster and articles and products made from them**					
10 13 01		waste preparation mixture before thermal processing			NL		
10 13 04		wastes from calcination and hydration of lime			NL		
10 13 06		particulates and dust (except 10 13 12 and 10 13 13)			NL		
10 13 07		sludges and filter cakes from gas treatment			NL		
10 13 09	*	wastes from asbestos-cement manufacture containing asbestos	asbestos, CaO	HP7, HP4	A2050	Y36	H11
10 13 10		wastes from asbestos-cement manufacture other than those mentioned in 10 13 09			NL		
10 13 11		wastes from cement-based composite materials other than those mentioned in 10 13 09 and 10 13 10			NL		
10 13 12	*	solid wastes from gas treatment containing hazardous substances	CaO, Tl, irritant chlorides	HP6, HP4	NL	**Art. 1 (1) b**	(H6.1)
10 13 13		solid wastes from gas treatment other than those mentioned in 10 13 12			B2070		
10 13 14		waste concrete and concrete sludge			B2040		
10 13 99		wastes not otherwise specified			NL		
10 14 00		**waste from crematoria**					
10 14 01	*	waste from gas cleaning containing mercury	Hg, activated carbon	HP6, HP15	A1030, A4160	Y29	H6.1, H13

Code	H	Designation	possible constituents/substances	EU		Basel	
				typical HP-Criteria	possible WSR-Codes	possible Y-Codes	typical H- Criteria
11 00 00		**WASTES FROM CHEMICAL SURFACE TREATMENT AND COATING OF METALS AND OTHER MATE-RIALS; NON-FERROUS HYDRO-METALLURGY**					
11 01 00		**wastes from chemical surface treatment and coating of metals and other materials (for example galvanic processes, zinc coating processes, pickling processes, etching, phosphating, alkaline degreasing, anodising)**					
11 01 05	*	pickling acids	acids, HM, psbl, HC	HP4, HP8, HP7, HP10	A1060	Y34	H8, H11
11 01 06	*	acids not otherwise specified	acids	HP4, HP8	A4090	Y34	H8
11 01 07	*	pickling bases	alkaline, HM, psbl, HC	HP4, HP8	A1060	Y35	H8
11 01 08	*	phosphatising sludges	Pb, Ni, Cr, Zn, etc.	HP4, HP8, HP7, HP10	A4090, A1060	Y23, Y34	H8; H11
11 01 09	*	sludges and filter cakes containing hazardous sub-stances	Pb, Ni, Cr, CN, B, F	HP4, HP5, HP8, HP7, HP10, HP6	A1050, A1020	Y7, Y31, Y32	**partly** (H8, H11, H6.1)
11 01 10		sludges and filter cakes other than those mentioned in 11 01 09			NL		
11 01 11	*	aqueous rinsing liquids containing hazardous substances	metal ions, cyanides, F, B	HP6, HP15	A1060	Y21, Y32	H6.1, H13
11 01 12		aqueous rinsing liquids other than those mentioned in 11 01 11			NL		
11 01 13	*	degreasing wastes containing hazardous substances	HC, alkaline, eventual sur-factants and metal ions	HP3, HP8, HP6, (HP14)	A4060, AC250	Y9	H8, H6.1, H3, (H12)
11 01 14		degreasing wastes other than those mentioned in 11 01 13			NL		
11 01 15	*	eluate and sludges from membrane systems or ion exchange systems containing hazardous substances	HM, acids, alkaline	HP4, HP7, HP10	A4090	Y34, Y35	H11
11 01 16	*	saturated or spent ion exchange resins	HM	HP7, HP10	AD120	**Art. 1 (1) b**	(H11)

Code	H	Designation	possible constituents/substances	EU		Basel	
				typical HP-Criteria	possible WSR-Codes	possible Y-Codes	typical H- Criteria
11 01 98	*	other wastes containing hazardous substances	CN-salts, Ba-salts, nitrate, tempering oil	HP6, HP8	NL	Art. 1 (1) b	(H6.1, H8)
11 01 99		wastes not otherwise specified			NL		
11 02 00		**wastes from non-ferrous hydrometallurgical processes**					
11 02 02	*	sludges from zinc hydrometallurgy (including jarosite, goethite)	Cd, Zn	HP4, HP7, HP14	A1070, A1020	Y23, Y26	H11, H12
11 02 03		wastes from the production of anodes for aqueous electrolytical processes			NL		
11 02 05	*	wastes from copper hydrometallurgical processes containing hazardous substances	Ni-, Pb-compounds	HP10, HP7	A1020, A1120, A1110	Y22, Y31	H11
11 02 06		wastes from copper hydrometallurgical processes other than those mentioned in 11 02 05			NL, GB040		
11 02 07	*	other wastes containing hazardous substances		HP4, HP6	NL	Art. 1 (1) b	(H6.1)
11 02 99		wastes not otherwise specified			NL		
11 03 00		**sludges and solids from tempering processes**					
11 03 01	*	wastes containing cyanide	CN-tempering salts	HP6, HP12	A4050	Y7, Y33, Y17	H6.1, H10
11 03 02	*	other wastes	barium salts, nitrates, HC (tempering oil) nitrates, etc.	HP6, HP12	AD100	Y17	H6.1, H10
11 05 00		**wastes from hot galvanising processes**					
11 05 01		hard zinc			B1100		
11 05 02		zinc ash			B1080		
11 05 03	*	solid wastes from gas treatment	ammonium salts, zinc compounds, Pb	HP4, HP6, HP10	A4100, A1080	Y23, Y26	H6.1, H11
11 05 04	*	spent flux	ammonium salts, acids	HP4, HP12	A4090	Y35	H12, (H8)
11 05 99		wastes not otherwise specified			NL		

Code	H	Designation	possible constituents/substances	EU		Basel	
				typical HP-Criteria	possible WSR-Codes	possible Y-Codes	typical H- Criteria
12 00 00		WASTES FROM SHAPING AND PHYSICAL AND ME-CHANICAL SURFACE TREATMENT OF METALS AND PLASTICS					
12 01 00		wastes from shaping and physical and mecha-nical surface treatment of metals and plastics					
12 01 01		ferrous metal filings and turnings			B1010		
12 01 02		ferrous metal dust and particles			B1010, NL		
12 01 03		non-ferrous metal filings and turnings			B1010, B1020, B1070		
12 01 04		non-ferrous metal dust and particles			B1031, B1070		
12 01 05		plastics shavings and turnings			B3011/EU3011, (B3010), B3130, (GH013)		
12 01 06	*	mineral-based machining oils containing halogens (except emulsions and solutions)	HC, R-CL, fine metals, HM	HP10, (HP7)	A4060, A3020	Y8, Y17, Y9	H11
12 01 07	*	mineral-based machining oils free of halogens (except emulsions and solutions)	HC, fine metals, HM	HP10, (HP7)	A4060, A3020	Y8, Y17, Y9	H11
12 01 08	*	machining emulsions and solutions containing halogens	HC, R-CL, fine metals, HM	HP10, (HP7)	A4060	Y17, Y9	H11
12 01 09	*	machining emulsions and solutions free of halogens	HC, fine metals, HM	HP10, (HP7)	A4060	Y17, Y9	H11
12 01 10	*	synthetic machining oils	HC, fine metals, HM	HP10, (HP7)	A3020, A4060	Y17, Y9	H11
12 01 12	*	spent waxes and fats	HC, PAH, Ni, Pb-com-pounds	HP10, (HP7)	A4060	Y17, Y9, Y31	H11
12 01 13		welding wastes			NL		
12 01 14	*	machining sludges containing hazardous substances	HC, PAH, Ni, Pb-com-pounds	HP7, HP10, HP5	A4060	Y17, Y9	H11
12 01 15		machining sludges other than those mentioned in 12 01 14			NL		

Code	H	Designation	possible constituents/substances	EU		Basel	
				typical HP-Criteria	possible WSR-Codes	possible Y-Codes	typical H- Criteria
12 01 16	*	waste blasting material containing hazardous substances	depending on the treated material	HP10, HP7, HP5	AB130	Y20 to Y31	H11
12 01 17		waste blasting material other than those mentioned in 12 01 16			AB130		
12 01 18	*	metal sludge (grinding, honing and lapping sludge) containing oil	HC, abraded material	HP7, HP10, HP5	A4060	Y17, Y9	H11
12 01 19	*	readily biodegradable machining oil	HC, stabilisers	HP15, HP10	A4060	Y17, Y9	H13, H11
12 01 20	*	spent grinding bodies and grinding materials containing hazardous substances	HC	HP15, HP10	A4060	Y17, Y9	H13, H11
12 01 21		spent grinding bodies and grinding materials other than those mentioned in 12 01 20			B2040		
12 01 99		wastes not otherwise specified			NL		
12 03 00		**wastes from water and steam degreasing processes (except 11)**					
12 03 01	*	aqueous washing liquids	HC, surfactants	HP15	A4060	Y9	H13
12 03 02	*	steam degreasing wastes	HC, surfactants	HP3, HP15	A4060	Y9	H3, H13
13 00 00		**OIL WASTES AND WASTES OF LIQUID FUELS (except edible oils, and those in chapters 05, 12 and 19)**					
13 01 00		**waste hydraulic oils**					
13 01 01	*	hydraulic oils, containing PCBs	PCB, take care of PCB substitutes (PCT, PCN, Ugilec)	HP5, HP6	A3180	Y10, Y9	H6.1, H11
13 01 04	*	chlorinated emulsions	PCB, PCT, PCN	HP5, HP6, HP7	A4060	Y8, Y9	H11, H6.1
13 01 05	*	non-chlorinated emulsions	PCB, PCT, PCN	HP5, HP6, HP7	A4060	Y8, Y9	H6.1, H11
13 01 09	*	mineral-based chlorinated hydraulic oils	PCB, PCT, PCN	HP5, HP6, HP7	A3020	Y8, Y9	H6.1, H11
13 01 10	*	mineral based non-chlorinated hydraulic oils	PCB, PCT, PCN	HP5, HP6, HP7	A3020	Y8, Y9	H6.1, H11
13 01 11	*	synthetic hydraulic oils	stabilizers	HP15, HP6, HP10	A3020	Y8, Y9	H13, H6.1, H11
13 01 12	*	readily biodegradable hydraulic oils	stabilizers	HP15, HP10	A3020	Y8, Y9	H13, H11

Code	H	Designation	possible constituents/substances	EU		Basel	
				typical HP-Criteria	possible WSR-Codes	possible Y-Codes	typical H- Criteria
13 01 13	*	other hydraulic oils	mainly esters (phosphoric acid esters, silicones, boric acid esters)	HP5, HP6	A3020	Y8, Y9	H6.1, H11
13 02 00		**waste engine, gear and lubricating oils**					
13 02 04	*	mineral-based chlorinated engine, gear and lubricating oils	AOX, HC, abraded metals	HP6, HP7, HP10, (HP3)	A3020	Y8, Y9	H11, H6.1, (H3)
13 02 05	*	mineral-based non-chlorinated engine, gear and lubricating oils	HC, abraded metals	HP3, HP7, HP10, HP6	A3020	Y8, Y9	H3, H6.1, H11
13 02 06	*	synthetic engine, gear and lubricating oils	AOX, HC, abraded metals, Mo-, Si-compounds	HP6, HP7, HP10, (HP3)	A3020	Y8, Y9	H11, H3, H6.1
13 02 07	*	readily biodegradable engine, gear and lubricating oils	AOX, HC, abraded metal, stabilizers (boric acid derivate) Mo-compounds	HP7, HP10, (HP3)	A3020	Y8, Y9	H11, (H3)
13 02 08	*	other engine, gear and lubricating oils	HC, abraded metal, stabilizers (boric acid derivate) Mo-compounds	HP7, HP10, HP6, (HP3)	A3020	Y8, Y9	H11, (H3), H6.1
13 03 00		**waste insulating and heat transmission oils**					
13 03 01	*	insulating or heat transmission oils containing PCBs	PCB, HC, take care of PCB substitutes (PCT, PCN, Ugilec)	HP5, HP7, HP10, HP11	A3180	Y10, Y8, Y9	H11
13 03 06	*	mineral-based chlorinated insulating and heat transmission oils other than those mentioned in 13 03 01	AOX, HC, psbl, PCN	HP5, HP6, HP7	A3020, A3040, A3180	Y8, Y9, Y10	H6.1, H11,
13 03 07	*	mineral-based non-chlorinated insulating and heat transmission oils	HC	HP5, HP6, HP7, HP3	A3020, A3040	Y8, Y9	H3, H6.1, H11
13 03 08	*	synthetic insulating and heat transmission oils	HC, silicones	HP5, HP6, HP7	A3020, A3040	Y8, Y9	H6.1, H11
13 03 09	*	readily biodegradable insulating and heat transmission oils	stabilisers (e.g. boric acid derivate)	HP15, HP7, HP10	A3040	Y8, Y9	H13, H11
13 03 10	*	other insulating and heat transmission oils		HP5, HP6, HP7	A3020	Y8, Y9	H6.1, H11
13 04 00		**bilge oils**					
13 04 01	*	bilge oils from inland navigation	HC, metal oxides sludge	HP3, HP14	A4060, A3020	Y8, Y9,	H3, H12

Code	H	Designation	possible constituents/substances	EU		Basel	
				typical HP-Criteria	possible WSR-Codes	possible Y-Codes	typical H- Criteria
13 04 02	*	bilge oils from jetty sewers	HC, metal oxides sludge	HP3, HP14	A4060, A3020	Y8, Y9,	H12, H3
13 04 03	*	bilge oils from other navigation	HC, metal oxides sludge	HP3, HP14	A4060, A3020	Y8, Y9,	H12, H3
13 05 00		**oil/water separator contents**					
13 05 01	*	solids from grit chambers and oil/water separators	HC, PAH	HP7, HP10		Y9	H11
13 05 02	*	sludges from oil/water separators	HC	HP15	A4060	Y9	H13
13 05 03	*	interceptor sludges	HC	HP15	A4060	Y9	H13
13 05 06	*	oil from oil/water separators	HC, contaminations (HM, sand)	HP3, HP6	A4060, A3020	Y8	H3, H6.1
13 05 07	*	oily water from oil/water separators	HC	HP15, HP8	A4060	Y8, Y9	H8, H13
13 05 08	*	mixtures of wastes from grit chambers and oil/water separators	HC	HP15	A4060	Y8, Y9	H13
13 07 00		**wastes of liquid fuels**					
13 07 01	*	fuel oil and diesel	HC	HP3	NL, A4060	**Art. 1 (1) b**	(H3)
13 07 02	*	Petrol	HC, additives, arom. HC	HP3, HP7	NL, A4060	**Art. 1 (1) b**	(H3, H11)
13 07 03	*	other fuels (including mixtures)	HC, additives, arom. HC	HP3, HP7	NL, A4060	**Art. 1 (1) b**	(H3, H11)
13 08 00		**oil wastes not otherwise specified**					
13 08 01	*	desalter sludges or emulsions	HC	HP15	A4060	Y8, Y9	H13
13 08 02	*	other emulsions	HC, PAH	HP7	A4060	Y8, Y9	H11
13 08 99	*	wastes not otherwise specified		HP3, HP7, HP10, HP15, HP11	A4060	Y8, Y9	H11, H13, H4.1
14 00 00		**WASTE ORGANIC SOLVENTS, REFRIGERANTS AND PROPELLANTS (except 07 and 08)**					
14 06 00		**waste organic solvents, refrigerants and foam/aerosol propellants**					
14 06 01	*	chlorofluorocarbons, HCFC, HFC	R-Cl	HP14	AC150	**Art. 1 (1) b**	
14 06 02	*	other halogenated solvents and solvent mixtures	R-X	HP6, HP7, HP3	A3150	Y6, Y41, Y12	H3, H11, H6.1

Code	H	Designation	possible constituents/substances	EU		Basel	
				typical HP-Criteria	possible WSR-Codes	possible Y-Codes	typical H- Criteria
14 06 03	*	other solvents and solvent mixtures	solvents	HP3, HP6, HP7	A3140	Y6, Y42, Y12	H3, H11, H6.1
14 06 04	*	sludges or solid wastes containing halogenated solvents	R-X	HP6, HP7, HP3	A3160, A3170	Y6, Y41, Y12	H3, H6.1, H11
14 06 05	*	sludges or solid wastes containing other solvents	R-X	HP6, HP7, HP3	A3160	Y6, Y42, Y12	H3, H11, H6.1
15 00 00		**WASTE PACKAGING; ABSORBENTS, WIPING CLOTHS, FILTER MATERIALS AND PROTECTIVE CLOTHING NOT OTHERWISE SPECIFIED**					
15 01 00		**packaging (including separately collected municipal packaging waste)**					
15 01 01		paper and cardboard packaging			B3020		
15 01 02		plastic packaging			Y48	Y48	
15 01 03		wooden packaging	all treated wood: AC170		B3050, B3130, AC170		
15 01 04		metallic packaging			B1010		
15 01 05		composite packaging			B3020, BEU04		
15 01 06		mixed packaging			NL		
15 01 07		glass packaging			B2020		
15 01 09		textile packaging			B3030		
15 01 10	*	packaging containing residues of or contaminated by hazardous substances		HP14, HP3	A3210, NL, A4130	**Art. 1 (1) b**	(H12, H3)
15 01 11	*	metallic packaging containing a hazardous solid porous matrix (for example asbestos), including empty pressure containers	acetylene containers	HP7, (HP3)	A2050, A4130	Y39	H11
15 02 00		**absorbents, filter materials, wiping cloths and protective clothing**					
15 02 02	*	absorbents, filter materials (including oil filters not otherwise specified), wiping cloths, protective clothing contaminated by hazardous substances	depending on origin/pro-cess	HP8, HP7, HP3	A4160, A4130	Y8, Y9, Y41, Y42	**partly** (H8, H11, H4.1)

208

Code	H	Designation	possible constituents/substances	EU		Basel	
				typical HP-Criteria	possible WSR-Codes	possible Y-Codes	typical H- Criteria
15 02 03		absorbents, filter materials, wiping cloths and protective clothing other than those mentioned in15 02 02			B2060		
16 00 00		**WASTES NOT OTHERWISE SPECIFIED IN THE LIST**					
16 01 00		**end-of-life vehicles from different means of transport (including off-road machinery) and wastes from dismantling of end-of-life vehicles and vehicle maintenance (except 13, 14, 16 06 and 16 08)**					
16 01 03		end-of-life tyres			B3140, (B3080)		
16 01 04	*	end-of-life vehicles	gasoline, batteries, oil, ex-plosives (air bag, belt stretcher), capacitors	HP3, HP10, HP8, HP1	NL	Y9, Y31, Y34, Y15	H11, H8, H3, H1
16 01 06		end-of-life vehicles, containing neither liquids nor other hazardous components			B1250, GC030		
16 01 07	*	oil filters	HC, abraded metal	HP7, HP3, HP14	A3020, NL	Y8	H3, H11, (H12)
16 01 08	*	components containing mercury	Hg	HP15, HP6	A1030	Y29	H13, H6.1
16 01 09	*	components containing PCBs	capacitors (old equipment)	HP5, HP7	A3180	Y10	H11
16 01 10	*	explosive components (for example air bags)	explosive belt stretcher	HP1	A4080	Y15	H1
16 01 11	*	brake pads containing asbestos	asbestos	HP7	A2050	Y36	H11
16 01 12		brake pads other than those mentioned in 16 01 11			NL		
16 01 13	*	brake fluids	boric acid derivate	HP5, HP6	AC070; NL	Y42	H6.1, H11
16 01 14	*	antifreeze fluids containing hazardous substances	alcohols	HP5, HP6	A3140, AC080	Y42	H6.1, H11
16 01 15		antifreeze fluids other than those mentioned in 16 01 14			NL		
16 01 16		tanks for liquefied gas			NL		
16 01 17		ferrous metal			B1010		
16 01 18		non-ferrous metal			B1010, B1020		

Code	H	Designation	possible constituents/substances	EU		Basel	
				typical HP-Criteria	possible WSR-Codes	possible Y-Codes	typical H- Criteria
16 01 19		plastic			B3011/EU3011, (B3010), B3140, (GH013)		
16 01 20		glass			B2020		
16 01 21	*	hazardous components other than those mentioned in 16 01 07 to 16 01 11 and 16 01 13 and16 01 14		HP3, HP6	NL	Art. 1 (1) b	(H4.1, H6.1)
16 01 22		components not otherwise specified			NL, B3040, B3080, B3090		
16 01 99		wastes not otherwise specified			NL		
16 02 00		**wastes from electrical and electronic equipment**					
16 02 09	*	transformers and capacitors containing PCBs	PCB, PCDD/PCDF (after thermal stress)	HP5, HP6, HP7	A1180, A3180	Y10, (Y43/44)	H11, H6.1
16 02 10	*	discarded equipment containing or contaminated by PCBs other than those mentioned in 16 02 09	PCB, PCDD/PCDF (after thermal stress)	HP5, HP6, HP7	A1180, A3180	Y10, (Y43/44)	H6.1, H11
16 02 11	*	discarded equipment containing chlorofluorocarbons, HCFC, HFC		HP14	AC150	Art. 1 (1) b	
16 02 12	*	discarded equipment containing free asbestos		HP7	A1180, A2050	Y36	H11
16 02 13	*	discarded equipment containing hazardous components*) other than those mentioned in 16 02 09 to16 02 12	batteries, x-ray tubes, back lighted LCDs (mercury containing backlight), ca-pacitors	HP4, HP8, HP5, HP6, HP7, HP10, HP14	NL, A1180	Art. 1 (1) b	(H11, H8, H6.1, H12)
16 02 14		discarded equipment other than those mentioned in 16 02 09 to 16 02 13			GC020, B1110		
16 02 15	*	hazardous components removed from discarded equipment	batteries, x-ray tubes, back lighted LCDs (mercury containing backlight), ca-pacitors	HP4, HP8, HP5, HP6, HP7, HP10, HP14	A1180, A1030, A2010, A3180	Y9, Y10, Y20, Y26, Y29, Y31, Y36, Y45	**partly** (H11, H8, H6.1, H12)

Code	H	Designation	possible constituents/substances	EU		Basel	
				typical HP-Criteria	possible WSR-Codes	possible Y-Codes	typical H- Criteria
16 02 16		components removed from discarded equipment other than those mentioned in 16 02 15			GC020, B1110, B1115		
16 03 00		**off-specification batches and unused products**					
16 03 03	*	inorganic wastes containing hazardous substances		HP4, HP5, HP6, HP7, HP10, HP14	A4130, A4140	Y according to composi-tion	**partly** (H6.1, H11, H3, H12)
16 03 04		inorganic wastes other than those mentioned in 16 03 03			NL		
16 03 05	*	organic wastes containing hazardous substances		HP3, HP4, HP5, HP6, HP7, HP10, HP14	A4130, A4140	Y according to composi-tion	**partly** (H6.1, H11, H3, H12)
16 03 06		organic wastes other than those mentioned in 16 03 05			B3130		
16 03 07	*	metallic mercury		HP6, HP14	A1030	Y29	H6.1, H12
16 04 00		**waste explosives**					
16 04 01	*	waste ammunition		HP1	A4080	Y15	H1
16 04 02	*	fireworks wastes		HP1	A4080	Y15	H1
16 04 03	*	other waste explosives		HP1	A4080	Y15	H1
16 05 00		**gases in pressure containers and discarded chemicals**					
16 05 04	*	gases in pressure containers (including halons) containing hazardous substances		HP3, HP4, HP6, HP8	AC150, AC160	**Art. 1 (1) b**	(H11, H5.1, H8, H6.1)
16 05 05		gases in pressure containers other than those mentioned in 16 05 04			NL		
16 05 06	*	laboratory chemicals, consisting of or containing hazardous substances, including mixtures of laboratory chemicals		HP6, HP8, HP11, HP14, HP3	A4140, A4130, A4150	(Y14)	**partly** (H6.1, H8, H3, H11, H12)
16 05 07	*	discarded inorganic chemicals consisting of or containing hazardous substances		HP6, HP7, HP10, HP11, HP8, HP14	A4140, A4130, A4150	(Y14)	**partly** (H6.1, H11, H8, H1, H12)

Code	H	Designation	possible constituents/substances	EU		Basel	
				typical HP-Criteria	possible WSR-Codes	possible Y-Codes	typical H- Criteria
16 05 08	*	discarded organic chemicals consisting of or containing hazardous substances		HP3, HP6, HP7, HP8, HP10, HP11, HP14	A4140, A4130, A4150	(Y14)	**partly** (H3, H6.1, H8, H11, H12)
16 05 09		discarded chemicals other than those mentioned in 16 05 06, 16 05 07 or 16 05 08			NL		
16 06 00		**batteries and accumulators**					
16 06 01	*	lead batteries	Pb, acids	HP8, HP10, HP6	A1160	Y31, Y34	H8, H6.1, H11
16 06 02	*	Ni-Cd batteries	Ni, Cd	HP6, HP7, HP8	A1170	Y26, Y35	H6.1, H8, H11
16 06 03	*	mercury-containing batteries	Hg	HP6, HP4, HP8	A1170	Y29, Y35	H6.1, H8
16 06 04		alkaline batteries (except 16 06 03)	alkaline electrolytes; mar-ganese oxides		B1090		
16 06 05		other batteries and accumulators			B1090		
16 06 06	*	separately collected electrolyte from batteries and accumulators	sulphuric acid, alkaline, or-ganic electrolytes	HP8	A4090	Y34, Y35	H8
16 07 00		**wastes from transport tank, storage tank and barrel cleaning (except 05 and 13)**					
16 07 08	*	wastes containing oil	HM (Pb Ni, Cr)	HP7, HP10	A4060	Y9	H11
16 07 09	*	wastes containing other hazardous substances	HM (Pb Ni, Cr)	HP4, HP5, HP6, HP7, HP8	NL, A4140	**Art. 1 (1) b**	(H11, H8, H6.1)
16 07 99		wastes not otherwise specified			NL		
16 08 00		**spent catalysts**					
16 08 01		spent catalysts containing gold, silver, rhenium, rhodium, palladium, iridium or platinum (except16 08 07)			B1130		
16 08 02	*	spent catalysts containing hazardous transition metals or hazardous transition metal compounds	V, Ni	HP7, HP3, HP6, HP5	B1120		

212

Code	H	Designation	possible constituents/substances	EU		Basel	
				typical HP-Criteria	possible WSR-Codes	possible Y-Codes	typical H- Criteria
16 08 03		spent catalysts containing transition metals or transition metal compounds not otherwise specified			B1120		
16 08 04		spent fluid catalytic cracking catalysts (except 16 08 07)			GC050		
16 08 05	*	spent catalysts containing phosphoric acid	HC, PAH, e.g. zeolites with immobilised phosphoric acid	HP4, HP8, HP3	A1140, A4090, A2030	Y34	H3, H8
16 08 06	*	spent liquids used as catalysts	acids, alkaline, salts (Ti-, Al-, P-compounds, amines, etc.	HP4, HP8, HP7, HP2	A1140, A4090, A2030	Y34, Y35	H5.1, H8, H11
16 08 07	*	spent catalysts contaminated with hazardous substances	e.g. HC (according to process)	HP7, (HP3)	NL	Art. 1 (1) b	H11, (H3)
16 09 00		**oxidising substances**					
16 09 01	*	permanganates, for example potassium permanganate		HP2, H5, HP8	NL	Art. 1 (1) b	(H5.1, H11, H8)
16 09 02	*	chromates, for example potassium chromate, potassium or sodium dichromate	Cr(VI)	HP2, HP8, HP7, HP14	A1040	Y21	H5.1, H11, H8, H12
16 09 03	*	peroxides, for example hydrogen peroxide		HP2, HP8, (HP1)	NL	Art. 1 (1) b	(H5.1, H1, H8)
16 09 04	*	oxidising substances, not otherwise specified		HP2, HP5, HP6, HP7	NL	Art. 1 (1) b	(H5.1, H11, H6.1)
16 10 00		**aqueous liquid wastes destined for off-site treatment**					
16 10 01	*	aqueous liquid wastes containing hazardous substances		HP6, HP7, HP14	NL	Art. 1 (1) b	(H6.1, H11, H12)
16 10 02		aqueous liquid wastes other than those mentioned in 16 10 01			NL		
16 10 03	*	aqueous concentrates containing hazardous substances		HP15, HP8, HP7	NL	Art. 1 (1) b	(H13, H8, H11)

Code	H	Designation	possible constituents/substances	EU		Basel	
				typical HP-Criteria	possible WSR-Codes	possible Y-Codes	typical H- Criteria
16 10 04		aqueous concentrates other than those mentioned in 16 10 03			NL		
16 11 00		**waste linings and refractories**					
16 11 01	*	carbon-based linings and refractories from metallurgical processes containing hazardous substances	cyanides, metal carbides, metal compounds, fluoride compounds	HP6, HP12, HP7, HP10	A1020, A4050	Y32	**partly** (H6.1, H10, H11)
16 11 02		carbon-based linings and refractories from metallurgical processes others than those mentioned in16 11 01			B1100		
16 11 03	*	other linings and refractories from metallurgical processes containing hazardous substances	Cr(VI), PAH, metal com-pounds	HP6, HP7, HP10	NL	**Art. 1 (1) b**	(H6.1, H11)
16 11 04		other linings and refractories from metallurgical processes other than those mentioned in 16 11 03			NL, B1100, GF010		
16 11 05	*	linings and refractories from non-metallurgical processes containing hazardous substances	Cr(VI), PAH, metal com-pounds	HP6, HP7, HP10	NL	**Art. 1 (1) b**	(H6.1, H11)
16 11 06		linings and refractories from non-metallurgical processes others than those mentioned in 16 11 05			NL, GF010		
17 00 00		**CONSTRUCTION AND DEMOLITION WASTES (IN-CLUDING EXCAVATED SOIL FROM CONTAMINA-TED SITES)**					
17 01 00		**concrete, bricks, tiles and ceramics**					
17 01 01		concrete			B2040		
17 01 02		bricks			GF010		
17 01 03		tiles and ceramics			GF010		
17 01 06	*	mixtures of, or separate fractions of concrete, bricks, tiles and ceramics containing hazardous substances	PAH	HP10, HP7	NL, A3180	**Art. 1 (1) b**	(H11)
17 01 07		mixtures of concrete, bricks, tiles and ceramics other than those mentioned in 17 01 06			NL		

Code	H	Designation	possible constituents/substances	EU		Basel	
				typical HP-Criteria	possible WSR-Codes	possible Y-Codes	typical H- Criteria
17 02 00		**wood, glass and plastic**					
17 02 01		wood			AC170		
17 02 02		glass			B2020		
17 02 03		plastic			B3011/EU3011, (B3010), B3130, (GH013)		
17 02 04	*	glass, plastic and wood containing or contaminated with hazardous substances	lead containing paints, PAH, e.g. wood with tar containing adhesives	HP10, HP7, H6	A3210, AC170; NL	**Art. 1 (1) b**	(H11, H6.1, H12)
17 03 00		**bituminous mixtures, coal tar and tarred products**					
17 03 01	*	bituminous mixtures containing coal tar	PAH, phenols	HP7	A3190, A3200	Y11	H11
17 03 02		bituminous mixtures other than those mentioned in 17 03 01			B2130		
17 03 03	*	coal tar and tarred products	PAH, phenols, HC	HP7, HP8, HP14	A3190	Y11	H11, H8, H12
17 04 00		**metals (including their alloys)**					
17 04 01		copper, bronze, brass			B1010, B1040		
17 04 02		aluminium			B1010, B1040		
17 04 03		lead			B1010, B1040		
17 04 04		zinc			B1010, B1040		
17 04 05		iron and steel			B1010, B1040		
17 04 06		tin			B1010, B1040		
17 04 07		mixed metals			B1010, B1040		
17 04 09	*	metal waste contaminated with hazardous substances	Ni, Cd, Pb, etc.	HP7, HP10	NL	**Art. 1 (1) b**	(H11)
17 04 10	*	cables containing oil, coal tar and other hazardous substances	PAH, HC, PCBs	HP7, HP14	A1190, (A3180)	Y11, Y10	(H11, H12)

Code	H	Designation	possible constituents/substances	EU		Basel	
				typical HP-Criteria	possible WSR-Codes	possible Y-Codes	typical H- Criteria
17 04 11		cables other than those mentioned in 17 04 10			B1115		
17 05 00		**soil (including excavated soil from contamina-ted sites), stones and dredging spoil**					
17 05 03	*	soil and stones containing hazardous substances	HC, PAH, HM, AOX/EOX, PCBs	HP7, HP6, HP14	NL	Art. 1 (1) b	(H11, H6.1, H12)
17 05 04		soil and stones other than those mentioned in 17 05 03			NL		
17 05 05	*	dredging spoil containing hazardous substances	HC, PAH, HM, organome-tallic compounds (TBT)	HP7, HP14	NL	Art. 1 (1) b	(H11, H12)
17 05 06		dredging spoil other than those mentioned in 17 05 05			NL		
17 05 07	*	track ballast containing hazardous substances	HC, PAH, psbl, herbicides	HP6, HP15, HP7	NL	Art. 1 (1) b	(H6.1, H13, H11)
17 05 08		track ballast other than those mentioned in 17 05 07			NL		
17 06 00		**insulation materials and asbestos-containing construction materials**					
17 06 01	*	insulation materials containing asbestos	free asbestos	HP7	A2050	Y36	H11
17 06 03	*	other insulation materials consisting of or containing hazardous substances	biocide or fire safe impreg-nation; PCBs, PAHs	HP5, HP7	RB020	Art. 1 (1) b	(H11)
17 06 04		insulation materials other than those mentioned in 17 06 01 and 17 06 03			B2030, B3010		
17 06 05	*	construction materials containing asbestos	weak bound asbestos, free asbestos	HP7	A2050	Y36	H11
17 08 00		**gypsum-based construction material**					
17 08 01	*	gypsum-based construction materials contaminated with hazardous substances	PCB-containing wastes (softener from paints)	HP7, HP6	NL	Art. 1 (1) b	(H11, H6.1)
17 08 02		gypsum-based construction materials other than those mentioned in 17 08 01			B2040		

Code	H	Designation	possible constituents/substances	EU		Basel	
				typical HP-Criteria	possible WSR-Codes	possible Y-Codes	typical H- Criteria
17 09 00		**other construction and demolition wastes**					
17 09 01	*	construction and demolition wastes containing mercury	Hg	HP6, HP7, HP14	A1030	Y29	H11, H6.1, H12
17 09 02	*	construction and demolition wastes containing PCB (for example PCB-containing sealants, PCB-containing resin-based floorings, PCB-containing sealed glazing units, PCB-containing capacitors)	PCB	HP5, HP6, HP7, HP14	A3180, A4110	Y10	H6.1, H11, H12
17 09 03	*	other construction and demolition wastes (including mixed wastes) containing hazardous substances		HP6, HP7, HP14	NL	Art. 1 (1) b	(H11, H6.1, H12)
17 09 04		mixed construction and demolition wastes other than those mentioned in 17 09 01, 17 09 02 and 17 09 03			NL		
18 00 00		**WASTES FROM HUMAN OR ANIMAL HEALTH CARE AND/OR RELATED RESEARCH (except kitchen and restaurant wastes not arising from immediate health care)**					
18 01 00		**wastes from natal care, diagnosis, treatment or prevention of disease in humans**					
18 01 01		sharps (except 18 01 03)			NL		
18 01 02		body parts and organs including blood bags and blood preserves (except 18 01 03)			NL		
18 01 03	*	wastes whose collection and disposal is subject to special requirements in order to prevent infection	infectious materials	HP9	A4020	Y1	H6.2
18 01 04		wastes whose collection and disposal is not subject to special requirements in order to prevent infection(for example dressings, plaster casts, linen, disposable clothing, diapers)			NL		
18 01 06	*	chemicals consisting of or containing hazardous substances	HC, solvents, etc.	HP6, HP8, HP11, HP14	NL	Art. 1 (1) b	(H6.1, H8, H11, H12)
18 01 07		chemicals other than those mentioned in 18 01 06			NL		

Code	H	Designation	possible constituents/substances	EU		Basel	
				typical HP-Criteria	possible WSR-Codes	possible Y-Codes	typical H- Criteria
18 01 08	*	cytotoxic and cytostatic medicines		HP5, HP11, HP6, HP7, HP3	A4010	Y3	H11, H6.1, H3
18 01 09		medicines other than those mentioned in 18 01 08			NL		
18 01 10	*	amalgam waste from dental care	Hg	HP6	A1030, A4020	Y29	H6.1
18 02 00		**wastes from research, diagnosis, treatment or prevention of disease involving animals**					
18 02 01		sharps (except 18 02 02)			NL		
18 02 02	*	wastes whose collection and disposal is subject to special requirements in order to prevent infection		HP9	A4020	Y1	H6.2
18 02 03		wastes whose collection and disposal is not subject to special requirements in order to prevent infection			NL		
18 02 05	*	chemicals consisting of or containing hazardous substances		HP6, HP8, HP3, HP11, HP14	A4140, A4130, A4150	Y1, Y14	H6.1, H8, H3, H11, H12
18 02 06		chemicals other than those mentioned in 18 02 05			NL		
18 02 07	*	cytotoxic and cytostatic medicines		HP11, HP6, HP5, HP3	A4010	Y3	H11, H6.1, H3
18 02 08		medicines other than those mentioned in 18 02 07			NL		
19 00 00		**WASTES FROM WASTE MANAGEMENT FACILITIES, OFF-SITE WASTE WATER TREATMENT PLANTS AND THE PREPARATION OF WATER INTENDED FOR HUMAN CONSUMPTION AND WATER FOR INDUSTRIAL USE**					
19 01 00		**wastes from incineration or pyrolysis of waste**					
19 01 02		ferrous materials removed from bottom ash			B1010, NL		
19 01 05	*	filter cake from gas treatment	HM (Hg, Pb, Ni, Cd), metal hydroxides, CaO, Ca-sul-phates, PCDD/PCDF	HP7, HP10, HP4	A4100, AB150	Y18, Y31, (Y22, Y43, Y44)	H11
19 01 06	*	aqueous liquid wastes from gas treatment and other aqueous liquid wastes	acidic/alkaline solutions, HCl, HF, HM-compounds	HP8, HP7, HP10, HP6	A4100	Y18 (Y22, Y34, Y35)	H8, H11, H6.1

Code	H	Designation	possible constituents/substances	EU		Basel	
				typical HP-Criteria	possible WSR-Codes	possible Y-Codes	typical H- Criteria
19 01 07	*	solid wastes from gas treatment	PAH, HC, HM, reactive metals (Al), PCDD/PCDF	HP10, HP7, HP6, (HP3)	A4100	Y18 (Y22, Y34, Y35)	H11, H6.1, (H3)
19 01 10	*	spent activated carbon from flue-gas treatment	PCDD/PCDF, PAH, Hg, other HM	HP3, HP7, HP10, HP6	A4160	Y18	H11, H4.1, H6.1
19 01 11	*	bottom ash and slag containing hazardous substances	PAH, HC, HM, reactive metals (e.g. Al)	HP10, HP7, HP14, (HP3)	A4100	Y18	H11, H12, (H3)
19 01 12		bottom ash and slag other than those mentioned in 19 01 11			NL		
19 01 13	*	fly ash containing hazardous substances	PAH, HC, HM, reactive metals, PCDD/PCDF,	HP10, HP7, (HP3)	A4100	Y18, (Y43, Y44)	H11, (H3)
19 01 14		fly ash other than those mentioned in 19 01 13			NL		
19 01 15	*	boiler dust containing hazardous substances	PAH, HM (Pb, etc.)	HP15, HP7, HP10	A4100	Y18	H11, H13
19 01 16		boiler dust other than those mentioned in 19 01 15			NL		
19 01 17	*	pyrolysis wastes containing hazardous substances	PCDD/PCDF, PAH, HM (Pb, Ni, Cr, Cd, etc.)	HP7, HP3	NL, A3190	**Art. 1 (1) b**	(H11, H4.2)
19 01 18		pyrolysis wastes other than those mentioned in 19 01 17			NL		
19 01 19		sands from fluidised beds			NL		
19 01 99		wastes not otherwise specified			NL		
19 02 00		**wastes from physico/chemical treatments of waste (including dechromatation, decyanidation, neutralisation)**					
19 02 03		premixed wastes composed only of non-hazardous wastes			NL		
19 02 04	*	premixed wastes composed of at least one hazardous waste		HP3, HP15, HP7, HP14, HP6, HP8, HP10, HP5	A3180, NL	Y18	(H11, H8, H3, H6.1, H13, H12)
19 02 05	*	sludges from physico/chemical treatment containing hazardous substances	HM- sludges, HC- containing sludges	HP5, HP6, HP7, HP10	A102C, A1030	Y18	H6.1, H11

Code	H	Designation	possible constituents/substances	EU		Basel	
				typical HP-Criteria	possible WSR-Codes	possible Y-Codes	typical H- Criteria
19 02 06		sludges from physico/chemical treatment other than those mentioned in 19 02 05			B2100, B1140, B2100		
19 02 07	*	oil and concentrates from separation	HC PAH, HM (Pb, Ni, Cr, psbl, Hg)	HP7, HP10, (HP3)	A3020, A4060	Y8, Y18	H11, (H3)
19 02 08	*	liquid combustible wastes containing hazardous substances	HC, PCB, AOX, psbl, Hg and other HM	HP3, HP7, HP10	A3180, NL	Y18	H3, H11
19 02 09	*	solid combustible wastes containing hazardous substances	HC, PCB, AOX, HM	HP3, HP7, HP10	A3180, NL	Y18	H3, H11
19 02 10		combustible wastes other than those mentioned in 19 02 08 and 19 02 09			NL		
19 02 11	*	other wastes containing hazardous substances		HP14, (HP3)	NL	Art. 1 (1) b	(H4.1, H12)
19 02 99		wastes not otherwise specified			NL		
19 03 00		**stabilised/solidified wastes**					
19 03 04	*	wastes marked as hazardous, partly stabilised other than 19 03 08		HP7, HP10, HP6, HP14, HP15	NL, A4140	Y18	H11, H6.1, H12, H13
19 03 05		stabilised wastes other than those mentioned in 19 03 04			NL		
19 03 06	*	wastes marked as hazardous, solidified		HP7, HP10, HP6, HP8	NL, A4140	Y18	H11, H6.1, H8
19 03 07		solidified wastes other than those mentioned in 19 03 06			NL		
19 03 08	*	partly stabilised mercury		HP6, HP14, HP13	A1030	Y29, Y18	H6.1, H12
19 04 00		**vitrified waste and wastes from vitrification**					
19 04 01		vitrified waste			NL		
19 04 02	*	fly ash and other flue-gas treatment wastes	HM (Ni, Pb, Cr, Cd, As, Zn), PCDD/PCDF	HP7, HP10, HP6	A4100	Y18, (Y43, Y44)	H11, H6.1
19 04 03	*	non-vitrified solid phase	HM (Ni, Pb, Cr, Cd, As)	HP7, HP10	NL, A4140	Y18	H11
19 04 04		aqueous liquid wastes from vitrified waste tempering			NL		

Code	H	Designation	possible constituents/substances	EU		Basel	
				typical HP-Criteria	possible WSR-Codes	possible Y-Codes	typical H- Criteria
19 05 00		**wastes from aerobic treatment of solid wastes**					
19 05 01		non-composted fraction of municipal and similar wastes			NL		
19 05 02		non-composted fraction of animal and vegetable waste			NL		
19 05 03		off-specification compost			NL		
19 05 99		wastes not otherwise specified			NL		
19 06 00		**wastes from anaerobic treatment of waste**					
19 06 03		liquor from anaerobic treatment of municipal waste			NL		
19 06 04		digestate from anaerobic treatment of municipal waste			NL		
19 06 05		liquor from anaerobic treatment of animal and vegetable waste			NL		
19 06 06		digestate from anaerobic treatment of animal and vegetable waste			NL		
19 06 99		wastes not otherwise specified			NL		
19 07 00		**landfill leachate**					
19 07 02	*	landfill leachate containing hazardous substances	HM, HC, AOX/EOX	HP15, HP14	NL	Y18	(H13, H12)
19 07 03		landfill leachate other than those mentioned in 19 07 02			NL		
19 08 00		**wastes from waste water treatment plants not otherwise specified**					
19 08 01		screenings			NL		
19 08 02		waste from desanding			NL		
19 08 05		sludges from treatment of urban waste water			AC270		
19 08 06	*	saturated or spent ion exchange resins	HM	HP7, HP10, HP4, HP8	AD120	**Art. 1 (1) b**	(H8, H11)
19 08 07	*	solutions and sludges from regeneration of ion exchangers	acids, alkaline, HM	HP4, HP8, HP7, HP10	A4090	Y34, Y35, Y18	H8, H11

Code	H	Designation	possible constituents/substances	EU		Basel	
				typical HP-Criteria	possible WSR-Codes	possible Y-Codes	typical H- Criteria
19 08 08	*	membrane system waste containing heavy metals	HM, HC	HP4, HP8, HP7, HP10, HP6	A4090	Y34, Y35, Y18	H8, H11, H6.1
19 08 09		grease and oil mixture from oil/water separation containing only edible oil and fats			B3065, NL		
19 08 10	*	grease and oil mixture from oil/water separation other than those mentioned in 19 08 09	HC, AOX, psbl, HM (Ni, Pb)	HP3, HP14, HP10, HP7	A4060	Y18, Y9	H3, H11, H12
19 08 11	*	sludges containing hazardous substances from biological treatment of industrial waste water	HM, HC, CaO	HP4, HP5, HP8	AC270	**Art. 1 (1) b**	(H8, H11)
19 08 12		sludges from biological treatment of industrial waste water other than those mentioned in 19 08 11			AC270		
19 08 13	*	sludges containing hazardous substances from other treatment of industrial waste water	HM, HC, CaO	HP8, HP7, HP10	AC270	**Art. 1 (1) b**	(H6.1, H8, H11)
19 08 14		sludges from other treatment of industrial waste water other than those mentioned in 19 08 13			AC270		
19 08 99		wastes not otherwise specified			NL		
19 09 00		**wastes from the preparation of water intended for human consumption or water for industrial use**					
19 09 01		solid waste from primary filtration and screenings			NL		
19 09 02		sludges from water clarification			NL		
19 09 03		sludges from decarbonation			NL		
19 09 04		spent activated carbon			B2060		
19 09 05		saturated or spent ion exchange resins			NL		
19 09 06		solutions and sludges from regeneration of ion exchangers			NL		
19 09 99		wastes not otherwise specified			NL		

Code	H	Designation	possible constituents/substances	EU		Basel	
				typical HP-Criteria	possible WSR-Codes	possible Y-Codes	typical H- Criteria
19 10 00		**wastes from shredding of metal-containing wastes**					
19 10 01		iron and steel waste			B1010		
19 10 02		non-ferrous waste			B1010, B1020, B1050		
19 10 03	*	fluff-light fraction and dust containing hazardous substances	HC, fine lead, PCB	HP10, HP15	A3120	Y31, Y9, Y10, Y18	H11, H13
19 10 04		fluff-light fraction and dust other than those mentioned in 19 10 03			NL		
19 10 05	*	other fractions containing hazardous substances	batteries, compensators, lead glass, etc.)	HP10	A4100	Y18	H11
19 10 06		other fractions other than those mentioned in 19 10 05			B1050, B1060		
19 11 00		**wastes from oil regeneration**					
19 11 01	*	spent filter clays	HC	HP14, HP15		Y18, Y9	H12, H13
19 11 02	*	acid tars	HC, PAH, sulphuric acids	HP8, HP7, HP10	A4090, A3190	Y18, Y34, Y11	H8, H11
19 11 03	*	aqueous liquid wastes	HC	HP14, HP15	A4060	Y18	H12, H13
19 11 04	*	wastes from cleaning of fuel with bases	HC, alkaline, HM (Ni, Pb)	HP8, HP14, HP10, HP7	A4090	Y18, Y35, Y9	H11, H8, H12
19 11 05	*	sludges from on-site effluent treatment containing hazardous substances	HC, HM	HP6, HP5	AC270	**Art. 1 (1) b**	(H6.1, H11)
19 11 06		sludges from on-site effluent treatment other than those mentioned in 19 11 05			AC270		
19 11 07	*	wastes from flue-gas cleaning		HP6, HP4	A4100, AB150	Y18	H6.1
19 11 99		wastes not otherwise specified			NL		
19 12 00		**wastes from the mechanical treatment of waste (for example sorting, crushing, compacting, pelletising) not otherwise specified**					
19 12 01		paper and cardboard			B3020		

Code	H	Designation	possible constituents/substances	EU		Basel	
				typical HP-Criteria	possible WSR-Codes	possible Y-Codes	typical H- Criteria
19 12 02		ferrous metal			B1010		
19 12 03		non-ferrous metal			B1010, B1020, B1050, B1060		
19 12 04		plastic and rubber			B3011/EU3011, (B3010), B3040, B3080, (GH013), B3130		
19 12 05		glass			B2020		
19 12 06	*	wood containing hazardous substances	PAH, Cr(VI), Cu- comp., organic pesticides, psbl, As-, B-, Pb- compounds	HP7, HP11, HP10	AC170, A3180	Y18	H11
19 12 07		wood other than that mentioned in 19 12 06			AC170, B3050		
19 12 08		textiles			B3030		
19 12 09		minerals (for example sand, stones)			NL		
19 12 10		combustible waste (refuse derived fuel)			NL		
19 12 11	*	other wastes (including mixtures of materials) from mechanical treatment of waste containing hazardous substances	batteries, containers with hazardous residues (HC, PAH, …)	HP3, HP6, HP14	A4130, A1180, (A1170)	Y18	H3, H6.1, H12
19 12 12		other wastes (including mixtures of materials) from mechanical treatment of wastes other than those mentioned in 19 12 11			NL		
19 13 00		**wastes from soil and groundwater remediation**					
19 13 01	*	solid wastes from soil remediation containing hazardous substances	PAH, detergents, phenols, HM,	HP6, HP7	NL, A3180	Y18	H6.1, H11
19 13 02		solid wastes from soil remediation other than those mentioned in 19 13 01			B2060		
19 13 03	*	sludges from soil remediation containing hazardous substances	PAH, detergents, phenols, HM,	HP6, HP7	NL, A3180	Y18	H6.1, H11
19 13 04		sludges from soil remediation other than those mentioned in 19 13 03			NL		

Code	H	Designation	possible constituents/substances	EU		Basel	
				typical HP-Criteria	possible WSR-Codes	possible Y-Codes	typical H- Criteria
19 13 05	*	sludges from groundwater remediation containing hazardous substances		HP6	NL, A3180	Y18	H6.1
19 13 06		sludges from groundwater remediation other than those mentioned in 19 13 05	e. g. Fe(OH)$_3$ sludge		NL		
19 13 07	*	aqueous liquid wastes and aqueous concentrates from groundwater remediation containing hazardous substances		HP6	NL	Art. 1 (1) b	(H6.1)
19 13 08		aqueous liquid wastes and aqueous concentrates from groundwater remediation other than those mentioned in 19 13 07			NL		
20 00 00		**MUNICIPAL WASTES (HOUSEHOLD WASTE AND SIMILAR COMMERCIAL, INDUSTRIAL AND INSTITU-TIONAL WASTES) INCLUDING SEPARATELY COL-LECTED FRACTIONS**					
20 01 00		**separately collected fractions (except 15 01)**					
20 01 01		paper and cardboard			B3020		
20 01 02		glass			B2020		
20 01 08		biodegradable kitchen and canteen waste			BEU05, B3060		
20 01 10		clothes			B3030		
20 01 11		textiles			B3030		
20 01 13	*	Solvents	HC, R-Cl, gasoline, petro-leum ether, ligroin, acetone	HP3, HP14, HP6, HP7	A3140, A3150	Y41, Y42	H3, H11, H12
20 01 14	*	acids	acids	HP4, HP8	A4090	Y34	H8
20 01 15	*	alkalines	ammonia, hypochlorites, NaOH	HP4, HP8	A4090	Y35	H8
20 01 17	*	photochemicals	quinones, amino phenols, KOH, glycine, borates; acetic acid; thiosulfates, amino-thiosulfates, Ag-complexes	HP4, HP8	AD090, A3070	Y16	H8

Code	H	Designation	possible constituents/substances	EU		Basel	
				typical HP-Criteria	possible WSR-Codes	possible Y-Codes	typical H- Criteria
20 01 19	*	pesticides	R-X, solvents, pyrethroids, phosphides (rodenticides)	HP6, HP3, HP5	A4030	Y4	H11, H6.1, H3
20 01 21	*	fluorescent tubes and other mercury-containing waste	Hg	HP6, HP14	A1030, A1180	Y29	H6.1, H12
20 01 23	*	discarded equipment containing chlorofluorocarbons	R-Cl	HP14, HP8	NL, A1180	**Art. 1 (1) b**	(H8, H12)
20 01 25		edible oil and fat			B3065		
20 01 26	*	oil and fat other than those mentioned in 20 01 25	HC	HP7, HP3	A3020	Y8, Y9	H11, H3
20 01 27	*	paint, inks, adhesives and resins containing hazardous substances	solvents, HM, peroxides, etc.	HP3, HP6, HP4, HP8	A4070	Y12	H3, (H8)
20 01 28		paint, inks, adhesives and resins other than those mentioned in 20 01 27			B4010, B4020		
20 01 29	*	detergents containing hazardous substances		HP4, HP8, HP5, HP14, HP3	AC250	**Art. 1 (1) b**	H3, H11, (H8)
20 01 30		detergents other than those mentioned in 20 01 29			AC250		
20 01 31	*	cytotoxic and cytostatic medicines		HP11, HP6, HP5	A4010	Y3	H11, H6.1
20 01 32		medicines other than those mentioned in 20 01 31			NL		
20 01 33	*	batteries and accumulators included in 16 06 01, 16 06 02 or 16 06 03 and unsorted batteries and accumulators containing these batteries	alkaline, MnO, etc.	HP8, HP5, HP7	A1170, A1160	Y31, Y26, Y34, Y35, Y42	H8, H11
20 01 34		batteries and accumulators other than those mentioned in 20 01 33			B1090		
20 01 35	*	discarded electrical and electronic equipment other than those mentioned in 20 01 21 and 20 01 23 containing hazardous components*⁾	waste oil, PCB, asbestos, Pb, Hg,	HP7, HP8, HP5	A1180	Y31, Y10, Y8, Y36, Y29	**partly** (H11, H8)
20 01 36		discarded electrical and electronic equipment other than those mentioned in 20 01 21, 20 01 23 and 20 01 35			B1110, GC020		
20 01 37	*	wood containing hazardous substances	PAH, Cr(VI) pesticides, Cu-compounds, psbl, As-, B-, Pb-compounds	HP7, HP10, HP11	AC170	**Art. 1 (1) b**	(H11)

Code	H	Designation	possible constituents/substances	EU		Basel	
				typical HP-Criteria	possible WSR-Codes	possible Y-Codes	typical H- Criteria
20 01 38		wood other than that mentioned in 20 01 37			AC170, B3050		
20 01 39		plastics			Y48, NL	Y48	
20 01 40		metals			B1010, B1020		
20 01 41		wastes from chimney sweeping			NL		
20 01 99		other fractions not otherwise specified			NL		
20 02 00		**garden and park wastes (including cemetery waste)**					
20 02 01		biodegradable waste			BEU05, B3060, B3050, B3070		
20 02 02		soil and stones			NL, (Y46)		
20 02 03		other non-biodegradable wastes			NL, (Y46)		
20 03 00		**other municipal wastes**					
20 03 01		mixed municipal waste			Y46	Y46	
20 03 02		waste from markets			B3060		
20 03 03		street-cleaning residues			Y46	Y46	
20 03 04		septic tank sludge			NL		
20 03 06		waste from sewage cleaning			NL		
20 03 07		bulky waste			Y46	Y46	
20 03 99		municipal wastes not otherwise specified			Y46	Y46	

*)	Hazardous components from electrical and electronic equipment may include accumulators and batteries mentioned in 16 06 and marked as hazardous; mercury switches, glass from cathode ray tubes and other activated glass, etc.

227

II.   Annexes

II.1   Basel Convention

BASEL CONVENTION[1] ON THE CONTROL OF TRANSBOUNDARY MOVEMENTS OF HAZARDOUS WASTES AND THEIR DISPOSAL

Preamble

The Parties to this Convention,

<u>Aware</u> of the risk of damage to human health and the environment caused by hazardous wastes and other wastes and the transboundary movement thereof,

<u>Mindful</u> of the growing threat to human health and the environment posed by the increased generation and complexity, and transboundary movement of hazardous wastes and other wastes,

<u>Mindful also</u> that the most effective way of protecting human health and the environment from the dangers posed by such wastes is the reduction of their generation to a minimum in terms of quantity and/or hazard potential,

<u>Convinced</u> that States should take necessary measures to ensure that the management of hazardous wastes and other wastes including their transboundary movement and disposal is consistent with the protection of human health and the environment whatever the place of disposal,

<u>Noting</u> that States should ensure that the generator should carry out duties with regard to the transport and disposal of hazardous wastes and other wastes in a manner that is consistent with the protection of the environment, whatever the place of disposal,

<u>Fully recognizing</u> that any State has the sovereign right to ban the entry or disposal of foreign hazardous wastes and other wastes in its territory,

<u>Recognizing also</u> the increasing desire for the prohibition of transboundary movements of hazardous wastes and their disposal in other States, especially developing countries,

<u>Recognizing</u> that transboundary movements of hazardous wastes, especially to developing countries, have a high risk of not constituting an environmentally sound

---

[1]   The present text incorporates amendments to the Convention adopted subsequent to its entry into force and that are in force as at 31 March 2020. Only the text of the Convention as kept in the custody of the Secretary-General of the United Nations in his capacity as Depositary constitutes the authentic version of the Convention, as modified by any amendments and/or corrections thereto. This publication is issued for information purposes only.

management of hazardous wastes as required by this Convention,[2]

<u>Convinced</u> that hazardous wastes and other wastes should, as far as is compatible with environmentally sound and efficient management, be disposed of in the State where they were generated,

<u>Aware also</u> that transboundary movements of such wastes from the State of their generation to any other State should be permitted only when conducted under conditions which do not endanger human health and the environment, and under conditions in conformity with the provisions of this Convention,

<u>Considering</u> that enhanced control of transboundary movement of hazardous wastes and other wastes will act as an incentive for their environmentally sound management and for the reduction of the volume of such transboundary movement,

<u>Convinced</u> that States should take measures for the proper exchange of information on and control of the transboundary movement of hazardous wastes and other wastes from and to those States,

<u>Noting</u> that a number of international and regional agreements have addressed the issue of protection and preservation of the environment with regard to the transit of dangerous goods,

<u>Taking into account</u> the Declaration of the United Nations Conference on the Human Environment (Stockholm, 1972), the Cairo Guidelines and Principles for the Environmentally Sound Management of Hazardous Wastes adopted by the Governing Council of the United Nations Environment Programme (UNEP) by decision 14/30 of 17 June 1987, the Recommendations of the United Nations Committee of Experts on the Transport of Dangerous Goods (formulated in 1957 and updated biennially), relevant recommendations, declarations, instruments and regulations adopted within the United Nations system and the work and studies done within other international and regional organizations,

<u>Mindful</u> of the spirit, principles, aims and functions of the World Charter for Nature adopted by the General Assembly of the United Nations at its thirty-seventh session (1982) as the rule of ethics in respect of the protection of the human environment and the conservation of natural resources,

<u>Affirming</u> that States are responsible for the fulfilment of their international obligations concerning the protection of human health and protection and preservation of the environment, and are liable in accordance with international law,

<u>Recognizing</u> that in the case of a material breach of the provisions of this Convention or any protocol thereto the relevant international law of treaties shall apply,

<u>Aware</u> of the need to continue the development and implementation of environmentally sound low-waste technologies, recycling options, good house-keeping and management systems with a view to reducing to a minimum the generation of hazardous wastes and other wastes,

---

[2] The Conference of the Parties adopted Decision III/1 at its third meeting to amend the Convention by adding, inter alia, a new preambular paragraph 7 bis. This amendment entered into force on 5 December 2019 (depositary notification C.N.420.2019)

<u>Aware also</u> of the growing international concern about the need for stringent control of transboundary movement of hazardous wastes and other wastes, and of the need as far as possible to reduce such movement to a minimum,

<u>Concerned</u> about the problem of illegal transboundary traffic in hazardous wastes and other wastes,

<u>Taking into account also</u> the limited capabilities of the developing countries to manage hazardous wastes and other wastes,

<u>Recognizing</u> the need to promote the transfer of technology for the sound management of hazardous wastes and other wastes produced locally, particularly to the developing countries in accordance with the spirit of the Cairo Guidelines and decision 14/16 of the Governing Council of UNEP on Promotion of the transfer of environmental protection technology,

<u>Recognizing also</u> that hazardous wastes and other wastes should be transported in accordance with relevant international conventions and recommendations,

<u>Convinced also</u> that the transboundary movement of hazardous wastes and other wastes should be permitted only when the transport and the ultimate disposal of such wastes is environmentally sound, and

<u>Determined</u> to protect, by strict control, human health and the environment against the adverse effects which may result from the generation and management of hazardous wastes and other wastes,

<u>Have agreed as follows:</u>

ARTICLE 1

Scope of the Convention

1. The following wastes that are subject to transboundary movement shall be "hazardous wastes" for the purposes of this Convention:

(a) Wastes that belong to any category contained in Annex I, unless they do not possess any of the characteristics contained in Annex III; and

(b) Wastes that are not covered under paragraph (a) but are defined as, or are considered to be, hazardous wastes by the domestic legislation of the Party of export, import or transit.

2. Wastes that belong to any category contained in Annex II that are subject to transboundary movement shall be "other wastes" for the purposes of this Convention.

3. Wastes which, as a result of being radioactive, are subject to other international control systems, including international instruments, applying specifically to radioactive materials, are excluded from the scope of this Convention.

4. Wastes which derive from the normal operations of a ship, the discharge of

which is covered by another international instrument, are excluded from the scope of this Convention.

ARTICLE 2

Definitions

For the purposes of this Convention:

1. "Wastes" are substances or objects which are disposed of or are intended to be disposed of or are required to be disposed of by the provisions of national law;

2. "Management" means the collection, transport and disposal of hazardous wastes or other wastes, including after-care of disposal sites;

3. "Transboundary movement" means any movement of hazardous wastes or other wastes from an area under the national jurisdiction of one State to or through an area under the national jurisdiction of another State or to or through an area not under the national jurisdiction of any State, provided at least two States are involved in the movement;

4. "Disposal" means any operation specified in Annex IV to this Convention;

5. "Approved site or facility" means a site or facility for the disposal of hazardous wastes or other wastes which is authorized or permitted to operate for this purpose by a relevant authority of the State where the site or facility is located;

6. "Competent authority" means one governmental authority designated by a Party to be responsible, within such geographical areas as the Party may think fit, for receiving the notification of a transboundary movement of hazardous wastes or other wastes, and any information related to it, and for responding to such a notification, as provided in Article 6;

7. "Focal point" means the entity of a Party referred to in Article 5 responsible for receiving and submitting information as provided for in Articles 13 and 16;

8. "Environmentally sound management of hazardous wastes or other wastes" means taking all practicable steps to ensure that hazardous wastes or other wastes are managed in a manner which will protect human health and the environment against the adverse effects which may result from such wastes;

9. "Area under the national jurisdiction of a State" means any land, marine area or airspace within which a State exercises administrative and regulatory response-bility in accordance with international law in regard to the protection of human health or the environment;

10. "State of export" means a Party from which a transboundary movement of hazardous wastes or other wastes is planned to be initiated or is initiated;

11. "State of import" means a Party to which a transboundary movement of hazardous wastes or other wastes is planned or takes place for the purpose of disposal therein or for the purpose of loading prior to disposal in an area not under the national jurisdiction of any State;

12.   "State of transit" means any State, other than the State of export or import, through which a movement of hazardous wastes or other wastes is planned or takes place;

13.   "States concerned" means Parties which are States of export or import, or transit States, whether or not Parties;

14.   "Person" means any natural or legal person;

15.   "Exporter" means any person under the jurisdiction of the State of export who arranges for hazardous wastes or other wastes to be exported;

16.   "Importer" means any person under the jurisdiction of the State of import who arranges for hazardous wastes or other wastes to be imported;

17.   "Carrier" means any person who carries out the transport of hazardous wastes or other wastes;

18.   "Generator" means any person whose activity produces hazardous wastes or other wastes or, if that person is not known, the person who is in possession and/or control of those wastes;

19.   "Disposer" means any person to whom hazardous wastes or other wastes are shipped and who carries out the disposal of such wastes;

20.   "Political and/or economic integration organization" means an organization constituted by sovereign States to which its Member States have transferred competence in respect of matters governed by this Convention and which has been duly authorized, in accordance with its internal procedures, to sign, ratify, accept, approve, formally confirm or accede to it;

21.   "Illegal traffic" means any transboundary movement of hazardous wastes or other wastes as specified in Article 9.

ARTICLE 3

National Definitions of Hazardous Wastes

1.   Each Party shall, within six months of becoming a Party to this Convention, inform the Secretariat of the Convention of the wastes, other than those listed in Annexes I and II, considered or defined as hazardous under its national legislation and of any requirements concerning transboundary movement procedures applicable to such wastes.

2.   Each Party shall subsequently inform the Secretariat of any significant changes to the information it has provided pursuant to paragraph 1.

3.   The Secretariat shall forthwith inform all Parties of the information it has received pursuant to paragraphs 1 and 2.

4.   Parties shall be responsible for making the information transmitted to them by the Secretariat under paragraph 3 available to their exporters.

ARTICLE 4

General Obligations

1.(a) Parties exercising their right to prohibit the import of hazardous wastes or other wastes for disposal shall inform the other Parties of their decision pursuant to Article 13.

(b)   Parties shall prohibit or shall not permit the export of hazardous wastes and other wastes to the Parties which have prohibited the import of such wastes, when notified pursuant to subparagraph (a) above.

(c)   Parties shall prohibit or shall not permit the export of hazardous wastes and other wastes if the State of import does not consent in writing to the specific import, in the case where that State of import has not prohibited the import of such wastes.

2.   Each Party shall take the appropriate measures to:

(a)   Ensure that the generation of hazardous wastes and other wastes within it is reduced to a minimum, taking into account social, technological and economic aspects;

(b)   Ensure the availability of adequate disposal facilities, for the environmentally sound management of hazardous wastes and other wastes, that shall be located, to the extent possible, within it, whatever the place of their disposal;

(c)   Ensure that persons involved in the management of hazardous wastes or other wastes within it take such steps as are necessary to prevent pollution due to hazardous wastes and other wastes arising from such management and, if such pollution occurs, to minimize the consequences thereof for human health and the environment;

(d)   Ensure that the transboundary movement of hazardous wastes and other wastes is reduced to the minimum consistent with the environmentally sound and efficient management of such wastes, and is conducted in a manner which will protect human health and the environment against the adverse effects which may result from such movement;

(e)   Not allow the export of hazardous wastes or other wastes to a State or group of States belonging to an economic and/or political integration organization that are Parties, particularly developing countries, which have prohibited by their legislation all imports, or if it has reason to believe that the wastes in question will not be managed in an environmentally sound manner, according to criteria to be decided on by the Parties at their first meeting;

(f)   Require that information about a proposed transboundary movement of hazardous wastes and other wastes be provided to the States concerned, according to Annex V A, to state clearly the effects of the proposed movement on human health and the environment;

(g)   Prevent the import of hazardous wastes and other wastes if it has reason to believe that the wastes in question will not be managed in an environmentally sound manner;

(h)   Co-operate in activities with other Parties and interested organizations, directly and through the Secretariat, including the dissemination of information on the trans-boundary movement of hazardous wastes and other wastes, in order to improve the environmentally sound management of such wastes and to achieve the prevention of illegal traffic.

3. The Parties consider that illegal traffic in hazardous wastes or other wastes is criminal.

4. Each Party shall take appropriate legal, administrative and other measures to implement and enforce the provisions of this Convention, including measures to pre-vent and punish conduct in contravention of the Convention.

5. A Party shall not permit hazardous wastes or other wastes to be exported to a non-Party or to be imported from a non-Party.

6. The Parties agree not to allow the export of hazardous wastes or other wastes for disposal within the area south of 60° South latitude, whether or not such wastes are subject to transboundary movement.

7. Furthermore, each Party shall:

(a) Prohibit all persons under its national jurisdiction from transporting or disposing of hazardous wastes or other wastes unless such persons are authorized or allowed to perform such types of operations;

(b) Require that hazardous wastes and other wastes that are to be the subject of a transboundary movement be packaged, labelled, and transported in conformity with generally accepted and recognized international rules and standards in the field of packaging, labelling, and transport, and that due account is taken of relevant internationally recognized practices;

(c) Require that hazardous wastes and other wastes be accompanied by a move-ment document from the point at which a transboundary movement commences to the point of disposal.

8. Each Party shall require that hazardous wastes or other wastes, to be exported, are managed in an environmentally sound manner in the State of import or else-where. Technical guidelines for the environmentally sound management of wastes subject to this Convention shall be decided by the Parties at their first meeting.

9. Parties shall take the appropriate measures to ensure that the transboundary movement of hazardous wastes and other wastes only be allowed if:

(a) The State of export does not have the technical capacity and the necessary facilities, capacity or suitable disposal sites in order to dispose of the wastes in ques-tion in an environmentally sound and efficient manner; or

(b) The wastes in question are required as a raw material for recycling or recovery industries in the State of import; or

(c) The transboundary movement in question is in accordance with other criteria to be decided by the Parties, provided those criteria do not differ from the objectives of this Convention.

10. The obligation under this Convention of States in which hazardous wastes and

other wastes are generated to require that those wastes are managed in an environmentally sound manner may not under any circumstances be transferred to the States of import or transit.

11. Nothing in this Convention shall prevent a Party from imposing additional requirements that are consistent with the provisions of this Convention, and are in accordance with the rules of international law, in order better to protect human health and the environment.

12. Nothing in this Convention shall affect in any way the sovereignty of States over their territorial sea established in accordance with international law, and the sovereign rights and the jurisdiction which States have in their exclusive economic zones and their continental shelves in accordance with international law, and the exercise by ships and aircraft of all States of navigational rights and freedoms as provided for in international law and as reflected in relevant international instruments.

13. Parties shall undertake to review periodically the possibilities for the reduction of the amount and/or the pollution potential of hazardous wastes and other wastes which are exported to other States, in particular to developing countries.

ARTICLE 4A[3]

General Obligations

1. Each Party listed in Annex VII shall prohibit all transboundary movements of hazardous wastes which are destined for operations according to Annex IV A to States not listed in Annex VII.

2. Each Party listed in Annex VII shall phase out by 31 December 1997, and prohibit as of that date, all transboundary movements of hazardous wastes under Article 1(1)(a) of the Convention which are destined for operations according to Annex IV B to States not listed in Annex VII. Such transboundary movement shall not be prohibited unless the wastes in question are characterized as hazardous under the Convention.

ARTICLE 5

Designation of Competent Authorities and Focal Point

To facilitate the implementation of this Convention, the Parties shall:

1. Designate or establish one or more competent authorities and one focal point. One competent authority shall be designated to receive the notification in case of a State of transit.

2. Inform the Secretariat, within three months of the date of the entry into force of

---

[3] The Conference of the Parties adopted Decision III/1 at its third meeting to amend the Convention by adding, inter alia, a new Article 4A. This amendment entered into force on 5 December 2019 (depositary notification C.N.420.2019). For information on the status of individual Parties in relation to the amendment/s, please see the Status of Ratifications page on the Basel Convention website.

this Convention for them, which agencies they have designated as their focal point and their competent authorities.

3.   Inform the Secretariat, within one month of the date of decision, of any changes regarding the designation made by them under paragraph 2 above.

ARTICLE 6

Transboundary Movement between Parties

1.   The State of export shall notify, or shall require the generator or exporter to notify, in writing, through the channel of the competent authority of the State of export, the competent authority of the States concerned of any proposed transboundary movement of hazardous wastes or other wastes. Such notification shall contain the declarations and information specified in Annex V A, written in a language acceptable to the State of import. Only one notification needs to be sent to each State concerned.

2.   The State of import shall respond to the notifier in writing, consenting to the movement with or without conditions, denying permission for the movement, or requesting additional information. A copy of the final response of the State of import shall be sent to the competent authorities of the States concerned which are Parties.

3.   The State of export shall not allow the generator or exporter to commence the transboundary movement until it has received written confirmation that:

(a)   The notifier has received the written consent of the State of import; and

(b)   The notifier has received from the State of import confirmation of the existence of a contract between the exporter and the disposer specifying environmentally sound management of the wastes in question.

4.   Each State of transit which is a Party shall promptly acknowledge to the notifier receipt of the notification. It may subsequently respond to the notifier in writing, within 60 days, consenting to the movement with or without conditions, denying permission for the movement, or requesting additional information. The State of export shall not allow the transboundary movement to commence until it has received the written consent of the State of transit. However, if at any time a Party decides not to require prior written consent, either generally or under specific conditions, for transit transboundary movements of hazardous wastes or other wastes, or modifies its requirements in this respect, it shall forthwith inform the other Parties of its decision pursuant to Article 13. In this latter case, if no response is received by the State of export within 60 days of the receipt of a given notification by the State of transit, the State of export may allow the export to proceed through the State of transit.

5.   In the case of a transboundary movement of wastes where the wastes are legally defined as or considered to be hazardous wastes only:

(a)   By the State of export, the requirements of paragraph 9 of this Article that apply to the importer or disposer and the State of import shall apply _mutatis mutandis_ to the exporter and State of export, respectively;

(b)   By the State of import, or by the States of import and transit which are Parties, the requirements of paragraphs 1, 3, 4 and 6 of this Article that apply to the exporter and State of export shall apply _mutatis mutandis_ to the importer or disposer and State of import, respectively; or

(c)   By any State of transit which is a Party, the provisions of paragraph 4 shall apply to such State.

6.   The State of export may, subject to the written consent of the States concerned, allow the generator or the exporter to use a general notification where hazardous wastes or other wastes having the same physical and chemical characteristics are shipped regularly to the same disposer via the same customs office of exit of the State of export via the same customs office of entry of the State of import, and, in the case of transit, via the same customs office of entry and exit of the State or States of transit.

7.   The States concerned may make their written consent to the use of the general notification referred to in paragraph 6 subject to the supply of certain information, such as the exact quantities or periodical lists of hazardous wastes or other wastes to be shipped.

8. The general notification and written consent referred to in paragraphs 6 and 7 may cover multiple shipments of hazardous wastes or other wastes during a maximum period of 12 months.

9.   The Parties shall require that each person who takes charge of a transboundary movement of hazardous wastes or other wastes sign the movement document either upon delivery or receipt of the wastes in question. They shall also require that the disposer inform both the exporter and the competent authority of the State of export of receipt by the disposer of the wastes in question and, in due course, of the completion of disposal as specified in the notification. If no such information is received within the State of export, the competent authority of the State of export or the exporter shall so notify the State of import.

10.   The notification and response required by this Article shall be transmitted to the competent authority of the Parties concerned or to such governmental authority as may be appropriate in the case of non-Parties.

11.   Any transboundary movement of hazardous wastes or other wastes shall be covered by insurance, bond or other guarantee as may be required by the State of import or any State of transit which is a Party.

ARTICLE 7

Transboundary Movement from a Party through States which are not Parties

Paragraph 1 of Article 6 of the Convention shall apply _mutatis mutandis_ to transboundary movement of hazardous wastes or other wastes from a Party through a State or States which are not Parties.

ARTICLE 8

Duty to Re-import

When a transboundary movement of hazardous wastes or other wastes to which the consent of the States concerned has been given, subject to the provisions of this Convention, cannot be completed in accordance with the terms of the contract, the State of export shall ensure that the wastes in question are taken back into the State of export, by the exporter, if alternative arrangements cannot be made for their disposal in an environmentally sound manner, within 90 days from the time that the importing State informed the State of export and the Secretariat, or such other period of time as the States concerned agree. To this end, the State of export and any Party of transit shall not oppose, hinder or prevent the return of those wastes to the State of export.

ARTICLE 9

Illegal Traffic

1.   For the purpose of this Convention, any transboundary movement of hazardous wastes or other wastes:

(a)   without notification pursuant to the provisions of this Convention to all States concerned; or

(b)   without the consent pursuant to the provisions of this Convention of a State concerned; or

(c)   with consent obtained from States concerned through falsification, misrepresentation or fraud; or

(d)   that does not conform in a material way with the documents; or

(e)   that results in deliberate disposal (e.g. dumping) of hazardous wastes or other wastes in contravention of this Convention and of general principles of international law,

shall be deemed to be illegal traffic.

2.   In case of a transboundary movement of hazardous wastes or other wastes deemed to be illegal traffic as the result of conduct on the part of the exporter or generator, the State of export shall ensure that the wastes in question are:

(a)   taken back by the exporter or the generator or, if necessary, by itself into the State of export, or, if impracticable,

(b)   are otherwise disposed of in accordance with the provisions of this Convention, within 30 days from the time the State of export has been informed about the illegal traffic or such other period of time as States concerned may agree. To this end the Parties concerned shall not oppose, hinder or prevent the return of those wastes to the State of export.

3.   In the case of a transboundary movement of hazardous wastes or other wastes deemed to be illegal traffic as the result of conduct on the part of the importer or

disposer, the State of import shall ensure that the wastes in question are disposed of in an environmentally sound manner by the importer or disposer or, if necessary, by itself within 30 days from the time the illegal traffic has come to the attention of the State of import or such other period of time as the States concerned may agree. To this end, the Parties concerned shall co-operate, as necessary, in the disposal of the wastes in an environmentally sound manner.

4.   In cases where the responsibility for the illegal traffic cannot be assigned either to the exporter or generator or to the importer or disposer, the Parties concerned or other Parties, as appropriate, shall ensure, through co-operation, that the wastes in question are disposed of as soon as possible in an environmentally sound manner either in the State of export or the State of import or elsewhere as appropriate.

5.   Each Party shall introduce appropriate national/domestic legislation to prevent and punish illegal traffic. The Parties shall co-operate with a view to achieving the objects of this Article.

ARTICLE 10

International Co-operation

1.   The Parties shall co-operate with each other in order to improve and achieve environmentally sound management of hazardous wastes and other wastes.

2.   To this end, the Parties shall:

(a)   Upon request, make available information, whether on a bilateral or multilateral basis, with a view to promoting the environmentally sound management of hazardous wastes and other wastes, including harmonization of technical standards and practices for the adequate management of hazardous wastes and other wastes;

(b)   Co-operate in monitoring the effects of the management of hazardous wastes on human health and the environment;

(c)   Co-operate, subject to their national laws, regulations and policies, in the development and implementation of new environmentally sound low-waste technologies and the improvement of existing technologies with a view to eliminating, as far as practicable, the generation of hazardous wastes and other wastes and achieving more effective and efficient methods of ensuring their management in an environmentally sound manner, including the study of the economic, social and environmental effects of the adoption of such new or improved technologies;

(d)   Co-operate actively, subject to their national laws, regulations and policies, in the transfer of technology and management systems related to the environmentally sound management of hazardous wastes and other wastes. They shall also co-operate in developing the technical capacity among Parties, especially those which may need and request technical assistance in this field;

(e)   Co-operate in developing appropriate technical guidelines and/or codes of practice.

3.   The Parties shall employ appropriate means to co-operate in order to assist

developing countries in the implementation of subparagraphs a, b, c and d of paragraph 2 of Article 4.

4.   Taking into account the needs of developing countries, co-operation between Parties and the competent international organizations is encouraged to promote, <u>inter alia</u>, public awareness, the development of sound management of hazardous wastes and other wastes and the adoption of new low-waste technologies.

ARTICLE 11

Bilateral, Multilateral and Regional Agreements

1.   Notwithstanding the provisions of Article 4 paragraph 5, Parties may enter into bilateral, multilateral, or regional agreements or arrangements regarding transboundary movement of hazardous wastes or other wastes with Parties or non-Parties provided that such agreements or arrangements do not derogate from the environmentally sound management of hazardous wastes and other wastes as required by this Convention. These agreements or arrangements shall stipulate provisions which are not less environmentally sound than those provided for by this Convention in particular taking into account the interests of developing countries.

2.   Parties shall notify the Secretariat of any bilateral, multilateral or regional agreements or arrangements referred to in paragraph 1 and those which they have entered into prior to the entry into force of this Convention for them, for the purpose of controlling transboundary movements of hazardous wastes and other wastes which take place entirely among the Parties to such agreements. The provisions of this Convention shall not affect transboundary movements which take place pursuant to such agreements provided that such agreements are compatible with the environmentally sound management of hazardous wastes and other wastes as required by this Convention.

ARTICLE 12

Consultations on Liability

The Parties shall co-operate with a view to adopting, as soon as practicable, a protocol setting out appropriate rules and procedures in the field of liability and compensation for damage resulting from the transboundary movement and disposal of hazardous wastes and other wastes.

ARTICLE 13

Transmission of Information

1.   The Parties shall, whenever it comes to their knowledge, ensure that, in the case of an accident occurring during the transboundary movement of hazardous wastes or other wastes or their disposal, which are likely to present risks to human health and the environment in other States, those States are immediately informed.

2.   The Parties shall inform each other, through the Secretariat, of:

(a)   Changes regarding the designation of competent authorities and/or focal points, pursuant to Article 5;

(b)   Changes in their national definition of hazardous wastes, pursuant to Article 3;

and, as soon as possible,

(c)   Decisions made by them not to consent totally or partially to the import of hazardous wastes or other wastes for disposal within the area under their national jurisdiction;

(d)   Decisions taken by them to limit or ban the export of hazardous wastes or other wastes;

(e)   Any other information required pursuant to paragraph 4 of this Article.

3.   The Parties, consistent with national laws and regulations, shall transmit, through the Secretariat, to the Conference of the Parties established under Article 15, before the end of each calendar year, a report on the previous calendar year, containing the following information:

(a)   Competent authorities and focal points that have been designated by them pursuant to Article 5;

(b)   Information regarding transboundary movements of hazardous wastes or other wastes in which they have been involved, including:

  (i)   The amount of hazardous wastes and other wastes exported, their category, characteristics, destination, any transit country and disposal method as stated on the response to notification;

  (ii)   The amount of hazardous wastes and other wastes imported, their category, characteristics, origin, and disposal methods;

  (iii)   Disposals which did not proceed as intended;

  (iv)   Efforts to achieve a reduction of the amount of hazardous wastes or other wastes subject to transboundary movement;

(c)   Information on the measures adopted by them in implementation of this Convention;

(d)   Information on available qualified statistics which have been compiled by them on the effects on human health and the environment of the generation, transportation and disposal of hazardous wastes or other wastes;

(e)   Information concerning bilateral, multilateral and regional agreements and arrangements entered into pursuant to Article 11 of this Convention;

(f)   Information on accidents occurring during the transboundary movement and disposal of hazardous wastes and other wastes and on the measures undertaken to deal with them;

(g)   Information on disposal options operated within the area of their national jurisdiction;

(h)   Information on measures undertaken for development of technologies for the reduction and/or elimination of production of hazardous wastes and other wastes; and

(i)   Such other matters as the Conference of the Parties shall deem relevant.

4.   The Parties, consistent with national laws and regulations, shall ensure that copies of each notification concerning any given transboundary movement of hazardous wastes or other wastes, and the response to it, are sent to the Secretariat when a Party considers that its environment may be affected by that transboundary movement has requested that this should be done.

ARTICLE 14

Financial Aspects

1.   The Parties agree that, according to the specific needs of different regions and subregions, regional or sub-regional centres for training and technology transfers regarding the management of hazardous wastes and other wastes and the minimization of their generation should be established. The Parties shall decide on the establishment of appropriate funding mechanisms of a voluntary nature.

2.   The Parties shall consider the establishment of a revolving fund to assist on an interim basis in case of emergency situations to minimize damage from accidents arising from transboundary movements of hazardous wastes and other wastes or during the disposal of those wastes.

ARTICLE 15

Conference of the Parties

1.   A Conference of the Parties is hereby established. The first meeting of the Conference of the Parties shall be convened by the Executive Director of UNEP not later than one year after the entry into force of this Convention. Thereafter, ordinary meetings of the Conference of the Parties shall be held at regular intervals to be determined by the Conference at its first meeting.

2.   Extraordinary meetings of the Conference of the Parties shall be held at such other times as may be deemed necessary by the Conference, or at the written request of any Party, provided that, within six months of the request being communicated to them by the Secretariat, it is supported by at least one third of the Parties.

3.   The Conference of the Parties shall by consensus agree upon and adopt rules of procedure for itself and for any subsidiary body it may establish, as well as financial rules to determine in particular the financial participation of the Parties under this Convention.

4.   The Parties at their first meeting shall consider any additional measures needed to assist them in fulfilling their responsibilities with respect to the protection and the preservation of the marine environment in the context of this Convention.

5.   The Conference of the Parties shall keep under continuous review and evalua-tion the effective implementation of this Convention, and, in addition, shall:

(a)   Promote the harmonization of appropriate policies, strategies and measures for minimizing harm to human health and the environment by hazardous wastes and other wastes;

(b)   Consider and adopt, as required, amendments to this Convention and its an-nexes, taking into consideration, _inter alia_, available scientific, technical, economic and environmental information;

(c)   Consider and undertake any additional action that may be required for the achievement of the purposes of this Convention in the light of experience gained in its operation and in the operation of the agreements and arrangements envisaged in Article 11;

(d)   Consider and adopt protocols as required; and

(e)   Establish such subsidiary bodies as are deemed necessary for the implemen-tation of this Convention.

6.   The United Nations, its specialized agencies, as well as any State not Party to this Convention, may be represented as observers at meetings of the Conference of the Parties. Any other body or agency, whether national or international, governmen-tal or non-governmental, qualified in fields relating to hazardous wastes or other wastes which has informed the Secretariat of its wish to be represented as an ob-server at a meeting of the Conference of the Parties, may be admitted unless at least one third of the Parties present object. The admission and participation of observers shall be subject to the rules of procedure adopted by the Conference of the Parties.

7.   The Conference of the Parties shall undertake three years after the entry into force of this Convention, and at least every six years thereafter, an evaluation of its effectiveness and, if deemed necessary, to consider the adoption of a complete or partial ban of transboundary movements of hazardous wastes and other wastes in light of the latest scientific, environmental, technical and economic information.

ARTICLE 16

Secretariat

1.   The functions of the Secretariat shall be:

(a)   To arrange for and service meetings provided for in Articles 15 and 17;

(b)   To prepare and transmit reports based upon information received in accor-dance with Articles 3, 4, 6, 11 and 13 as well as upon information derived from mee-tings of subsidiary bodies established under Article 15 as well as upon, as appropri-ate, information provided by relevant intergovernmental and non-governmental entities;

(c)   To prepare reports on its activities carried out in implementation of its functions under this Convention and present them to the Conference of the Parties;

(d)   To ensure the necessary coordination with relevant international bodies, and in particular to enter into such administrative and contractual arrangements as may be required for the effective discharge of its function;

(e)   To communicate with focal points and competent authorities established by the Parties in accordance with Article 5 of this Convention;

(f)   To compile information concerning authorized national sites and facilities of Parties available for the disposal of their hazardous wastes and other wastes and to circulate this information among Parties;

(g)   To receive and convey information from and to Parties on:

- sources of technical assistance and training;

- available technical and scientific know-how;

- sources of advice and expertise; and

- availability of resources

with a view to assisting them, upon request, in such areas as:

- the handling of the notification system of this Convention;

- the management of hazardous wastes and other wastes;

- environmentally sound technologies relating to hazardous wastes and other wastes, such as low- and non-waste technology;

- the assessment of disposal capabilities and sites;

- the monitoring of hazardous wastes and other wastes; and

- emergency responses;

(h)   To provide Parties, upon request, with information on consultants or consulting firms having the necessary technical competence in the field, which can assist them to examine a notification for a transboundary movement, the concurrence of a shipment of hazardous wastes or other wastes with the relevant notification, and/or the fact that the proposed disposal facilities for hazardous wastes or other wastes are environmentally sound, when they have reason to believe that the wastes in question will not be managed in an environmentally sound manner. Any such examination would not be at the expense of the Secretariat;

(i)   To assist Parties upon request in their identification of cases of illegal traffic and to circulate immediately to the Parties concerned any information it has received regarding illegal traffic;

(j)   To co-operate with Parties and with relevant and competent international organizations and agencies in the provision of experts and equipment for the purpose of rapid assistance to States in the event of an emergency situation; and

(k)   To perform such other functions relevant to the purposes of this Convention as may be determined by the Conference of the Parties.

2.   The secretariat functions will be carried out on an interim basis by UNEP until

the completion of the first meeting of the Conference of the Parties held pursuant to Article 15.

3.   At its first meeting, the Conference of the Parties shall designate the Secretariat from among those existing competent intergovernmental organizations which have signified their willingness to carry out the secretariat functions under this Convention. At this meeting, the Conference of the Parties shall also evaluate the implementation by the interim Secretariat of the functions assigned to it, in particular under paragraph 1 above, and decide upon the structures appropriate for those functions.

ARTICLE 17

Amendment of the Convention

1.   Any Party may propose amendments to this Convention and any Party to a protocol may propose amendments to that protocol. Such amendments shall take due account, _inter alia_, of relevant scientific and technical considerations.

2.   Amendments to this Convention shall be adopted at a meeting of the Conference of the Parties. Amendments to any protocol shall be adopted at a meeting of the Parties to the protocol in question. The text of any proposed amendment to this Convention or to any protocol, except as may otherwise be provided in such protocol, shall be communicated to the Parties by the Secretariat at least six months before the meeting at which it is proposed for adoption. The Secretariat shall also communicate proposed amendments to the Signatories to this Convention for information.

3.   The Parties shall make every effort to reach agreement on any proposed amendment to this Convention by consensus. If all efforts at consensus have been exhausted, and no agreement reached, the amendment shall as a last resort be adopted by a three-fourths majority vote of the Parties present and voting at the meeting, and shall be submitted by the Depositary to all Parties for ratification, approval, formal confirmation or acceptance.

4.   The procedure mentioned in paragraph 3 above shall apply to amendments to any protocol, except that a two-thirds majority of the Parties to that protocol present and voting at the meeting shall suffice for their adoption.

5.   Instruments of ratification, approval, formal confirmation or acceptance of amendments shall be deposited with the Depositary. Amendments adopted in accordance with paragraphs 3 or 4 above shall enter into force between Parties having accepted them on the ninetieth day after the receipt by the Depositary of their instrument of ratification, approval, formal confirmation or acceptance by at least three-fourths of the Parties who accepted them or by at least two thirds of the Parties to the protocol concerned who accepted them, except as may otherwise be provided in such protocol. The amendments shall enter into force for any other Party on the ninetieth day after that Party deposits its instrument of ratification, approval, formal confirmation or acceptance of the amendments.

6.   For the purpose of this Article, "Parties present and voting" means Parties present and casting an affirmative or negative vote.

ARTICLE 18

Adoption and Amendment of Annexes

1.   The annexes to this Convention or to any protocol shall form an integral part of this Convention or of such protocol, as the case may be and, unless expressly provided otherwise, a reference to this Convention or its protocols constitutes at the same time a reference to any annexes thereto. Such annexes shall be restricted to scientific, technical and administrative matters.

2.   Except as may be otherwise provided in any protocol with respect to its annexes, the following procedure shall apply to the proposal, adoption and entry into force of additional annexes to this Convention or of annexes to a protocol:

(a)   Annexes to this Convention and its protocols shall be proposed and adopted according to the procedure laid down in Article 17, paragraphs 2, 3 and 4;

(b)   Any Party that is unable to accept an additional annex to this Convention or an annex to any protocol to which it is party shall so notify the Depositary, in writing, within six months from the date of the communication of the adoption by the Depositary. The Depositary shall without delay notify all Parties of any such notification received. A Party may at any time substitute an acceptance for a previous declaration of objection and the annexes shall thereupon enter into force for that Party;

(c)   On the expiry of six months from the date of the circulation of the communication by the Depositary, the annex shall become effective for all Parties to this Convention or to any protocol concerned, which have not submitted a notification in accordance with the provision of subparagraph (b) above.

3.   The proposal, adoption and entry into force of amendments to annexes to this Convention or to any protocol shall be subject to the same procedure as for the proposal, adoption and entry into force of annexes to the Convention or annexes to a protocol. Annexes and amendments thereto shall take due account, _inter alia_, of relevant scientific and technical considerations.

4.   If an additional annex or an amendment to an annex involves an amendment to this Convention or to any protocol, the additional annex or amended annex shall not enter into force until such time the amendment to this Convention or to the protocol enters into force.

ARTICLE 19

Verification

Any Party which has reason to believe that another Party is acting or has acted in breach of its obligations under this Convention may inform the Secretariat thereof, and in such an event, shall simultaneously and immediately inform, directly or through the Secretariat, the Party against whom the allegations are made. All relevant information should be submitted by the Secretariat to the Parties.

ARTICLE 20

Settlement of Disputes

1.   In case of a dispute between Parties as to the interpretation or application of, or compliance with, this Convention or any protocol thereto, they shall seek a settlement of the dispute through negotiation or any other peaceful means of their own choice.

2.   If the Parties concerned cannot settle their dispute through the means mentioned in the preceding paragraph, the dispute, if the Parties to the dispute agree, shall be submitted to the International Court of Justice or to arbitration under the conditions set out in Annex VI on Arbitration. However, failure to reach common agreement on submission of the dispute to the International Court of Justice or to arbitration shall not absolve the Parties from the responsibility of continuing to seek to resolve it by the means referred to in paragraph 1.

3.   When ratifying, accepting, approving, formally confirming or acceding to this Convention, or at any time thereafter, a State or political and/or economic integration organization may declare that it recognizes as compulsory _ipso facto_ and without special agreement, in relation to any Party accepting the same obligation:

(a)   submission of the dispute to the International Court of Justice; and/or

(b)   arbitration in accordance with the procedures set out in Annex VI.

Such declaration shall be notified in writing to the Secretariat which shall communicate it to the Parties.

ARTICLE 21

Signature

This Convention shall be open for signature by States, by Namibia, represented by the United Nations Council for Namibia, and by political and/or economic integration organizations, in Basel on 22 March 1989, at the Federal Department of Foreign Affairs of Switzerland in Berne from 23 March 1989 to 30 June 1989 and at United Nations Headquarters in New York from 1 July 1989 to 22 March 1990.

ARTICLE 22

Ratification, Acceptance, Formal Confirmation or Approval

1.   This Convention shall be subject to ratification, acceptance or approval by States and by Namibia, represented by the United Nations Council for Namibia, and to formal confirmation or approval by political and/or economic integration organizations. Instruments of ratification, acceptance, formal confirmation, or approval shall be deposited with the Depositary.

2.   Any organization referred to in paragraph 1 above which becomes a Party to this Convention without any of its Member States being a Party shall be bound by all the obligations under the Convention. In the case of such organizations, one or more

of whose Member States is a Party to the Convention, the organization and its Member States shall decide on their respective responsibilities for the performance of their obligations under the Convention. In such cases, the organization and the Member States shall not be entitled to exercise rights under the Convention concurrently.

3. In their instruments of formal confirmation or approval, the organizations referred to in paragraph 1 above shall declare the extent of their competence with respect to the matters governed by the Convention. These organizations shall also inform the Depositary, who will inform the Parties of any substantial modification in the extent of their competence.

ARTICLE 23

Accession

1. This Convention shall be open for accession by States, by Namibia, represented by the United Nations Council for Namibia, and by political and/or economic integration organizations from the day after the date on which the Convention is closed for signature. The instruments of accession shall be deposited with the Depositary.

2. In their instruments of accession, the organizations referred to in paragraph 1 above shall declare the extent of their competence with respect to the matters governed by the Convention. These organizations shall also inform the Depositary of any substantial modification in the extent of their competence.

3. The provisions of Article 22, paragraph 2, shall apply to political and/or economic integration organizations which accede to this Convention.

ARTICLE 24

Right to Vote

1. Except as provided for in paragraph 2 below, each Contracting Party to this Convention shall have one vote.

2. Political and/or economic integration organizations, in matters within their competence, in accordance with Article 22, paragraph 3, and Article 23, paragraph 2, shall exercise their right to vote with a number of votes equal to the number of their Member States which are Parties to the Convention or the relevant protocol. Such organizations shall not exercise their right to vote if their Member States exercise theirs, and vice versa.

ARTICLE 25

Entry into Force

1. This Convention shall enter into force on the ninetieth day after the date of deposit of the twentieth instrument of ratification, acceptance, formal confirmation, approval or accession.

2.   For each State or political and/or economic integration organization which ratifies, accepts, approves or formally confirms this Convention or accedes thereto after the date of the deposit of the twentieth instrument of ratification, acceptance, approval, formal confirmation or accession, it shall enter into force on the ninetieth day after the date of deposit by such State or political and/or economic integration organization of its instrument of ratification, acceptance, approval, formal confirmation or accession.

3.   For the purpose of paragraphs 1 and 2 above, any instrument deposited by a political and/or economic integration organization shall not be counted as additional to those deposited by Member States of such organization.

ARTICLE 26

Reservations and Declarations

1.   No reservation or exception may be made to this Convention.

2.   Paragraph 1 of this Article does not preclude a State or political and/or economic integration organization, when signing, ratifying, accepting, approving, formally confirming or acceding to this Convention, from making declarations or statements, however phrased or named, with a view, _inter alia_, to the harmonization of its laws and regulations with the provisions of this Convention, provided that such declarations or statements do not purport to exclude or to modify the legal effects of the provisions of the Convention in their application to that State.

ARTICLE 27

Withdrawal

1.   At any time after three years from the date on which this Convention has entered into force for a Party, that Party may withdraw from the Convention by giving written notification to the Depositary.

2.   Withdrawal shall be effective one year from receipt of notification by the Depositary, or on such later date as may be specified in the notification.

ARTICLE 28

Depository

The Secretary-General of the United Nations shall be the Depository of this Convention and of any protocol thereto.

ARTICLE 29

Authentic texts

The original Arabic, Chinese, English, French, Russian and Spanish texts of this Convention are equally authentic.

IN WITNESS WHEREOF the undersigned, being duly authorized to that effect, have signed this Convention.

Done at Basel on the 22 day of March 1989.

ANNEX I

CATEGORIES OF WASTES TO BE CONTROLLED

Waste Streams

Y1	Clinical wastes from medical care in hospitals, medical centers and clinics
Y2	Wastes from the production and preparation of pharmaceutical products
Y3	Waste pharmaceuticals, drugs and medicines
Y4	Wastes from the production, formulation and use of biocides and phyto-pharmaceuticals
Y5	Wastes from the manufacture, formulation and use of wood preserving chemicals
Y6	Wastes from the production, formulation and use of organic solvents
Y7	Wastes from heat treatment and tempering operations containing cyanides
Y8	Waste mineral oils unfit for their originally intended use
Y9	Waste oils/water, hydrocarbons/water mixtures, emulsions
Y10	Waste substances and articles containing or contaminated with polychlorinated biphenyls (PCBs) and/or polychlorinated terphenyls (PCTs) and/or polybrominated biphenyls (PBBs)
Y11	Waste tarry residues arising from refining, distillation and any pyrolytic treatment
Y12	Wastes from production, formulation and use of inks, dyes, pigments, paints, lacquers, varnish
Y13	Wastes from production, formulation and use of resins, latex, plasticizers, glues/adhesives
Y14	Waste chemical substances arising from research and development or teaching activities which are not identified and/or are new and whose effects on man and/or the environment are not known
Y15	Wastes of an explosive nature not subject to other legislation
Y16	Wastes from production, formulation and use of photographic chemicals and processing materials
Y17	Wastes resulting from surface treatment of metals and plastics
Y18	Residues arising from industrial waste disposal operations

Wastes having as constituents:

Y19	Metal carbonyls
Y20	Beryllium; beryllium compounds
Y21	Hexavalent chromium compounds
Y22	Copper compounds
Y23	Zinc compounds
Y24	Arsenic; arsenic compounds

Y25   Selenium; selenium compounds

Y26   Cadmium; cadmium compounds

Y27   Antimony; antimony compounds

Y28   Tellurium; tellurium compounds

Y29   Mercury; mercury compounds

Y30   Thallium; thallium compounds

Y31   Lead; lead compounds

Y32   Inorganic fluorine compounds excluding calcium fluoride

Y33   Inorganic cyanides

Y34   Acidic solutions or acids in solid form

Y35   Basic solutions or bases in solid form

Y36   Asbestos (dust and fibers)

Y37   Organic phosphorus compounds

Y38   Organic cyanides

Y39   Phenols; phenol compounds including chlorophenols

Y40   Ethers

Y41   Halogenated organic solvents

Y42   Organic solvents excluding halogenated solvents

Y43   Any congenor of polychlorinated dibenzo-furan

Y44   Any congenor of polychlorinated dibenzo-p-dioxin

Y45   Organohalogen compounds other than substances referred to in this Annex (e.g. Y39, Y41, Y42, Y43, Y44)

(a)   To facilitate the application of this Convention, and subject to paragraphs (b), (c) and (d), wastes listed in Annex VIII are characterized as hazardous pursuant to Article 1, paragraph 1 (a), of this Convention, and wastes listed in Annex IX are not covered by Article 1, paragraph 1 (a), of this Convention.

(b)   Designation of a waste on Annex VIII does not preclude, in a particular case, the use of Annex III to demonstrate that a waste is not hazardous pursuant to Article 1, paragraph 1 (a), of this Convention.

(c)   Designation of a waste on Annex IX does not preclude, in a particular case, characterization of such a waste as hazardous pursuant to Article 1, paragraph 1 (a), of this Convention if it contains Annex I material to an extent causing it to exhibit an Annex III characteristic.

(d)   Annexes VIII and IX do not affect the application of Article 1, paragraph 1 (a), of this Convention for the purpose of characterization of wastes.[4]

---

[4]   The amendment whereby paragraphs (a), (b), (c) and (d) were added to at the end of Annex I entered into force on 6 November 1998, six months following the issuance of depositary notification C.N.77.1998 of 6 May 1998 (reflecting Decision IV/9, adopted by the Conference of the Parties at its fourth meeting).

ANNEX II[5]

CATEGORIES OF WASTES REQUIRING SPECIAL CONSIDERATION

Y46   Wastes collected from households

Y47   Residues arising from the incineration of household wastes

Y48[6/7]   Plastic waste, including mixtures of such waste, with the exception of the following:

- Plastic waste that is hazardous waste pursuant to paragraph 1 (a) of Article 1[8]

- Plastic waste listed below, provided it is destined for recycling[9] in an environmentally sound manner and almost free from contamination and other types of wastes:[10]

   — Plastic waste almost exclusively[11] consisting of one non-halogenated polymer, including but not limited to the following polymers:

   - Polyethylene (PE)
   - Polypropylene (PP)
   - Polystyrene (PS)
   - Acrylonitrile butadiene styrene (ABS)
   - Polyethylene terephthalate (PET)
   - Polycarbonates (PC)
   - Polyethers

   — Plastic waste almost exclusively[11] consisting of one cured resin or condensation product, including but not limited to the following resins:

   - Urea formaldehyde resins
   - Phenol formaldehyde resins
   - Melamine formaldehyde resins

---

5   This amendment to Annex II whereby one new entry was added entered into force on 24 March 2020 (depositary notification C.N. 432.2019), reflecting decision BC-14/12 adopted by the Conference of the Parties at its fourteenth meeting. For information on the status of individual Parties in relation to the amendment/s, please see the Status of Ratifications page on the Basel Convention website.

6   This entry becomes effective as of 1 January 2021.

7   Parties can impose stricter requirements in relation to this entry.

8   Note the related entry on list A A3210 in Annex VIII.

9   Recycling/reclamation of organic substances that are not used as solvents (R3 in Annex IV, sect. B) or, if needed, temporary storage limited to one instance, provided that it is followed by operation R3 and evidenced by contractual or relevant official documentation.

10   In relation to "almost free from contamination and other types of wastes", international and national specifications may offer a point of reference.

11   In relation to "almost exclusively", international and national specifications may offer a point of reference.

  - Epoxy resins

  - Alkyd resins

— Plastic waste almost exclusively[11] consisting of one of the following fluorinated polymers:[12]

  - Perfluoroethylene/propylene (FEP)

  - Perfluoroalkoxy alkanes:

    ▪ Tetrafluoroethylene/perfluoroalkyl vinyl ether (PFA)

    ▪ Tetrafluoroethylene/perfluoromethyl vinyl ether (MFA)

  - Polyvinylfluoride (PVF)

  - Polyvinylidenefluoride (PVDF)

• Mixtures of plastic waste, consisting of polyethylene (PE), polypropylene (PP) and/or polyethylene terephthalate (PET), provided they are destined for separate recycling[13] of each material and in an environmentally sound manner and almost free from contamination and other types of wastes.[10]

---

[12] Post-consumer wastes are excluded.

[13] Recycling/reclamation of organic substances that are not used as solvents (R3 in Annex IV, sect. B), with prior sorting and, if needed, temporary storage limited to one instance, provided that it is followed by operation R3 and evidenced by contractual or relevant official documentation.

ANNEX III

LIST OF HAZARDOUS CHARACTERISTICS

UN Class[14]	Code	Characteristics
1	H1	Explosive

An explosive substance or waste is a solid or liquid substance or waste (or mixture of substances or wastes) which is in itself capable by chemical reaction of producing gas at such a temperature and pressure and at such a speed as to cause damage to the surroundings.

| 3 | H3 | Flammable liquids |

The word "flammable" has the same meaning as "inflammable". Flammable liquids are liquids, or mixtures of liquids, or liquids containing solids in solution or suspension (for example, paints, varnishes, lacquers, etc., but not including substances or wastes otherwise classified on account of their dangerous characteristics) which give off a flammable vapour at temperatures of not more than 60.5°C, closed-cup test, or not more than 65.6°C, open-cup test. (Since the results of open-cup tests and of closed-cup tests are not strictly comparable and even individual results by the same test are often variable, regulations varying from the above figures to make allowance for such differences would be within the spirit of this definition.)

| 4.1 | H4.1 | Flammable solids |

Solids, or waste solids, other than those classed as explosives, which under conditions encountered in transport are readily combustible, or may cause or contribute to fire through friction.

| 4.2 | H4.2 | Substances or wastes liable to spontaneous combustion |

Substances or wastes which are liable to spontaneous heating under normal conditions encountered in transport, or to heating up on contact with air, and being then liable to catch fire.

| 4.3 | H4.3 | Substances or wastes which, in contact with water emit flammable gases |

Substances or wastes, which, by interaction with water, are liable to become spontaneously flammable or to give off flammable gases in dangerous quantities.

| 5.1 | H5.1 | Oxidizing |

Substances or wastes which, while in themselves not necessarily combustible, may, generally by yielding oxygen cause, or contribute to, the combustion of other materials.

| 5.2 | H5.2 | Organic Peroxides |

Organic substances or wastes, which contain the bivalent-o-o-struc-

---

[14] Corresponds to the hazard classification system included in the United Nations Recommendations on the Transport of Dangerous Goods (ST/SG/AC.10/1Rev.5, United Nations, New York, 1988).

UN Class[14]	Code	Characteristics
		ture, are thermally unstable substances, which may undergo exothermic self-accelerating decomposition.
6.1	H6.1	Poisonous (Acute)
		Substances or wastes liable either to cause death or serious injury or to harm human health if swallowed or inhaled or by skin contact.
6.2	H6.2	Infectious substances
		Substances or wastes containing viable microorganisms or their toxins, which are known or suspected to cause disease in animals or humans.
8	H8	Corrosives
		Substances or wastes which, by chemical action, will cause severe damage when in contact with living tissue, or, in the case of leakage, will materially damage, or even destroy, other goods or the means of transport; they may also cause other hazards-
9	H10	Liberation of toxic gases in contact with air or water
		Substances or wastes, which, by interaction with air or water, are liable to give off toxic gases in dangerous quantities.
9	H11	Toxic (Delayed or chronic)
		Substances or wastes, which, if they are inhaled or ingested or if they penetrate the skin, may involve delayed or chronic effects, including carcinogenicity.
9	H12	Ecotoxic
		Substances or wastes which if released present or may present immediate or delayed adverse impacts to the environment by means of bioaccumulation and/or toxic effects upon biotic systems.
9	H13	Capable, by any means, after disposal, of yielding another material, e.g., leachate, which possesses any of the characteristics listed above.

Tests

The potential hazards posed by certain types of wastes are not yet fully documented; tests to define quantitatively these hazards do not exist. Further research is necessary in order to develop means to characterise potential hazards posed to man and/or the environment by these wastes. Standardized tests have been derived with respect to pure substances and materials. Many countries have developed national tests, which can be applied to materials listed in Annex I, in order to decide if these materials exhibit any of the characteristics listed in this Annex.

ANNEX IV

DISPOSAL OPERATIONS

A. Operations which do not lead to the possibility of resource recovery, recycling, reclamation, direct re-use or alternative uses

Section A encompasses all such disposal operations which occur in practice.

D1	Deposit into or onto land, (e.g., landfill, etc.)
D2	Land treatment, (e.g., biodegradation of liquid or sludgy discards in soils, etc.)
D3	Deep injection, (e.g., injection of pumpable discards into wells, salt domes of naturally occurring repositories, etc.)
D4	Surface impoundment, (e.g., placement of liquid or sludge discards into pits, ponds or lagoons, etc.)
D5	Specially engineered landfill, (e.g., placement into lined discrete cells which are capped and isolated from one another and the environment, etc.)
D6	Release into a water body except seas/oceans
D7	Release into seas/oceans including sea-bed insertion
D8	Biological treatment not specified elsewhere in this Annex which results in final compounds or mixtures which are discarded by means of any of the operations in Section A
D9	Physico-chemical treatment not specified elsewhere in this Annex which results in final compounds or mixtures which are discarded by means of any of the operations in Section A, (e.g., evaporation, drying, calcination, neutralization, precipitation, etc.)
D10	Incineration on land
D11	Incineration at sea
D12	Permanent storage (e.g., emplacement of containers in a mine, etc.)
D13	Blending or mixing prior to submission to any of the operations in Section A
D14	Repackaging prior to submission to any of the operations in Section A
D15	Storage pending any of the operations in Section A

B. Operations which may lead to resource recovery, recycling reclamation, direct re-use or alternative uses

Section B encompasses all such operations with respect to materials legally defined as or considered to be hazardous wastes and which otherwise would have been destined for operations included in Section A

R1   Use as a fuel (other than in direct incineration) or other means to generate energy

R2   Solvent reclamation/regeneration

R3   Recycling/reclamation of organic substances which are not used as solvents

R4   Recycling/reclamation of metals and metal compounds

R5   Recycling/reclamation of other inorganic materials

R6   Regeneration of acids or bases

R7   Recovery of components used for pollution abatement

R8   Recovery of components from catalysts

R9   Used oil re-refining or other reuses of previously used oil

R10   Land treatment resulting in benefit to agriculture or ecological improvement

R11   Uses of residual materials obtained from any of the operations numbered R1-R10

R12   Exchange of wastes for submission to any of the operations numbered R1-R11

R13   Accumulation of material intended for any operation in Section B

ANNEX V A

Information to be provided on notification

1. Reason for waste export

2. Exporter of the waste 1/

3. Generator(s) of the waste and site of generation 1/

4. Disposer of the waste and actual site of disposal 1/

5. Intended carrier(s) of the waste or their agents, if known 1/

6. Country of export of the waste

    Competent authority 2/

7. Expected countries of transit

    Competent authority 2/

8. Country of import of the waste

    Competent authority 2/

9. General or single notification

10. Projected date(s) of shipment(s) and period of time over which waste is to be exported and proposed itinerary (including point of entry and exit) 3/

11. Means of transport envisaged (road, rail, sea, air, inland waters)

12. Information relating to insurance 4/

13. Designation and physical description of the waste including Y number and UN number and its composition 5/ and information on any special handling requirements including emergency provisions in case of accidents

14. Type of packaging envisaged (e.g. bulk, drummed, tanker)

15. Estimated quantity in weight/volume 6/

16. Process by which the waste is generated 7/

17. For wastes listed in Annex I, classifications from Annex III: hazardous characteristic, H number, and UN class

18. Method of disposal as per Annex IV

19. Declaration by the generator and exporter that the information is correct

20. Information transmitted (including technical description of the plant) to the exporter or generator from the disposer of the waste upon which the latter has based his assessment that there was no reason to believe that the wastes will not be managed in an environmentally sound manner in accordance with the laws and regulations of the country of import

21. Information concerning the contract between the exporter and disposer.

Basel Convention

<u>Notes</u>

1/   Full name and address, telephone, telex or telefax number and the name, address, telephone, telex or telefax number of the person to be contacted.

2/   Full name and address, telephone, telex or telefax number.

3/   In the case of a general notification covering several shipments, either the expected dates of each shipment or, if this is not known, the expected frequency of the shipments will be required.

4   Information to be provided on relevant insurance requirements and how they are met by exporter, carrier and disposer.

5/   The nature and the concentration of the most hazardous components, in terms of toxicity and other dangers presented by the waste both in handling and in relation to the proposed disposal method.

6/   In the case of a general notification covering several shipments, both the estimated total quantity and the estimated quantities for each individual shipment will be required.

7/   Insofar as this is necessary to assess the hazard and determine the appropriateness of the proposed disposal operation.

ANNEX V B

Information to be provided on the movement document

1. Exporter of the waste 1/

2. Generator(s) of the waste and site of generation 1/

3. Disposer of the waste and actual site of disposal 1/

4. Carrier(s) of the waste 1/ or his agent(s)

5. Subject of general or single notification

6. The date the transboundary movement started and date(s) and signature on receipt by each person who takes charge of the waste

7. Means of transport (road, rail, inland waterway, sea, air) including countries of export, transit and import, also point of entry and exit where these have been designated

8. General description of the waste (physical state, proper UN shipping name and class, UN number, Y number and H number as applicable)

9. Information on special handling requirements including emergency provision in case of accidents

10. Type and number of packages

11. Quantity in weight/volume

12. Declaration by the generator or exporter that the information is correct

13. Declaration by the generator or exporter indicating no objection from the competent authorities of all States concerned which are Parties

14. Certification by disposer of receipt at designated disposal facility and indication of method of disposal and of the approximate date of disposal.

<u>Notes</u>

The information required on the movement document shall where possible be integrated in one document with that required under transport rules. Where this is not possible the information should complement rather than duplicate that required under the transport rules. The movement document shall carry instructions as to who is to provide information and fill-out any form.

1/ Full name and address, telephone, telex or telefax number and the name, address, telephone, telex or telefax number of the person to be contacted in case of emergency.

ANNEX VI

ARBITRATION

Article 1

Unless the agreement referred to in Article 20 of the Convention provides otherwise, the arbitration procedure shall be conducted in accordance with Articles 2 to 10 below.

Article 2

The claimant Party shall notify the Secretariat that the Parties have agreed to submit the dispute to arbitration pursuant to paragraph 2 or paragraph 3 of Article 20 and include, in particular, the Articles of the Convention the interpretation or application of which are at issue. The Secretariat shall forward the information thus received to all Parties to the Convention.

Article 3

The arbitral tribunal shall consist of three members. Each of the Parties to the dispute shall appoint an arbitrator, and the two arbitrators so appointed shall designate by common agreement the third arbitrator, who shall be the chairman of the tribunal. The latter shall not be a national of one of the Parties to the dispute, nor have his usual place of residence in the territory of one of these Parties, nor be employed by any of them, nor have dealt with the case in any other capacity.

Article 4

1.   If the chairman of the arbitral tribunal has not been designated within two months of the appointment of the second arbitrator, the Secretary-General of the United Nations shall, at the request of either Party, designate him within a further two months period.

2.   If one of the Parties to the dispute does not appoint an arbitrator within two months of the receipt of the request, the other Party may inform the Secretary-General of the United Nations who shall designate the chairman of the arbitral tribunal within a further two months' period. Upon designation, the chairman of the arbitral tribunal shall request the Party which has not appointed an arbitrator to do so within two months. After such period, he shall inform the Secretary-General of the United Nations, who shall make this appointment within a further two months' period.

Article 5

1.   The arbitral tribunal shall render its decision in accordance with international law and in accordance with the provisions of this Convention.

2.   Any arbitral tribunal constituted under the provisions of this Annex shall draw up its own rules of procedure.

Article 6

1.   The decisions of the arbitral tribunal both on procedure and on substance, shall be taken by majority vote of its members.

2.   The tribunal may take all appropriate measures in order to establish the facts.

It may, at the request of one of the Parties, recommend essential interim measures of protection.

3.   The Parties to the dispute shall provide all facilities necessary for the effective conduct of the proceedings.

4.   The absence or default of a Party in the dispute shall not constitute an impediment to the proceedings.

Article 7

The tribunal may hear and determine counter-claims arising directly out of the subject-matter of the dispute.

Article 8

Unless the arbitral tribunal determines otherwise because of the particular circumstances of the case, the expenses of the tribunal, including the remuneration of its members, shall be borne by the Parties to the dispute in equal shares. The tribunal shall keep a record of all its expenses, and shall furnish a final statement thereof to the Parties.

Article 9

Any Party that has an interest of a legal nature in the subject-matter of the dispute which may be affected by the decision in the case, may intervene in the proceedings with the consent of the tribunal.

Article 10

1.   The tribunal shall render its award within five months of the date on which it is established unless it finds it necessary to extend the time-limit for a period which should not exceed five months.

2.   The award of the arbitral tribunal shall be accompanied by a statement of reasons. It shall be final and binding upon the Parties to the dispute.

3.   Any dispute which may arise between the Parties concerning the interpretation or execution of the award may be submitted by either Party to the arbitral tribunal which made the award or, if the latter cannot be seized thereof, to another tribunal constituted for this purpose in the same manner as the first.

ANNEX VII[15]

Parties and other States which are members of OECD, EC, Liechtenstein

---

[15] The amendment whereby Annex VII was added to the Convention entered into force on 5 December 2019, in accordance with article 17(5) of the Convention (depositary notification C.N.420.2019), reflecting decision III/1 adopted by the Conference of the Parties at its third meeting.

ANNEX VIII[16]

List A

Wastes contained in this Annex are characterized as hazardous under Article 1, paragraph 1 (a), of this Convention, and their designation on this Annex does not preclude the use of Annex III to demonstrate that a waste is not hazardous.

A1 Metal and metal-bearing wastes

A1010   Metal wastes and waste consisting of alloys of any of the following:

— Antimony

— Arsenic

— Beryllium

— Cadmium

— Lead

— Mercury

— Selenium

— Tellurium

— Thallium

but excluding such wastes specifically listed on list B.

A1020   Waste having as constituents or contaminants, excluding metal waste in massive form, any of the following:

— Antimony; antimony compounds

— Beryllium; beryllium compounds

— Cadmium; cadmium compounds

— Lead; lead compounds

— Selenium; selenium compounds

— Tellurium; tellurium compounds

---

[16]   The amendment whereby Annex VIII was added to the Convention entered into force on 6 November 1998, six months following the issuance of depositary notification C.N.77.1998 of 6 May 1998 (reflecting Decision IV/9 adopted by the Conference of the Parties at its fourth meeting). The amendment to Annex VIII whereby new entries were added entered into force on 20 November 2003 (depositary notification C.N.1314.2003), six months following the issuance of depositary notification C.N.399.2003 of 20 May 2003 (reflecting Decision VI/35 adopted by the Conference of the Parties at its sixth meeting). The amendment to Annex VIII whereby one new entry was added entered into force on 8 October 2005 (depositary notification C.N.1044.2005), six months following the issuance of depositary notification C.N.263.2005 of 8 April 2005 (re-issued on 13 June 2005, reflecting Decision VII/19 adopted by the Conference of the Parties at its seventh meeting). The amendment to Annex VIII whereby one new entry was added entered into force on 24 March 2020, six months following the issuance of depositary notification C.N.432.2019 of 24 September 2019 (reflecting decision BC-14/12 adopted by the Conference of the Parties at its fourteenth meeting). The present text includes all amendments. For information on the status of individual Parties in relation to the amendment/s, please see the Status of Ratifications page on the Basel Convention website.

A1030   Wastes having as constituents or contaminants any of the following:

— Arsenic; arsenic compounds

— Mercury; mercury compounds

— Thallium; thallium compounds

A1040   Wastes having as constituents any of the following:

— Metal carbonyls

— Hexavalent chromium compounds

A1050   Galvanic sludges

A1060   Waste liquors from the pickling of metals

A1070   Leaching residues from zinc processing, dust and sludges such as jarosite, hematite, etc.

A1080   Waste zinc residues not included on list B, containing lead and cadmium in concentrations sufficient to exhibit Annex III characteristics

A1090   Ashes from the incineration of insulated copper wire

A1100   Dusts and residues from gas cleaning systems of copper smelters

A1110   Spent electrolytic solutions from copper electrorefining and electrowinning operations

A1120   Waste sludges, excluding anode slimes, from electrolyte purification systems in copper electrorefining and electrowinning operations

A1130   Spent etching solutions containing dissolved copper

A1140   Waste cupric chloride and copper cyanide catalysts

A1150   Precious metal ash from incineration of printed circuit boards not included on list B[17]

A1160   Waste lead-acid batteries, whole or crushed

A1170   Unsorted waste batteries excluding mixtures of only list B batteries. Waste batteries not specified on list B containing Annex I constituents to an extent to render them hazardous

A1180   Waste electrical and electronic assemblies or scrap[18] containing components such as accumulators and other batteries included on list A, mercury-switches, glass from cathode-ray tubes and other activated glass and PCB-capacitors, or contaminated with Annex I constituents (e.g., cadmium, mercury, lead, polychlorinated biphenyl) to an extent that they possess any of the characteristics contained in Annex III (note the related entry on list B B1110)[19]

A1190   Waste metal cables coated or insulated with plastics containing or contaminated with coal tar, PCB[20], lead, cadmium, other organohalogen compounds or other Annex I constituents to an extent that they exhibit Annex III characteristics.

---

[17]   Note that mirror entry on list B (B1160) does not specify exceptions

[18]   This entry does not include scrap assemblies from electric power generation.

[19]   PCBs are at concentration level of 50 mg/kg or more.

[20]   PCBs are at concentration level of 50 mg/kg or more.

A2 Wastes containing principally inorganic constituents, which may contain metals and organic materials

A2010	Glass waste from cathode-ray tubes and other activated glasses
A2020	Waste inorganic fluorine compounds in the form of liquids or sludges but excluding such wastes specified on list B
A2030	Waste catalysts but excluding such wastes specified on list B
A2040	Waste gypsum arising from chemical industry processes, when containing Annex I constituents to the extent that it exhibits an Annex III hazardous characteristic (note the related entry on list B B2080)
A2050	Waste asbestos (dusts and fibers)
A2060	Coal-fired power plant fly-ash containing Annex I substances in concentrations sufficient to exhibit Annex III characteristics (note the related entry on list B B2050)

A3 Wastes containing principally organic constituents, which may contain metals and inorganic materials

A3010	Waste from the production or processing of petroleum coke and bitumen
A3020	Waste mineral oils unfit for their originally intended use
A3030	Wastes that contain, consist of or are contaminated with leaded anti-knock compound sludges
A3040	Waste thermal (heat transfer) fluids
A3050	Wastes from production, formulation and use of resins, latex, plasticizers, glues/adhesives excluding such wastes specified on list B (note the related entry on list B B4020)
A3060	Waste nitrocellulose
A3070	Waste phenols, phenol compounds including chlorophenol in the form of liquids or sludges
A3080	Waste ethers not including those specified on list B
A3090	Waste leather dust, ash, sludges and flours when containing hexavalent chromium compounds or biocides (note the related entry on list B B3100)
A3100	Waste paring and other waste of leather or of composition leather not suitable for the manufacture of leather articles containing hexavalent chromium compounds or biocides (note the related entry on list B B3090)
A3110	Fellmongery wastes containing hexavalent chromium compounds or biocides or infectious substances (note the related entry on list B B3110)
A3120	Fluff - light fraction from shredding
A3130	Waste organic phosphorous compounds
A3140	Waste non-halogenated organic solvents but excluding such wastes specified on list B
A3150	Waste halogenated organic solvents

A3160   Waste halogenated or unhalogenated non-aqueous distillation residues arising from organic solvent recovery operations

A3170   Wastes arising from the production of aliphatic halogenated hydrocarbons (such as chloromethane, dichloro-ethane, vinyl chloride, vinylidene chloride, allyl chloride and epichlorhydrin)

A3180   Wastes, substances and articles containing, consisting of or contaminated with polychlorinated biphenyl (PCB), polychlorinated terphenyl (PCT), polychlorinated naphthalene (PCN) or polybrominated biphenyl (PBB), or any other polybrominated analogues of these compounds, at a concentration level of 50 mg/kg or more[21]

A3190   Waste tarry residues (excluding asphalt cements) arising from refining, distillation and any pyrolitic treatment of organic materials

A3200   Bituminous material (asphalt waste) from road construction and maintenance, containing tar (note the related entry on list B, B2130)

A3210[22]   Plastic waste, including mixtures of such waste, containing or contaminated with Annex I constituents, to an extent that it exhibits an Annex III characteristic (note the related entries Y48 in Annex II and on list B B3011).

A4 Wastes which may contain either inorganic or organic constituents

A4010   Wastes from the production, preparation and use of pharmaceutical products but excluding such wastes specified on list B

A4020   Clinical and related wastes; that is wastes arising from medical, nursing, dental, veterinary, or similar practices, and wastes generated in hospitals or other facilities during the investigation or treatment of patients, or research projects

A4030   Wastes from the production, formulation and use of biocides and phytopharmaceuticals, including waste pesticides and herbicides which are off-specification, outdated,[23] or unfit for their originally intended use

A4040   Wastes from the manufacture, formulation and use of wood-preserving chemicals[24]

A4050   Wastes that contain, consist of or are contaminated with any of the following:

—   Inorganic cyanides, excepting precious-metal-bearing residues in solid form containing traces of inorganic cyanides

—   Organic cyanides

A4060   Waste oils/water, hydrocarbons/water mixtures, emulsions

A4070   Wastes from the production, formulation and use of inks, dyes, pigments, paints, lacquers, varnish excluding any such waste specified on list B (note the related entry on list B B4010)

A4080   Wastes of an explosive nature (but excluding such wastes specified on list B)

---

[21]   The 50 mg/kg level is considered to be an internationally practical level for all wastes. However, many individual countries have established lower regulatory levels (e.g., 20 mg/kg) for specific wastes.

[22]   This entry becomes effective as of 1 January 2021.

[23]   "Outdated" means unused within the period recommended by the manufacturer.

[24]   This entry does not include wood treated with wood preserving chemicals.

A4090 Waste acidic or basic solutions, other than those specified in the corresponding entry on list B (note the related entry on list B B2120)

A4100 Wastes from industrial pollution control devices for cleaning of industrial off-gases but excluding such wastes specified on list B

A4110 Wastes that contain, consist of or are contaminated with any of the following:

— Any congenor of polychlorinated dibenzo-furan

— Any congenor of polychlorinated dibenzo-p-dioxin

A4120 Wastes that contain, consist of or are contaminated with peroxides

A4130 Waste packages and containers containing Annex I substances in concentrations sufficient to exhibit Annex III hazard characteristics

A4140 Waste consisting of or containing off specification or outdated[25] chemicals corresponding to Annex I categories and exhibiting Annex III hazard characteristics

A4150 Waste chemical substances arising from research and development or teaching activities which are not identified and/or are new and whose effects on human health and/or the environment are not known

A4160 Spent activated carbon not included on list B (note the related entry on list B B2060)

---

[25] "Outdated" means unused within the period recommended by the manufacture

ANNEX IX[26]

List B

Wastes contained in the Annex will not be wastes covered by Article 1, paragraph 1 (a), of this Convention unless they contain Annex I material to an extent causing them to exhibit an Annex III characteristic.

B1 Metal and metal-bearing wastes

B1010   Metal and metal-alloy wastes in metallic, non-dispersible form:
- Precious metals (gold, silver, the platinum group, but not mercury)
- Iron and steel scrap
- Copper scrap
- Nickel scrap
- Aluminium scrap
- Zinc scrap
- Tin scrap
- Tungsten scrap
- Molybdenum scrap
- Tantalum scrap
- Magnesium scrap
- Cobalt scrap
- Bismuth scrap
- Titanium scrap
- Zirconium scrap
- Manganese scrap
- Germanium scrap
- Vanadium scrap
- Scrap of hafnium, indium, niobium, rhenium and gallium
- Thorium scrap
- Rare earths scrap

---

[26] The amendment whereby Annex IX was added to the Convention entered into force on 6 November 1998, six months following the issuance of depositary notification C.N.77.1998 (reflecting Decision IV/9 adopted by the Conference of the Parties at its fourth meeting). The amendment to Annex IX whereby new entries were added entered into force on 20 November 2003 (depositary notification C.N.1314.2003), six months following the issuance of depositary notification C.N.399.2003 of 20 May 2003 (reflecting Decision VI/35 adopted by the Conference of the Parties at its sixth meeting). The amendment to Annex IX whereby one entry was added entered into force on 8 October 2005 (depositary notification C.N.1044.2005) six months following the issuance of depositary notification C.N.263.2005 of 8 April 2005 (re-issued on 13 June 2005, reflecting Decision VII/19 adopted by the Conference of the Parties at its seventh meeting). The amendment to Annex IX whereby new entries where added entered into force on 27 May 2014 (depositary notification C.N. 304.2014) six months following the issuance of depositary notification C.N. 965.2013 of 26 November 2013 (reflecting decision BC-11/6 adopted by the Conference of the Parties at its eleventh meeting). Amendments to Annex IX entered into force on 24 March 2020, six months following the issuance of depositary notification C.N.432.2019 of 24 September 2019 (reflecting decision BC-14/12 adopted by the Conference of the Parties at its fourteenth meeting). The present text includes all amendments. For information on the status of individual Parties in relation to the amendment/s, please see the Status of Ratifications page on the Basel Convention website.

        — Chromium scrap

B1020   Clean, uncontaminated metal scrap, including alloys, in bulk finished form (sheet, plate, beams, rods, etc.), of:

        — Antimony scrap

        — Beryllium scrap

        — Cadmium scrap

        — Lead scrap (but excluding lead-acid batteries)

        — Selenium scrap

        — Tellurium scrap

B1030   Refractory metals containing residues

B1031   Molybdenum, tungsten, titanium, tantalum, niobium and rhenium metal and metal alloy wastes in metallic dispersible form (metal powder), excluding such wastes as specified in list A under entry A1050, Galvanic sludges

B1040   Scrap assemblies from electrical power generation not contaminated with lubricating oil, PCB or PCT to an extent to render them hazardous

B1050   Mixed non-ferrous metal, heavy fraction scrap, not containing Annex I materials in concentrations sufficient to exhibit Annex III characteristics[27]

B1060   Waste selenium and tellurium in metallic elemental form including powder

B1070   Waste of copper and copper alloys in dispersible form, unless they contain Annex I constituents to an extent that they exhibit Annex III characteristics

B1080   Zinc ash and residues including zinc alloys residues in dispersible form unless containing Annex I constituents in concentration such as to exhibit Annex III characteristics[28]

B1090   Waste batteries conforming to a specification, excluding those made with lead, cadmium or mercury

B1100   Metal-bearing wastes arising from melting, smelting and refining of metals:

        — Hard zinc spelter

        — Zinc-containing drosses:

            - Galvanizing slab zinc top dross (>90% Zn)

            - Galvanizing slab zinc bottom dross (>92% Zn)

            - Zinc die casting dross (>85% Zn)

            - Hot dip galvanizers slab zinc dross (batch)(>92% Zn)

            - Zinc skimmings

        — Aluminium skimmings (or skims) excluding salt slag

        — Slags from copper processing for further processing or refining not containing arsenic, lead or cadmium to an extent that they exhibit Annex III hazard characteristics

---

[27]   Note that even where low level contamination with Annex I materials initially exists, subsequent processes, including recycling processes, may result in separated fractions containing significantly enhanced concentrations of those Annex I materials.

[28]   The status of zinc ash is currently under review and there is a recommendation with the United Nations Conference on Trade and Development (UNCTAD) that zinc ashes should not be dangerous goods.

  — Wastes of refractory linings, including crucibles, originating from copper smelting

  — Slags from precious metals processing for further refining

  — Tantalum-bearing tin slags with less than 0.5% tin

B1110 Electrical and electronic assemblies:

  — Electronic assemblies consisting only of metals or alloys

  — Waste electrical and electronic assemblies or scrap[29] (including printed circuit boards) not containing components such as accumulators and other batteries included on list A, mercury-switches, glass from cathode-ray tubes and other activated glass and PCB-capacitors, or not contaminated with Annex I constituents (e.g., cadmium, mercury, lead, polychlorinated biphenyl) or from which these have been removed, to an extent that they do not possess any of the characteristics contained in Annex III (note the related entry on list A A1180)

  — Electrical and electronic assemblies (including printed circuit boards, electronic components and wires) destined for direct reuse,[30] and not for recycling or final disposal[31]

B1115 Waste metal cables coated or insulated with plastics, not included in list A A1190, excluding those destined for Annex IVA operations or any other disposal operations involving, at any stage, uncontrolled thermal processes, such as open-burning.

B1120 Spent catalysts excluding liquids used as catalysts, containing any of:

Transition metals, excluding waste catalysts (spent catalysts, liquid used catalysts or other catalysts) on list A:

	Lanthanides (rare earth metals):
— Scandium — Zirconium	— Lanthanum
— Vanadium — Molybdenum	— Praseodymium
— Manganese — Tantalum	— Samarium
— Cobalt — Rhenium	— Gadolinium
— Copper	— Dysprosium
— Yttrium	— Erbium
— Niobium	— Ytterbium
— Hafnium	— Cerium
— Tungsten	— Neodymium
— Titanium	— Europium
— Chromium	— Terbium
— Iron	— Holmium
— Nickel	— Thulium
— Zinc	— Lutetium

---

[29] This entry does not include scrap from electrical power generation.

[30] Reuse can include repair, refurbishment or upgrading, but not major reassembly.

[31] In some countries these materials destined for direct re-use are not considered wastes.

B1130   Cleaned spent precious-metal-bearing catalysts

B1140   Precious-metal-bearing residues in solid form which contain traces of inorganic cyanides

B1150   Precious metals and alloy wastes (gold, silver, the platinum group, but not mercury) in a dispersible, non-liquid form with appropriate packaging and labelling

B1160   Precious-metal ash from the incineration of printed circuit boards (note the related entry on list A A1150)

B1170   Precious-metal ash from the incineration of photographic film

B1180   Waste photographic film containing silver halides and metallic silver

B1190   Waste photographic paper containing silver halides and metallic silver

B1200   Granulated slag arising from the manufacture of iron and steel

B1210   Slag arising from the manufacture of iron and steel including slags as a source of TiO_2 and vanadium

B1220   Slag from zinc production, chemically stabilized, having a high iron content (above 20%) and processed according to industrial specifications (e.g., DIN 4301) mainly for construction

B1230   Mill scaling arising from the manufacture of iron and steel

B1240   Copper oxide mill-scale

B1250   Waste end-of-life motor vehicles, containing neither liquids nor other hazardous components

B2 Wastes containing principally inorganic constituents, which may contain metals and organic materials

B2010   Wastes from mining operations in non-dispersible form:

— Natural graphite waste

— Slate waste, whether or not roughly trimmed or merely cut, by sawing or otherwise

— Mica waste

— Leucite, nepheline and nepheline syenite waste

— Feldspar waste

— Fluorspar waste

— Silica wastes in solid form excluding those used in foundry operations

B2020   Glass waste in non-dispersible form:

— Cullet and other waste and scrap of glass except for glass from cathode-ray tubes and other activated glasses

B2030   Ceramic wastes in non-dispersible form:

— Cermet wastes and scrap (metal ceramic composites)

— Ceramic based fibers not elsewhere specified or included

B2040   Other wastes containing principally inorganic constituents:

— Partially refined calcium sulphate produced from flue-gas desulphurization

    (FGD)

— Waste gypsum wallboard or plasterboard arising from the demolition of buildings

— Slag from copper production, chemically stabilized, having a high iron content (above 20%) and processed according to industrial specifications (e.g., DIN 4301 and DIN 8201) mainly for construction and abrasive applications

— Sulphur in solid form

— Limestone from the production of calcium cyanamide (having a pH less than 9)

— Sodium, potassium, calcium chlorides

— Carborundum (silicon carbide)

— Broken concrete

— Lithium-tantalum and lithium-niobium containing glass scraps

B2050   Coal-fired power plant fly-ash, not included on list A (note the related entry on list A A2060)

B2060   Spent activated carbon not containing any Annex I constituents to the extent they exhibit Annex III characteristics, for example, carbon resulting from the treatment of potable water and processes of the food industry and vitamin production (note the related entry on list A A4160)

B2070   Calcium fluoride sludge

B2080   Waste gypsum arising from chemical industry processes not included on list A (note the related entry on list A A2040)

B2090   Waste anode butts from steel or aluminium production made of petroleum coke or bitumen and cleaned to normal industry specifications (excluding anode butts from chlor-alkali electrolysis and from metallurgical industry)

B2100   Waste hydrates of aluminium and waste alumina and residues from alumina production excluding such materials used for gas cleaning, flocculation or filtration processes

B2110   Bauxite residue ("red mud") (pH moderated to less than 11.5)

B2120   Waste acidic or basic solutions with a pH greater than 2 and less than 11.5, which are not corrosive or otherwise hazardous (note the related entry on list A A4090)

B2130   Bituminous material (asphalt waste) from road construction and maintenance, not containing tar[32] (note the related entry on list A, A3200)

---

[32]   The concentration level of Benzo [a] pyrene should not be 50mg/kg or more.

B3 Wastes containing principally organic constituents, which may contain metals and inorganic materials

B3011[33]   Plastic waste (note the related entries Y48 in Annex II and on list A A3210):

- Plastic waste listed below, provided it is destined for recycling[34] in an environmentally sound manner and almost free from contamination and other types of wastes:[35]

- Plastic waste almost exclusively[36] consisting of one non-halogenated polymer, including but not limited to the following polymers:
    - Polyethylene (PE)
    - Polypropylene (PP)
    - Polystyrene (PS)
    - Acrylonitrile butadiene styrene (ABS)
    - Polyethylene terephthalate (PET)
    - Polycarbonates (PC)
    - Polyethers

- Plastic waste almost exclusively[36] consisting of one cured resin or condensation product, including but not limited to the following resins:
    - Urea formaldehyde resins
    - Phenol formaldehyde resins
    - Melamine formaldehyde resins
    - Epoxy resins
    - Alkyd resins

- Plastic waste almost exclusively[36] consisting of one of the following fluorinated polymers:[37]
    - Perfluoroethylene/propylene (FEP)
    - Perfluoroalkoxy alkanes:
        - Tetrafluoroethylene/perfluoroalkyl vinyl ether (PFA)
        - Tetrafluoroethylene/perfluoromethyl vinyl ether (MFA)
    - Polyvinylfluoride (PVF)
    - Polyvinylidenefluoride (PVDF)

- Mixtures of plastic waste, consisting of polyethylene (PE), polypropylene (PP) and/or polyethylene terephthalate (PET), provided they are destined for separate recycling[38] of each material and in an environmentally sound manner, and almost free from contamination and other types of wastes.[39]

---

[33]   This entry becomes effective as of 1 January 2021.

[34]   Recycling/reclamation of organic substances that are not used as solvents (R3 in Annex IV, sect. B) or, if needed, temporary storage limited to one instance, provided that it is followed by operation R3 and evidenced by contractual or relevant official documentation.

[35]   In relation to "almost free from contamination and other types of wastes", international and national specifications may offer a point of reference.

[36]   In relation to "almost exclusively", international and national specifications may offer a point of reference.

[37]   Post-consumer wastes are excluded

[38]   Recycling/reclamation of organic substances that are not used as solvents (R3 in Annex IV, sect. B), with prior sorting and, if needed, temporary storage limited to one instance, provided that it is followed by operation R3 and evidenced by contractual or relevant official documentation.

[39]   In relation to "almost free from contamination and other types of wastes", international and national specifications may offer a point of reference.

B3020   Paper, paperboard and paper product wastes

The following materials, provided they are not mixed with hazardous wastes:

Waste and scrap of paper or paperboard of:

— unbleached paper or paperboard or of corrugated paper or paperboard

— other paper or paperboard, made mainly of bleached chemical pulp, not coloured in the mass

— paper or paperboard made mainly of mechanical pulp (for example, newspapers, journals and similar printed matter)

— other, including but not limited to

    1)   laminated paperboard

    2)   unsorted scrap

B3026   The following waste from the pre-treatment of composite packaging for liquids, not containing Annex I materials in concentrations sufficient to exhibit Annex III characteristics:

— Non-separable plastic fraction

— Non-separable plastic-aluminium fraction

B3027   Self-adhesive label laminate waste containing raw materials used in label material

B3030   Textile wastes

The following materials, provided they are not mixed with other wastes and are prepared to a specification:

— Silk waste (including cocoons unsuitable for reeling, yarn waste and garnetted stock)

    - not carded or combed

    - other

— Waste of wool or of fine or coarse animal hair, including yarn waste but excluding garnetted stock

    - noils of wool or of fine animal hair

    - other waste of wool or of fine animal hair

    - waste of coarse animal hair

— Cotton waste (including yarn waste and garnetted stock)

    - yarn waste (including thread waste)

    - garnetted stock

    - other

— Flax tow and waste

— Tow and waste (including yarn waste and garnetted stock) of true hemp (Cannabis sativa L.)

— Tow and waste (including yarn waste and garnetted stock) of jute and other textile baste fibres (excluding flax, true hemp and ramie)

— Tow and waste (including yarn waste and garnetted stock) of sisal and other textile fibers of the genus Agave

— Tow, noils and waste (including yarn waste and garnetted stock) of coconut

— Tow, noils and waste (including yarn waste and garnetted stock) of abaca (Manila hemp or Musa textilis Nee)

— Tow, noils and waste (including yarn waste and garnetted stock) of ramie and other vegetable textile fibers, not elsewhere specified or included

— Waste (including noils, yarn waste and garnetted stock) of man-made fibers

- of synthetic fibers

- of artificial fibers

— Worn clothing and other worn textile articles

— Used rags, scrap twine, cordage, rope and cables and worn out articles of twine, cordage, rope or cables of textile materials

- sorted

- other

B3035   Waste textile floor coverings, carpets

B3040   Rubber wastes

The following materials, provided they are not mixed with other wastes:

— Waste and scrap of hard rubber (e.g., ebonite)

— Other rubber wastes (excluding such wastes specified elsewhere)

B3050   Untreated cork and wood waste:

— Wood waste and scrap, whether or not agglomerated in logs, briquettes, pellets or similar forms

— Cork waste: crushed, granulated or ground cork

B3060   Wastes arising from agro-food industries provided it is not infectious:

— Wine lees

— Dried and sterilized vegetable waste, residues and byproducts, whether or not in the form of pellets, of a kind used in animal feeding, not elsewhere specified or included

— Degras: residues resulting from the treatment of fatty substances or animal or vegetable waxes

— Waste of bones and horn-cores, unworked, defatted, simply prepared (but not cut to shape), treated with acid or degelatinised

— Fish waste

— Cocoa shells, husks, skins and other cocoa waste

— Other wastes from the agro-food industry excluding by-products which meet national and international requirements and standards for human or animal consumption

B3065   Waste edible fats and oils of animal or vegetable origin (e.g. frying oils), provided they do not exhibit an Annex III characteristic

B3070   The following wastes:

— Waste of human hair

— Waste straw

— Deactivated fungus mycelium from penicillin production to be used as animal feed

B3080 Waste parings and scrap of rubber

B3090 Paring and other wastes of leather or of composition leather not suitable for the manufacture of leather articles, excluding leather sludges, not containing hexavalent chromium compounds and biocides (note the related entry on list A A3100)

B3100 Leather dust, ash, sludges or flours not containing hexavalent chromium compounds or biocides (note the related entry on list A A3090)

B3110 Fellmongery wastes not containing hexavalent chromium compounds or biocides or infectious substances (note the related entry on list A A3110)

B3120 Wastes consisting of food dyes

B3130 Waste polymer ethers and waste non-hazardous monomer ethers incapable of forming peroxides

B3140 Waste pneumatic tyres, excluding those destined for Annex IVA operations

B4 Wastes which may contain either inorganic or organic constituents

B4010 Wastes consisting mainly of water-based/latex paints, inks and hardened varnishes not containing organic solvents, heavy metals or biocides to an extent to render them hazardous (note the related entry on list A A4070)

B4020 Wastes from production, formulation and use of resins, latex, plasticizers, glues/adhesives, not listed on list A, free of solvents and other contaminants to an extent that they do not exhibit Annex III characteristics, e.g., water-based, or glues based on casein starch, dextrin, cellulose ethers, polyvinyl alcohols (note the related entry on list A A3050)

B4030 Used single-use cameras, with batteries not included on list A

II.2 OECD-Council Decision C(2001)107/FINAL

DECISION OF THE COUNCIL C(2001)107/FINAL ON THE CONTROL OF TRANSBOUNDARY MOVEMENTS OF WASTES DESTINED FOR RECOVERY OPERATIONS[1]

THE COUNCIL,

1. Having regard to Article 5a) of the Convention on the Organisation for Economic Co-operation and Development of 14 December 1960;

2. Having regard to the Decision of the Council of 30 March 1992 concerning the control of transfrontier movements of wastes destined for recovery operations C(92)39/FINAL, as amended, which establishes an operational control system for transboundary movements of wastes destined for recovery operations;

3. Having regard to the Basel Convention on the Control of Transboundary Movements of Hazardous Wastes and their Disposal, which entered into force on 5 May 1992, as amended on 6 November 1998 with Annexes VIII and IX listing respectively wastes characterised as hazardous pursuant to Article 1(1)(a) of the Convention and wastes not covered by Article 1(1)(a) of the Convention;

4. Noting that most OECD Member countries (hereafter Member countries) and the European Community have become Parties to the Basel Convention;

5. Noting that Member countries agreed at the Working Group on Waste Management Policy (WGWMP) meeting in Vienna in October 1998 to further harmonisation of procedures and requirements of OECD Decision C(92)39/FINAL with those of the Basel Convention;

6. Noting that recovery of valuable materials and energy from wastes is an integral part of the international economic system and that well established international markets exist for the collection and processing of such materials within Member countries;

7. Noting further that many industrial sectors in Member countries have already implemented waste recovery techniques in an environmentally sound and economically efficient manner, thus increasing resource efficiency and contributing to sustainable development, and convinced that further efforts to promote and facilitate waste recovery are necessary and should be encouraged;

8. Recognising that the environmentally sound and economically efficient reco-

---

[1] Decision of the Council C(2001)107/Final of 22. May 2001 on the Revision of Council Decision C(92)39/FINAL on the Control of Transboundary Movements of Wastes Destined for Recovery Operations, as amended
State: 1. January 2021

very of wastes may justify transboundary movements of wastes between Member countries;

9. Recognising that the operational Control System established by Decision C(92)39/FINAL has provided a valuable framework for Member countries to control transboundary movements of wastes destined for recovery operations in an environmentally sound and economically efficient manner;

10. Desiring, therefore, to continue this agreement or arrangement under Article 11.2 of the Basel Convention;

11. Recognising that Member countries may, within their jurisdiction, impose requirements consistent with this Decision and in accordance with the rules of international law, in order to better protect human health and the environment; and

12. Recognising the need to revise Decision C(92)39/FINAL in order to improve certain elements of the Control System and to enhance harmonisation with the Basel Convention,

on the proposal of the Environment Policy Committee:

DECIDES that the text of Decision C(92)39/FINAL is revised as follows:

CHAPTER I:

1. DECIDES that Member countries shall control transboundary movements of wastes destined for recovery operations within the OECD area in accordance with the provisions set out in Chapter II of this Decision and in the appendices to it.

2. INSTRUCTS the Environment Policy Committee in co-operation with other relevant OECD bodies, in particular the Trade Committee, to ensure that the provisions of this Control System remain compatible with the needs of Member countries to recover wastes in an environmentally sound and economically efficient manner.

3. RECOMMENDS Member countries to use for the Notification Document and Movement Document the forms contained in Appendix 8 to this Decision.

4. INSTRUCTS the Environment Policy Committee to amend the forms for the Notification Document and Movement Document as necessary.

5. INSTRUCTS the Environment Policy Committee to review the procedure for amending the waste lists under Chapter II. B, (3) at the latest seven (7) years after the adoption of the present Decision.

6. REQUESTS Member countries to provide the information that is necessary for the implementation of this Decision and is listed in Appendix 7 to this Decision.

7. REQUESTS the Secretary General to transmit this Decision to the United Nations Environment Programme and the Secretariat of the Basel Convention.

CHAPTER II

A. DEFINITIONS

For the purposes of this Decision:

1. WASTES are substances or objects, other than radioactive materials covered by other international agreements, which:

    (i)   are disposed of or are being recovered; or

    (ii) are intended to be disposed of or recovered; or

    (iii) are required, by the provisions of national law, to be disposed of or recovered.

2. HAZARDOUS WASTES are:

    (i)   Wastes that belong to any category contained in Appendix 1 to this Decision unless they do not possess any of the characteristics contained in Appendix 2 to this Decision; and

    (ii) Wastes that are not covered under sub-paragraph 2.(i) but are defined as, or are considered to be, hazardous wastes by the domestic legislation of the Member country of export, import or transit. Member countries shall not be required to enforce laws other than their own.

3. DISPOSAL means any of the operations specified in Appendix 5.A to this Decision.

4. RECOVERY means any of the operations specified in Appendix 5.B to this Decision.

5. TRANSBOUNDARY MOVEMENT means any movement of wastes from an area under the national jurisdiction of a Member country to an area under the national jurisdiction of another Member country.

6. RECOVERY FACILITY means a facility which, under applicable domestic law, is operating or is authorised or permitted to operate in the country of import to receive wastes and to perform recovery operations on them.

7. COUNTRY OF EXPORT means a Member country from which a transboundary movement of wastes is planned to be initiated or is initiated.

8. COUNTRY OF IMPORT means a Member country to which a transboundary movement of wastes is planned or takes place.

9. COUNTRY OF TRANSIT means a Member country other than the country of export or import through which a transboundary movement of wastes is planned or takes place.

10. COUNTRIES CONCERNED means the countries of export and import and any country of transit, as defined above.

11. OECD AREA means all land and marine areas, under the national jurisdiction of any Member country.

12. COMPETENT AUTHORITIES means the regulatory authorities of countries concerned having jurisdiction over transboundary movements of wastes covered by this Decision.

13. PERSON means any natural or legal person.

14. EXPORTER means any person under the jurisdiction of the country of export who initiates the transboundary movement of wastes or who has, at the time the planned transboundary movement commences, possession or other forms of legal control of the wastes.

15. IMPORTER means any person under the jurisdiction of the country of import to whom possession or other form of legal control of the waste is assigned at the time the waste is received in the country of import.

16. RECOGNISED TRADER means any person under the jurisdiction of a Member country who, with appropriate authorisation of countries concerned, acts in the role of principal to purchase and subsequently sell wastes; such a person may act to arrange and facilitate transboundary movements of wastes destined for recovery operations.

17. GENERATOR means any person whose activities create wastes.

18. A MIXTURE OF WASTES means a waste that results from an intentional or unintentional mixing of two or more different wastes. A single shipment of wastes, consisting of two or more wastes, where each waste is separated, is not a mixture of wastes.

B.   GENERAL PROVISIONS

(1) Conditions

The following conditions shall apply to transboundary movements of wastes subject to this Decision:

(a) The wastes shall be destined for recovery operations within a recovery facility which will recover the wastes in an environmentally sound manner according to national laws, regulations and practices to which the facility is subject.

(b) All persons involved in any contracts or arrangements for transboundary movements of wastes destined for recovery operations should have the appropriate legal status, in accordance with domestic legislation and regulations.

(c) The transboundary movements shall be carried out under the terms of applicable international transport agreements.

(d) Any transit of wastes through a non-member country shall be subject to international law and to all applicable national laws and regulations.

(2) Control Procedures

A two-tiered system serves to delineate controls to be applied to such transboundary movements of wastes:

a) Green Control Procedure:

Wastes falling under the Green control procedure are those wastes in Appendix 3 to this Decision. This Appendix has two parts:

- Part I contains the wastes in Annex IX of the Basel Convention, some of which are subject to a note for the purposes of this Decision;

- Part II contains additional wastes that OECD Member countries agreed to be subject to the Green control procedure, in accordance with criteria referred to in Appendix 6 to this Decision.

The Green control procedure is described in Section C.

b) Amber Control Procedure:

Wastes falling under the Amber control procedure are those wastes in Appendix 4 to this Decision. This Appendix has two parts:

- Part I contains the wastes in Annexes II and VIII of the Basel Convention, some of which are subject to a note for the purposes of this Decision;

- Part II contains additional wastes that OECD Member countries agreed to be subject to the Amber control procedure, in accordance with criteria referred to in Appendix 6 to this Decision.

The Amber control procedure is described in Section D.

(3) Procedure for Amendments to the Lists of Wastes in Appendices 3 and 4

Normally, and without any other formal decision, amendments made to Annex IX under the Basel Convention will be incorporated into Part I of Appendix 3 to this Decision and amendments made to Annexes II and VIII under the Basel Convention will be incorporated into Part I of Appendix 4 to this Decision, entering into effect from the date on which the amendment to the Basel Convention (hereafter the Amendment) becomes effective for the Parties to the Convention. On that same date any relevant change will be automatically made to Part II of Appendices 3 or 4.

In exceptional cases:

(a) A Member country that determines, in accordance with the criteria referred to in Appendix 6, that a different level of control is justified for one or more wastes covered by the Amendment, may object in writing to the OECD Secretariat within sixty (60) days following the adoption of the Amendment by the Conference of the Parties to the Basel Convention. Such an objection, which shall provide an alternative proposal for inclusion into the relevant appendix or appendices to this Decision, will be immediately disseminated by the OECD Secretariat to all Member countries.

(b) The notification of an objection to the OECD Secretariat suspends the incorporation of the waste(s) concerned into the relevant appendix to this Decision. Pending examination of the objection by the appropriate OECD body, the waste(s) concerned shall be subject to the provisions of Section 6 (b) and 6 (c) hereunder.

(c) The appropriate OECD body shall promptly examine the objection and the related alternative proposal and shall reach a conclusion one month before the Amendment becomes effective for the Parties to the Basel Convention.

(d) If consensus is reached within the appropriate OECD body during that period, the relevant Appendix to this Decision will be modified as appropriate. Any modification becomes effective on the same date on which the amendment to the Basel Convention becomes effective for the Parties to the Convention.

(e) If no consensus is reached within the appropriate OECD body during that period, the Amendment will not be applied within the OECD Control System. With respect to the waste(s) concerned, the relevant appendix to this Decision will be modified as appropriate. Each Member country retains its right to control such waste(s) in conformity with its domestic legislation and international law.

(4) Provision for Specific National Control

(a) This Decision does not prejudice the right of a Member country to control, on an exceptional basis, certain wastes differently, in conformity with domestic legislation and the rules of international law, in order to protect human health and the environment.

(b) Thus, a Member country may control wastes subject to the Green control procedure as if those wastes had been subject to the Amber control procedure.

(c) A Member country may, in conformity with domestic legislation, legally define or consider a waste subject to the Amber control procedure as subject to the Green control procedure because it does not exhibit any of the hazardous characteristics listed in Appendix 2 of this Decision, as determined using national procedures[2].

(d) In the case of a transboundary movement of wastes where the wastes are legally defined as, or considered to be, wastes subject to the Amber control procedure only by the country of import, the requirements of section D that apply to the exporter and the country of export, shall apply mutatis mutandis to the importer and the country of import, respectively.

---

[2] In addition, certain Member countries have developed regulations used to determine whether or not wastes are controlled as hazardous wastes.

(5) Information Requirements

Any Member country exercising the right to apply a different level of control shall immediately inform the OECD secretariat citing the specific waste(s) and applicable legislative requirements.

Member countries which prescribe the use of certain tests and testing procedures in order to determine whether a waste exhibits one or more of the hazardous characteristics listed in Appendix 2 of this Decision shall also inform the OECD secretariat concerning which tests and testing procedures are being so utilised; and, if possible, which wastes would or would not be legally defined or considered to be hazardous wastes based upon application of these national procedures. All the above information requirements are specified in Appendix 7 to this Decision.

(6) Wastes not Listed in Appendices 3 or 4 to this Decision

Wastes which are destined for recovery operations but have not yet been assigned to Appendices 3 or 4 of this Decision shall be eligible for transboundary movements pursuant to this Decision subject to the following conditions:

(a) Member countries shall identify such wastes and, if appropriate, make applications to the Technical Working Group of the Basel Convention in order to amend the relevant Annexes of the Basel Convention;

(b) Pending assignment to a list, such wastes shall be subject to the controls required for the transboundary movements of wastes by the domestic legislation of the countries concerned in order that no country is obliged to enforce laws other than its own;

(c) However, if such wastes exhibit a hazardous characteristic listed in Appendix 2 to this Decision as determined by using national procedures[3] and any applicable international agreements, such wastes shall be subject to the Amber control procedure

(7) Generator of Mixed or Transformed Waste

If two or more lots of wastes are mixed and/or otherwise subjected to physical or chemical transformation operations, the person who performs these operations shall be deemed to be the generator of the new wastes resulting from these operations.

(8) Procedures for Mixtures of Wastes

Having regard to paragraph 11 of the preamble of this Decision, a mixture of wastes, for which no individual entry exists, shall be subject to the following control procedure:

(i) a mixture of two or more Green wastes shall be subject to the Green control procedure, provided the composition of this mixture does not impair

---

[3] In addition, certain Member countries have developed regulations used to determine whether or not wastes are controlled as hazardous wastes

its environmentally sound recovery;

(ii)   a mixture of a Green waste and more than a de minimis amount of an Amber waste or a mixture of two or more Amber wastes shall be subject to the Amber control procedure, provided the composition of this mixture does not impair its environmentally sound recovery.

C.   GREEN CONTROL PROCEDURE

Transboundary movements of wastes subject to the Green control procedure shall be subject to all existing controls normally applied in commercial transactions.

Regardless of whether or not wastes are included on the list of wastes subject to the Green Control Procedure (Appendix 3), they may not be subject to the Green control procedure if they are contaminated by other materials to an extent which (a) increases the risks associated with the wastes sufficiently to render them appropriate for submission to the amber control procedure, when taking into account the criteria in Appendix 6 to this Decision, or (b) prevents the recovery of the wastes in an environmentally sound manner.

D.   AMBER CONTROL PROCEDURE

(1)   Conditions

(a)   Contracts

Transboundary movements of wastes under the Amber control procedure may only occur under the terms of a valid written contract, or chain of contracts, or equivalent arrangements between facilities controlled by the same legal entity, starting with the exporter and terminating at the recovery facility. All persons involved in the contracts, or arrangements shall have appropriate legal status.

The contracts shall:

(i)   clearly identify: the generator of each type of waste, each person who shall have legal control of the wastes and the recovery facility;

(ii)   provide that relevant requirements of this Decision are taken into account and are binding on all parties to the contracts.

(iii)   specify which party to the contract (i) shall assume responsibility for an alternative management of the wastes in compliance with applicable laws and regulations including, if necessary, the return of the wastes in accordance with section D. (3) (a) below and (ii), as the case may be, shall provide the notification for re-export in accordance with section D.(3) (b) below.

Upon the request of the competent authorities of the countries of export or import, the exporter shall provide copies of such contracts or portions thereof.

Any information contained in the contracts provided under terms of the above paragraph shall be held strictly confidential in accordance with and to the extent required by domestic laws.

(b) Financial Guarantees

Where applicable, the exporter or the importer shall provide financial guarantees in accordance with national or international law requirements, for alternative recycling, disposal or other means of environmentally sound management of the wastes in cases where arrangements for the transboundary movement and the recovery operations cannot be carried out as foreseen.

(c) Transboundary Movements of Amber Wastes for Laboratory Analysis

Member countries may exempt a transboundary movement of a waste from the Amber control procedure, if it is explicitly destined for laboratory analysis to assess its physical or chemical characteristics or to determine its suitability for recovery operations. The amount of such waste so exempted shall be determined by the minimum quantity reasonably needed to adequately perform the analysis in each particular case, but not more than 25 kg. Analytical samples must be appropriately packaged and labelled and they remain subject to the conditions set out in Chapter II Section B. (1)(c) and (d) of this Decision. Where a competent authority of a country of import or country of export is required to be informed under its domestic legislation, the exporter shall inform that authority of a transboundary movement of a laboratory sample.

(2) Functioning of the Amber Control Procedure:

Procedures are provided under the Amber control procedure for the following two cases:

Case 1: individual transboundary movements or multiple shipments to a recovery facility;

Case 2: transboundary movements to pre-consented recovery facilities

Case 1: Individual transboundary movements of wastes or multiple shipments to a recovery facility.

(a) Prior to commencement of each transboundary movement of wastes, the exporter shall provide written notification ("**single notification**") to the competent authorities of the countries concerned. The notification document shall include all of the information listed in Appendix 8.A to this Decision. In accordance with domestic laws, the competent authorities of the country of export, instead of the exporter, may themselves transmit this notification.

(b) In instances where competent authorities acting under the terms of their domestic laws are required to review the contracts referred to in section D. (1) above, the contract(s) or portions thereof to be reviewed must be sent together with the notification document in order that such review may be appropriately performed.

(c) The competent authorities of the countries concerned may request additional information if the notification is not complete. Upon receipt of the complete notification document referred to in paragraph (a) above, the competent authorities of the country of import and, if applicable, of the country of export shall transmit an **acknowledgement** to the exporter with a copy to the competent authorities of all other countries concerned within three (3) working days of the receipt of the notification.

(d) The competent authorities of the countries concerned shall have **thirty (30) days to object**, according to their domestic laws, to the proposed transboundary movement of wastes. The thirty (30)-day period for possible objection shall commence upon issuance of the acknowledgement of the competent authority of the country of import.

(e) **Any objection** by any of the competent authorities of the countries concerned must be provided in writing to the exporter and to the competent authorities of all other countries concerned within the thirty (30)-day period.

(f) If no objection has been lodged (**tacit consent**), the transboundary movement of wastes may commence after this thirty (30)-day period has passed. Tacit consent expires within one (1) calendar year from the end of the thirty (30)-day period.

(g) In cases where the competent authorities of the countries concerned do not object and decide to provide **written consent**, it shall be issued within the thirty (30)-day period commencing upon issuance of the acknowledgement of the receipt of notification by the competent authority of the country of import. The transboundary movement of wastes may commence after all consents are received. Copies of the written consent(s) shall be sent to competent authorities of all countries concerned. Written consent is valid for up to one (1) calendar year from the date of its issuance.

(h) Objection or written consent may be provided by post, e-mail with a digital signature, e-mail without digital signature followed by post, or telefax followed by post.

(i) The transboundary movement of wastes may only take place during the period when the consents of all competent authorities (tacit or written consent) are valid.

(j) Each transboundary movement of wastes shall be accompanied by a **movement document** which includes the information listed in Appendix 8.B to this Decision.

(k) Within three (3) days of the receipt of the wastes by the recovery facility, the recovery facility shall return **a signed copy of the movement document** to the exporter and to the competent authorities of the countries of export, transit and import. Those countries of transit that do not wish to receive a signed copy of the movement document shall inform the OECD Secretariat. The recovery facility shall retain the original of the movement document for three (3) years.

(l) As soon as possible, but no later than thirty (30) days after the completion of recovery and no later than one (1) calendar year following the receipt of the waste, the recovery facility shall send **a certificate of recovery** to the exporter and to the competent authorities of the countries of export and import by post, e-mail with a digital signature, email without digital signature followed by post, or telefax followed by post.

(m) In cases where essentially similar wastes (e.g. those having essentially similar physical and chemical characteristics) are to be sent periodically to the same recovery facility by the same exporter, the competent authorities of the countries concerned may elect to accept one "**general notification**" for such multiple shipments for a period of up to one year. Each shipment must be accompanied by its own movement document, which includes the information listed in Appendix 8.B to this Decision.

(n) Revocation of the acceptance in (m) above may be accomplished by means of an official notice to the exporter from any of the competent authorities of the countries concerned. Notice of revocation of acceptance for transboundary movements previously granted under this provision shall be given to the competent authorities of all countries concerned by the competent authorities of the country that revokes such acceptance.

Case 2: *Transboundary movements of wastes to pre-consented recovery facilities*

(a) Competent authorities having jurisdiction over specific recovery facilities in the country of import may decide not to raise objections concerning transboundary movements of certain types of wastes to a specific recovery facility (**pre-consented recovery facility**). Such decisions can be limited to a specified period of time and can be revoked at any time.

(b) Competent authorities that elect this option shall inform the OECD secretariat of the recovery facility name, address, technologies employed, waste types to which the pre-consent applies, and the period covered. The OECD secretariat must also be notified of any revocations.

(c) For all transboundary movements of wastes to such facilities paragraphs (a), (b) and (c) of Case 1 shall apply.

(d) The competent authorities of the countries of export and transit shall have seven (7) working days to object, according to their domestic laws, to the proposed transboundary movement of wastes. The seven (7) working days period for possible objection shall commence upon issuance of the acknowledgement of the competent authority of the country of import. In exceptional cases where the competent authority of the country of export needs more than seven (7) working days in order to receive additional information from the exporter as necessary to meet the requirements of its domestic law, it may inform the exporter within the seven (7) working days that additional time is needed. This additional time may be up to thirty (30) days starting from the day of the issuance of the acknowledgement of the competent authority of the country of import.

(e) Paragraphs (e), (f) and (g) of Case 1 shall apply with a period of seven (7) working days instead of thirty (30) days but for the exceptional cases mentioned in paragraph (d) above, in which case the period shall remain thirty (30) days.

(f)   Paragraphs (h), (i), (j), (k) and (l) of Case 1 shall apply.

(g)   In the case of the acceptance of a general notification, paragraph (m) of Case I shall apply with the exception that the shipments can cover a period of up to three (3) years. For the revocation of this acceptance, paragraph (n) in Case 1 shall apply.

(3)   Duty to Return or Re-export Wastes Subject to the Amber Control Procedure

When a transboundary movement of wastes subject to the Amber control procedure, to which countries concerned have given consent, cannot be completed in accordance with the terms of the contract, for any reason such as illegal shipments, the competent authority of the country of import shall immediately inform the competent authority of the country of export. If alternative arrangements cannot be made to recover these wastes in an environmentally sound manner in the country of import, the following provisions shall apply as the case may be:

(a)   Return from a country of import to the country of export:

The competent authority of the country of import shall inform the competent authorities of the countries of export and transit, mentioning in particular the reason for returning the waste. The competent authority of the country of export shall admit the return of those wastes. In addition, the competent authorities of the countries of export and transit shall not oppose or prevent the return of these wastes. The return should take place within ninety (90) days from the time the country of import informs the country of export or such other period of time as the concerned Member countries agree. Any new transit country would require a new notification.

(b)   Re-export from a country of import to a country other than the initial country of export:

Re-export from a country of import of wastes subject to the Amber control procedure may only occur following notification by an exporter in the country of import to the countries concerned, as well as to the initial country of export. The notification and control procedure shall follow the provisions set out in Case 1 of Section D. (2) with the addition that the provisions concerning the competent authorities of countries concerned shall also apply to the competent authority of the initial country of export.

(4)   Duty to Return Wastes Subject to the Amber Control Procedure from a Country of Transit

When the competent authority of the country of transit observes that a transboundary movement of wastes subject to the Amber control procedure, to which countries concerned have given consent, does not comply with the requirements of the notification and movement documents or otherwise constitutes illegal shipment, the competent authority of the country of transit shall immediately inform the competent authorities of the countries of export and import and any other countries of transit.

If alternative arrangements cannot be made to recover these wastes in an environmentally sound manner, the competent authority of the country of export shall admit the return of the shipment of these wastes. In addition, the competent authorities of the country of export and other countries of transit shall not oppose or prevent the return of the wastes. The return should take place within ninety (90) days from the time the country of transit informs the country of export or such other period of time as the concerned countries agree.

(5) Provisions Relating to Recognised Traders

(a) A recognised trader may act as an exporter or importer for wastes with all the responsibilities associated with being an exporter or importer.

(b) The notification document called for in Chapter II section D (2), case 1, a) above shall include a signed declaration by the exporter that the appropriate contracts referred to in Chapter II section D (1) (a) are in place and are legally enforceable in all countries concerned.

(6) Provisions Relating to Exchange (R12) and Accumulation (R13) Operations

For transboundary movements of wastes destined for exchange (R12) or accumulation (R13) operations paragraphs (a) to (j), (m) and (n) of Case 1 shall apply. In addition:

(a) If wastes are destined for a facility or facilities where a R12 or R13 recovery operation as designated in Appendix 5.B to this Decision takes place, the recovery facility or facilities where the subsequent R1-R11 recovery operation as designated in Appendix 5.B takes place or may take place, shall also be indicated in the notification document.

(b) Within three (3) days of the receipt of the wastes by the R12/R13 recovery facility or facilities, the facilit(y)ies shall return a signed copy of the movement document to the exporter and to the competent authorities of the countries of export and import. The facilit(y)ies shall retain the original of the movement document for three (3) years.

(c) As soon as possible but no later than thirty (30) days after the completion of the R12/R13 recovery operation and no later than one (1) calendar year following the receipt of the waste, the R12 or R13 facilit(y)ies shall send a certificate of recovery to the exporter and to the competent authorities of the countries of export and import by post, e-mail with a digital signature, email without digital signature followed by post, or telefax followed by post.

(d) When an R12/R13 recovery facility delivers wastes for recovery to an R1-R11 recovery facility located in the country of import, it shall obtain as soon as possible but no later than one calendar year following delivery of the waste, a certification from the R1-R11 facility that recovery of the wastes at that facility has been completed. The R12/R13 facility shall promptly transmit the applicable certification(s) to the competent authorities of the countries of import and export, identifying the transboundary movements to which the certification(s) pertain.

(e) When an R12/R13 recovery facility delivers wastes for recovery to an R1-R11 recovery facility located:

(f) in the initial country of export, a new notification is required in accordance with Section D. (2); or

(g) in a third country other than the initial country of export, a new notification is required in accordance with Section D. (3)(b).

APPENDIX 1:

CATEGORIES OF WASTES TO BE CONTROLLED[4]

Waste streams:

Y1   Clinical wastes from medical care in hospitals, medical centres and clinics

Y2   Wastes from the production and preparation of pharmaceutical products

Y3   Waste pharmaceuticals, drugs and medicines

Y4   Wastes from the production, formulation and use of biocides and phyto-pharmaceuticals

Y5   Wastes from the manufacture, formulation and use of wood preserving chemicals

Y6   Wastes from the production, formulation and use of organic solvents

Y7   Wastes from heat treatment and tempering operations containing cyanides

Y8   Waste mineral oils unfit for their originally intended use

Y9   Waste oil/water, hydrocarbon/water mixtures, emulsions

Y10   Waste substances and articles containing or contaminated with polychlorinated biphenyls (PCB's) and/or polychlorinated terphenyls (PCT's) and/or polybrominated biphenyls (PBB's)

Y11   Waste tarry residues arising from refining, distillation and any pyrolytic treatment

Y12   Wastes from production, formulation and use of inks, dyes, pigments, paints, laquers, varnish

Y13   Wastes from production, formulation and use of resins, latex, plasticizers, glues/adhesives

Y14   Waste chemical substances arising from research and development or teaching activities which are not identified and/or are new and whose effects on man and/or the environment are not known

Y15   Wastes of an explosive nature not subject to other legislation

Y16   Wastes from production, formulation and use of photographic chemicals and processing materials

Y17   Wastes resulting from surface treatment of metals and plastics

Y18   Residues arising from industrial waste disposal operations

---

[4]   This Appendix is identical to Annex I of the Basel Convention.

Wastes having as constituents:

Y19 Metal carbonyls

Y20 Beryllium; beryllium compounds

Y21 Hexavalent chromium compounds

Y22 Copper compounds

Y23 Zinc compounds

Y24 Arsenic; arsenic compounds

Y25 Selenium; selenium compounds

Y26 Cadmium; cadmium compounds

Y27 Antimony; antimony compounds

Y28 Tellurium; tellurium compounds

Y29 Mercury; mercury compounds

Y30 Thallium; thallium compounds

Y31 Lead; lead compounds

Y32 Inorganic fluorine compounds excluding calcium fluoride

Y33 Inorganic cyanides

Y34 Acidic solutions or acids in solid form

Y35 Basic solutions or bases in solid form

Y36 Asbestos (dust and fibres)

Y37 Organic phosphorous compounds

Y38 Organic cyanides

Y39 Phenols; phenol compounds including chlorophenols

Y40 Ethers

Y41 Halogenated organic solvents

Y42 Organic solvents excluding halogenated solvents

Y43 Any congener of polychlorinated dibenzo-furan

Y44 Any congener of polychlorinated dibenzo-p-dioxin

Y45 Organohalogen compounds other than substances referred to in this Appendix (e.g. Y39, Y41, Y42 Y43, Y44)

APPENDIX 2:

LIST OF HAZARDOUS CHARACTERISTICS[5]

Code[6]   Characteristics

H1:   **Explosive**

An explosive substance or waste is a solid or liquid substance or waste (or mixture of substances or wastes) which is in itself capable by chemical reaction of producing gas at such a temperature and pressure and at such a speed as to cause damage to the surroundings.

H3:   **Flammable Liquids**

The word "flammable" has the same meaning as "inflammable". Flammable liquids are liquids, or mixtures of liquids, or liquids containing solids in solution or suspension (for example, paints, varnishes, lacquers, etc. but not including substances or wastes otherwise classified on account of their dangerous characteristics) which give off a flammable vapour at temperatures of not more than 60.5 C, closed-cup test, or not more than 65.6 C, open-cup test. (Since the results of open-cup tests and of closed-cup tests are not strictly comparable and even individual results by the same test are often variable, regulations varying from the above figures to make allowance for such differences would be within the spirit of this definition.)

H4.1:   **Flammable Solids.**

Solids, or waste solids, other than those classed as explosives, which under conditions encountered in transport are readily combustible, or may cause or contribute to fire through friction.

H4.2:   **Substances or Wastes Liable to Spontaneous Combustion.**

Substances or wastes which are liable to spontaneous heating under normal conditions encountered in transport, or to heating up in contact with air, and being liable to catch fire.

H4.3:   **Substances or Wastes which, in Contact with Water, Emit Flammable Gases.**

Substances or wastes which, by interaction with water, are liable to become spontaneously flammable or to give off flammable gases in dangerous quantities.

H5.1:   **Oxidising.**

Substances or wastes which, while in themselves not necessarily combustible, may, generally by yielding oxygen cause, or contribute to, the

---

[5]   Codes and hazardous characteristics are identical to those in Annex III of the Basel Convention.

[6]   Corresponds to hazard classification system included in the United Nations Recommendations on the Transport of Dangerous Goods (11th Revised Edition, UN, New York, October 1999) for H1 through H9; omissions of H2, H7 and H9 are deliberate. Codes H10-H13 correspond to UN class 9.

Code[6]   Characteristics

combustion of other materials.

H5.2:   Organic Peroxides.

Organic substances or wastes that contain the bivalent-0-0-structure are thermally unstable substances which may undergo exothermic self-accelerating decomposition.

H6.1:   Poisonous (Acute)

Substances or wastes liable either to cause death or serious injury or to harm human health if swallowed or inhaled or by skin contact.

H6.2:   Infectious Substances

Substances or wastes containing viable micro-organisms or their toxins which are known or suspected to cause disease in animals or humans.

H8:   Corrosives

Substances or wastes that, by chemical action, will cause severe damage when in contact with living tissue, or, in the case of leakage, will materially damage, or even destroy, other goods or the means of transport; they may also cause other hazards.

H10:   Liberation of Toxic Gases in Contact with Air or Water

Substances or wastes which, by interaction with air or water, are liable to give off toxic gases in dangerous quantities.

H11:   Toxic (Delayed or Chronic)

Substances or wastes which, if they are inhaled or ingested or if they penetrate the skin, may involve delayed or chronic effects, including carcinogenicity.

H12:   Ecotoxic

Substances or wastes which if released present or may present immediate or delayed adverse impacts to the environment by means of bioaccumulation and/or toxic effects upon biotic systems.

H13:

Capable, by any means, after disposal, of yielding another material, e.g., leachate, which possesses any of the characteristics listed above.

Tests

The potential hazards posed by certain types of wastes are not yet fully documented; objective tests to define quantitatively these hazards do not exist. Further research is necessary in order to develop means to characterise potential hazards posed to man and/or the environment by these wastes. Standardised tests have been derived with respect to pure substances and materials. Many Member countries have developed tests which can be applied to materials destined for disposal or recovery by means of operations listed in Appendices 5.A or 5.B in order to decide if these materials exhibit any of the characteristics listed in this Appendix.

APPENDIX 3:

LIST OF WASTES SUBJECT TO THE GREEN CONTROL PROCEDURE

Regardless of whether or not wastes are included on this list, they may not be subject to the Green control procedure if they are contaminated by other materials to an extent which (a) increases the risks associated with the wastes sufficiently to render them appropriate for submission to the amber control procedure, when taking into account the criteria in Appendix 6, or (b) prevents the recovery of the wastes in an environmentally sound manner.

Part I:

Wastes listed in Annex IX of the Basel Convention.

For the purposes of this Decision:

(a)   Any reference in Annex IX of the Basel Convention to list A shall be understood as a reference to Appendix 4 of this Decision.

(a bis) Any reference to Annex I or III of the Basel Convention shall be understood as a reference to Appendix 1 or 2 of this Decision, respectively.

(b)   In Basel entry B1020 the term "bulk finished form" includes all metallic non-dispersible[7] forms of the scrap listed therein.

(c)   Pending approval by the Basel Convention, Basel entry B1030 shall read: "Residues containing Refractory metals".

(d)   The part of Basel entry B1100 that refers to "Slags from copper processing" etc. does not apply and OECD entry GB040 in Part II applies instead.

(e)   Basel entry B1110 does not apply and OECD entries GC010 and GC020 in Part II apply instead.

(f)   Basel entry B2050 does not apply and OECD entry GG040 in Part II applies instead.

(g)   Basel entry B3011 does not apply as no consensus has been reached among OECD Member countries to incorporate this entry into this Decision.[8] Also, no consensus has been reached among OECD Member countries on whether or not the prior Basel entry B3010[9] continues to apply in this Decision. As

---

[7]   "Non-dispersible" does not include any wastes in the form of powder, sludge, dust or solid items containing encased hazardous waste liquids.

[8]   The appropriate OECD body should review whether consensus could be reached to incorporate this entry or an alternative into the Decision before the end of 2024.

[9]   The text of prior entry B3010 is provided below:
Solid plastic waste:
The following plastic or mixed plastic materials, provided they are not mixed with other wastes and are prepared to a specification:
- Scrap plastic of non-halogenated polymers and co-polymers, including but not limited to the following:*
  - ethylene. – styrene, - polypropylene, - polyethylene terephthalate, - acrylonitrile, - butadiene, - polyacetals, - polyamides, - polybutylene terephthalate, - polycarbonates, - polyethers, - polyphenylene sulphides, - acrylic polymers, - alkanes C10-C13, (plasticiser), - polyurethane (not containing CFCs), - polysiloxanes, - polymethyl methacrylate, - polyvinyl alcohol, - polyvinyl butyral, - polyvinyl acetate
- Cured waste resins or condensation products including the following:
  - urea formaldehyde resins, - phenol formaldehyde resins, - melamine formaldehyde resins, - epoxy resins, - alkyd

a result of this situation, each Member country retains its right to control the plastic waste covered by Basel entry B3011 in conformity with its domestic legislation and international law. Member countries should inform the OECD Secretariat about their controls for plastic waste covered by Basel entry B3011 by 15 January 2021, as well as about any future changes of such controls, in a timely manner. The OECD Secretariat should publish the information received on the OECD website[10].

Part II:

The following wastes will also be subject to the Green control procedure:

<u>Metal Bearing Wastes Arising from Melting, Smelting and Refining of Metals</u>

GB040	7112 262030 262090	Slags from precious metals and copper processing for further refining

<u>Other Wastes Containing Metals</u>

GC010		Electrical assemblies consisting only of metals or alloys.
GC020		Electronic scrap (e.g. printed circuit boards, electronic components, wire, etc.) and reclaimed electronic components suitable for base and precious metal recovery.
GC030	ex 890800	Vessels and other floating structures for breaking up, properly emptied of any cargo and other materials arising from the operation of the vessel which may have been classified as a dangerous substance or waste
GC040	ex 8701-05 ex 8709-11	Motor vehicle wrecks, drained of liquids
GC050		Spent Fluid Catalytic Cracking (FCC) Catalysts (e.g.: aluminium oxide, zeolites)

<u>Glass Waste in Non-dispersible Form</u>

GE020	ex 7001 ex 701939	Glass Fibre Waste

---

resins, - polyamides
- The following fluorinated polymer wastes**
  - perfluoroethylene/propylene (FEP),
  - perfluoro alkoxyl alkane
    - tetrafluoroethylene/per fluoro vinyl ether (PFA)
    - tetrafluoroethylene/per fluoro methylvinyl ether (MFA)
    - polyvinylfluoride (PVF), - polyvinylidenefluoride (PVDF)

* It is understood that such scraps are completely polymerized.
** - Post-consumer wastes are excluded from this entry
- Wastes shall not be mixed
- Problems arising from open-burning practices to be considered

[10] https://www.oecd.org/environment/waste/Reporting-of-controls-non-hazardous-waste.pdf

Ceramic Wastes in Non-Dispersible Form

GF010		Ceramic wastes which have been fired after shaping, including ceramic vessels (before and/or after use)

Other Wastes Containing Principally Inorganic Constituents, Which May Contain Metals and Organic Materials

GG030	ex 2621	Bottom ash and slag tap from coal fired power plants
GG040	ex 2621	Coal fired power plants fly ash

Solid Plastic Wastes

GH013[11]	391530 ex 390410-40	Polymers of vinyl chloride

Wastes Arising from Tanning and Fellmongery Operations and Leather Use

GN010	ex 050200	Waste of pigs', hogs' or boars' bristles and hair or of badger hair and other brush making hair
GN020	ex 050300	Horsehair waste, whether or not put up as a layer with or without supporting material
GN030	ex 050590	Waste of skins and other parts of birds, with their feathers or down, of feathers and parts of feathers (whether or not with trimmed edges) and down, not further worked than cleaned, disinfected or treated for preservation

---

[11] No consensus has been reached among OECD Member countries to incorporate Basel entry Y48 into this Decision. Also, no consensus has been reached among OECD Member countries on whether or not GH013 continues to apply in this Decision. As a result of this situation, each Member country retains its right to control waste of polymers of vinyl chloride in conformity with its domestic legislation and international law.

APPENDIX 4:

LIST OF WASTES SUBJECT TO THE AMBER CONTROL PROCEDURE

Part I:

Wastes listed in Annexes II and VIII of the Basel Convention.

For the purposes of this Decision:

(a)   Any reference in Annex VIII of the Basel Convention to List B shall be understood as a reference to Appendix 3 of this Decision.

(a bis) Any reference to Annex I or III of the Basel Convention shall be understood as a reference to Appendix 1 or 2 of this Decision, respectively.

(b)   In Basel entry A1010, the term "excluding such wastes specifically listed on List B (Annex IX)" is a reference both to Basel entry B1020 and the note on B1020 in Appendix 3 to this Decision, Part I (b).

(c)   Basel entries A1180 and A2060 do not apply and OECD entries GC010, GC020 and GG040 in Appendix 3 Part II apply instead when appropriate. Member countries may control these wastes differently in accordance with Chapter II B 6 of this Decision concerning wastes not listed in Appendices 3 or 4, and the chapeau of Appendix 3.

(d)   Basel entry A4050 Includes spent potlinings from aluminium smelting because they contain Y33 inorganic cyanides. If the cyanides have been destroyed, spent potlinings are assigned to Part II entry AB120 because they contain Y32, inorganic fluorine compounds excluding calcium fluoride.

(e)   Basel entry A3210 does not apply and AC300 in Part II applies instead.

(f)   Basel entry Y48[12] does not apply as no consensus has been reached among OECD countries to incorporate this entry into this Decision.[13] As a result of this situation, each Member country retains its right to control the plastic waste covered by Basel entry Y48 in conformity with its domestic legislation and international law. Member countries should inform the OECD Secretariat about their controls for plastic waste covered by Basel entry Y48 by 15 January 2021, as well as about any future changes of such controls, in a timely manner. The OECD Secretariat should publish the information received on the OECD website.

Part II:

The following wastes will also be subject to the Amber control procedure:

<u>Metal Bearing Wastes</u>

AA010	261900	Dross, scalings and other wastes from the manufacture of iron and steel[14]

---

[12]   Basel entry Y48 covers the waste covered by entry GH013.

[13]   The appropriate OECD body should review whether consensus could be reached to incorporate this entry or an alternative into the Decision before the end of 2024

[14]   This listing includes wastes in the form of ash, residue, slag, dross, skimming, scaling, dust, powder, sludge and cake, unless a material is expressly listed elsewhere.

AA060	262050	Vanadium ashes and residues
AA190	810420 ex 810430	Magnesium waste and scrap that is flammable, pyrophoric or emits, upon contact with water, flammable gases in dangerous quantities

<u>Wastes Containing Principally Inorganic Constituents, Which May Contain Metals and Organic Materials</u>

AB030		Wastes from non-cyanide based systems which arise from surface treatment of metals
AB070		Sands used in foundry operations
AB120	ex 281290 ex 3824	Inorganic halide compounds, not elsewhere specified or included
AB130		Used blasting grit
AB150	ex 382490	Unrefined calcium sulphite and calcium sulphate from flue gas desulphurisation (FGD)

<u>Wastes Containing Principally Organic Constituents, Which May Contain Metals and Inorganic Materials</u>

AC060	ex 381900	Hydraulic fluids
AC070	ex 381900	Brake fluids
AC080	ex 382000	Antifreeze fluids
AC150		Chlorofluorocarbons
AC160		Halons
AC170	ex 440310	Treated cork and wood wastes
AC250		Surface active agents (surfactants)
AC260	ex 3101	Liquid pig manure; faeces
AC270		Sewage sludge
AC300		Plastic waste, including mixtures of such wastes, containing or contaminated with Appendix 1 constituents, to an extent that it exhibits an Appendix 2 characteristic

<u>Wastes Which May Contain either Inorganic or Organic Constituents</u>

AD090	ex 382490	Wastes from production, formulation and use of reprographic and photographic chemicals and materials not elsewhere specified or included
AD100		Wastes from non-cyanide based systems which arise from surface treatment of plastics
AD120	ex 391400 ex 3915	Ion exchange resins
AD150		Naturally occurring organic material used as a filter medium (such as bio-filters)

<u>Wastes Containing Principally Inorganic Constituents, Which May Contain Metals and Organic Materials</u>

RB020	ex 6815	Ceramic based fibres of physico chemical characteristics similar to those of asbestos

APPENDIX 5.A:

DISPOSAL OPERATIONS[15]

Appendix 5.A is meant to encompass all such disposal operations that occur in practice, whether or not they are adequate from the point of view of environmental protection.

D1   Deposit into or onto land, (e.g., landfill, etc.)

D2   Land treatment, (e.g., biodegradation of liquid or sludgy discards in soils, etc.)

D3   Deep injection, (e.g., injection of pumpable discards into wells, salt domes or naturally occurring repositories, etc.)

D4   Surface impoundment, (e.g., placement of liquid or sludge discards into pits, ponds or lagoons, etc.)

D5   Specially engineered landfill, (e.g., placement into lined discrete cells which are capped and isolated from one another and the environment, etc.)

D6   Release into a water body except seas/oceans

D7   Release into seas/oceans including sea-bed insertion

D8   Biological treatment not specified elsewhere in this Appendix which results in final compounds or mixtures which are discarded by means of any of the operations in Appendix 5.A

D9   Physico chemical treatment not specified elsewhere in this Appendix which results in final compounds or mixtures which are discarded by means of any of the operations in Appendix 5.A, (e.g., evaporation, drying, calcination, etc.)

D10   Incineration on land

D11   Incineration at sea

D12   Permanent storage (e.g., emplacement of containers in a mine, etc.)

D13   Blending or mixing prior to submission to any of the operations in Appendix 5.A

D14   Repackaging prior to submission to any of the operations in Appendix 5.A

D15   Storage pending any of the operations in Appendix 5.A

---

[15]   The wording of D1 to D15 in Appendix 5.A is identical to that of Annex IV.A of the Basel Convention.

APPENDIX 5.B:

RECOVERY OPERATIONS[16]

Appendix 5.B is meant to encompass all such operations with respect to materials considered to be or legally defined as wastes and which otherwise would have been destined for operations included in Appendix 5.A.

R1	Use as a fuel (other than in direct incineration) or other means to generate energy
R2	Solvent reclamation/regeneration
R3	Recycling/reclamation of organic substances which are not used as solvents
R4	Recycling/reclamation of metals and metal compounds
R5	Recycling/reclamation of other inorganic materials
R6	Regeneration of acids or bases
R7	Recovery of components used for pollution abatement
R8	Recovery of components from catalysts
R9	Used oil re-refining or other reuses of previously used oil
R10	Land treatment resulting in benefit to agriculture or ecological improvement
R11	Uses of residual materials obtained from any of the operations numbered R1-R10
R12	Exchange of wastes for submission to any of the operations numbered R1-R11
R13	Accumulation of material intended for any operation in Appendix 5.B

---

[16] The wording of R1 to R13 in Appendix 5.B is identical to that of Annex IV.B of the Basel Convention.

APPENDIX 6:

CRITERIA FOR THE OECD RISK-BASED APPROACH

A)   Properties

1) Does the waste normally exhibit any of the hazardous characteristics listed in Appendix 2 to this Decision? Furthermore, it is useful to know if the waste is legally defined as or considered to be a hazardous waste in one or more Member countries.

2) Is the waste typically contaminated?

3) What is the physical state of the waste?

4) What is the degree of difficulty of cleanup in the case of accidental spillage or mismanagement?

5) What is the economic value of the waste bearing in mind historical price fluctuations?

B)   Management

6) Does the technological capability to recover the waste exist?

7) Is there a history of adverse environmental incidents arising from trans-boundary movements of the waste or associated recovery operations?

8) Is the waste routinely traded through established channels and is that evidenced by commercial classification?

9) Is the waste usually moved internationally under the terms of a valid contract or chain of contracts?

10) What is the extent of reuse and recovery of the waste and how is any portion separated from the waste but not subject to recovery managed?

11) What are the overall environmental benefits arising from the recovery operations?

APPENDIX 7:

PRACTICAL INFORMATION TO BE PROVIDED BY MEMBER COUNTRIES

(1) **Competent Authority:** indicates the address, telephone, e-mail and fax numbers of the regulatory authority having jurisdiction over transboundary movements of wastes destined for recovery operations. If separate competent authorities are known to exist for different types of movements (e.g. different authorities for transit than for import/export), this is also indicated. When applicable, indicate the code number of the national competent authorities.

(2) **Contact Point:** provides the point of correspondence, including the address, telephone e-mail and fax numbers, through which individuals can, if desired, obtain additional or complementary information.

(3) **Acceptable Languages:** indicates the languages that can be used by the exporter so that the notification document is understandable for the competent authority receiving it.

(4) **Required Points of Entry/Exit:** notes if and when national regulations prescribe that shipments of recoverable wastes must enter or exit the territory through specific customs offices.

(5) **Pre-consented Recovery Facilities:** indicates if a Member country has granted pre-consent for certain wastes to be accepted by one or more pre-consented recovery facilities within its jurisdiction, in conformity with Chapter II, D, (2), Case 2. Details on the company, the location, the expiry of pre-consent, the relevant waste types, and total quantity pre-consented is also indicated when known.

(6) **Classification Differences:** this item is meant to indicate when divergent classifications exist between the OECD Appendices 3 and 4 and national waste lists, according to provisions of Section B(4) of this Decision. When known specific wastes and associated controls are cited.

(7) **Prohibitions:** provides information on wastes specifically banned or prohibited for import or export under the Member country's pertinent national laws or regulations.

(8) **Contractual Requirements:** notes requirements concerning contracts between the exporter and the importer, including whether the competent authority shall review the contract.

(9) **Written Consent:** indicates if Member countries require written consent for exports or imports of wastes.

(10) **Information related to Environmentally Sound Management:** Indicates additional information under the terms of domestic legislation on environmentally sound management of wastes.

(11) **Notification for Export:** Indicates whether notifications for export are transmitted by the competent authorities instead of the exporter.

(12) **Movement Document:** Indicates if a country of transit does not wish to receive a signed copy of the movement document, indicating the receipt of wastes by

the recovery facility in the country of import.

(13) <u>Financial Requirements:</u> If Member countries require financial guarantees for transboundary movements of recoverable wastes, such requirements would be specified under this entry. Information provided may inter alia include: the types of guarantee (e.g. insurance statement, bank letters, bonds, etc.), the amount of guarantee (minimum and maximum, if any), whether the guarantee varies according to amount and/or hazardousness of the waste, the damages to be covered.

(14) <u>Pertinent National Laws/Regulations:</u> provides citations to relevant domestic laws and regulations containing provisions that relate to the conditions of this Decision.

(15) <u>Other is used to indicate:</u>

- additional differences between this Decision and national provisions;

- pending amendments to pertinent national laws/regulations; and

- other requirements or issues deemed relevant by the Member country.

APPENDIX 8:

NOTIFICATION AND MOVEMENT DOCUMENTS

A. Information to be Included in the Notification Document:

1) Serial number or other accepted identifier of notification document.

2) Exporter name, address, telephone, telefax, e-mail and contact person.

3) Recovery facility name, address, telephone, telefax, e-mail and technologies employed.

4) Importer name, address, telephone, telefax, e-mail.

5) Address, telephone, telefax, e-mail of any intended carrier(s) and/or their agents.

6) Country of export and relevant competent authority.

7) Countries of transit and relevant competent authorities.

8) Country of import and relevant competent authority.

9) Single notification or general notification. If general, period of validity requested.

10) Date(s) foreseen for commencement of transboundary movement(s).

11) Means of transport envisaged.

12) Certification that any applicable insurance or other financial guarantee is or shall be in force covering the transboundary movement.

13) Designation of waste type(s) on the appropriate list (Part I or II of Appendix 3 or 4) and their description(s), probable total quantity of each, and any hazardous characteristics.

14) Specification of the recovery operation(s) according to Appendix 5.B to this Decision.

15) Certification of the existence of written contract or chain of contracts or equivalent arrangement as required by this Decision.

16) Certification by the exporter that the information is complete and correct to the best of his knowledge.

B. Information to be Included in the Movement Document:

Include all information at A. above plus:

(a) Date shipment has commenced.

(b) Carrier(s) name, address, telephone, telefax, e-mail.

(c) Type of packaging envisaged.

(d) Any special precautions to be taken by carrier(s).

(e) Declaration by exporter that no objection has been lodged by the competent authorities of all countries concerned. This declaration requires signature of the exporter.

(f)   Appropriate signatures for each custody transfer.

C.   Recommended forms for the notification and movement documents (see Chapter I paragraph 3) for transboundary movements of wastes destined for recovery operations within the OECD area and instructions for completing those forms:

[Not published herein]

Cf. explanatory notes sect. I.4.2.2 fig. 4 and 5 and sect. I.4.5.1 fig. 7 and 8

II.3 EU Waste Shipment Regulation

REGULATION (EC) No 1013/2006 OF THE EUROPEAN PARLIAMENT AND OF THE COUNCIL

of 14 June 2006[1]

on shipments of waste

THE EUROPEAN PARLIAMENT AND THE COUNCIL OF THE EUROPEAN UNION,

Having regard to the Treaty establishing the European Community, and in particular Article 175(1) thereof,

Having regard to the proposal from the Commission,

Having regard to the opinion of the European Economic and Social Committee[2],

After consulting the Committee of the Regions,

Acting in accordance with the procedure laid down in Article 251 of the Treaty[3],

Whereas:

(1)   The main and predominant objective and component of this Regulation is the protection of the environment, its effects on international trade being only incidental.

(2)   Council Regulation (EEC) No 259/93 of 1 February 1993 on the supervision and control of shipments of waste within, into and out of the European Community[4] has already been significantly amended on several occasions and requires further amendment. It is necessary, in particular, to incorporate in that Regulation the content of Commission Decision 94/774/EC of 24 November 1994 concerning the standard consignment note referred to in Council Regulation (EEC) No 259/93[5] and of Commission Decision 1999/412/EC of 3 June 1999 concerning a questionnaire for the reporting obligation of Member States pursuant to Article 41(2) of Council Regulation (EEC) No 259/93[6]. Regulation (EEC) No 259/93 should therefore be replaced in the interests of clarity.

---

[1]   OJ L 190 12.7.2006, p. 1, as amended by Commission Delegated Regulation (EU) 2020/2174 of 19 October 2020 OJ L 433, p. 11 of 22 December 2020

[2]   OJ C 108, 30.4.2004, p. 58.

[3]   Opinion of the European Parliament of 19 November 2003 (OJ C 87 E, 7.4.2004, p. 281), Council Common Position of 24 June 2005 (OJ C 206 E, 23.8.2005, p. 1) and position of the European Parliament of 25 October 2005 (not yet published in the Official Journal). Council Decision of 29 May 2006.

[4]   OJ L 30, 6.2.1993, p. 1. Regulation as last amended by Commission Regulation (EC) No 2557/2001 (OJ L 349, 31.12.2001, p. 1).

[5]   OJ L 310, 3.12.1994, p. 70.

[6]   OJ L 156, 23.6.1999, p. 37.

(3)   Council Decision 93/98/EEC[7] concerned the conclusion, on behalf of the Community, of the Basel Convention of 22 March 1989 on the control of transboundary movements of hazardous wastes and their disposal[8], to which the Community has been a Party since 1994. By adopting Regulation (EEC) No 259/93, the Council has established rules to curtail and to control such movements designed, inter alia, to make the existing Community system for the supervision and control of waste movements comply with the requirements of the Basel Convention.

(4)   Council Decision 97/640/EC[9] concerned the approval, on behalf of the Community, of the amendment to the Basel Convention, as laid down in Decision III/1 of the Conference of the Parties. By that amendment, all exports of hazardous waste destined for disposal from countries listed in Annex VII to the Convention to countries not listed therein were prohibited, as were, with effect from 1 January 1998, all such exports of the hazardous waste referred to in Article 1(1)(a) of the Convention and destined for recovery. Regulation (EEC) No 259/93 was amended accordingly by Council Regulation (EC) No 120/97[10].

(5)   In view of the fact that the Community has approved Decision C(2001)107/Final of the OECD Council concerning the revision of Decision C(92)39/Final on the control of transboundary movements of wastes destined for recovery operations (OECD Decision), in order to harmonise waste lists with the Basel Convention and revise certain other requirements, it is necessary to incorporate the content of that Decision in Community legislation.

(6)   The Community has signed the Stockholm Convention of 22 May 2001 on persistent organic pollutants.

(7)   It is important to organise and regulate the supervision and control of shipments of waste in a way which takes account of the need to preserve, protect and improve the quality of the environment and human health and which promotes a more uniform application of the Regulation throughout the Community.

(8)   It is also important to bear in mind the requirement laid down in Article 4(2)(d) of the Basel Convention that shipments of hazardous waste are to be reduced to a minimum, consistent with environmentally sound and efficient management of such waste.

(9)   Furthermore, it is important to bear in mind the right of each Party to the Basel Convention, pursuant to Article 4(1) thereof, to prohibit the import of hazardous waste or of waste listed in Annex II to that Convention.

(10) Shipments of waste generated by armed forces or relief organisations should be excluded from the scope of this Regulation when imported into the Community in certain situations (including transit within the Community when the waste enters the Community). The requirements of international law and international agreements should be respected in relation to such shipments. In such cases, any competent authority of transit and the competent authority of destination in the Community

---

[7]   OJ L 39, 16.2.1993, p. 1.
[8]   OJ L 39, 16.2.1993, p. 3.
[9]   OJ L 272, 4.10.1997, p. 45.
[10]   OJ L 22, 24.1.1997, p. 14.

should be informed in advance concerning the shipment and its destination.

(11) It is necessary to avoid duplication with Regulation (EC) No 1774/2002 of the European Parliament and of the Council of 3 October 2002 laying down health rules concerning animal by-products not intended for human consumption[11], which already contains provisions covering the overall consignment, channelling and movement (collection, transport, handling, processing, use, recovery or disposal, record keeping, accompanying documents and traceability) of animal by-products within, into and out of the Community.

(12) The Commission should report by the date of entry into force of this Regulation on the relationship between the existing sectoral legislation on animal and public health and the provisions of this Regulation, and should submit by that date any proposals needed to bring such legislation into line with this Regulation in order to achieve an equivalent level of control.

(13) Although the supervision and control of shipments of waste within a Member State is a matter for that Member State, national systems concerning shipments of waste should take account of the need for coherence with the Community system in order to ensure a high level of protection of the environment and human health.

(14) In the case of shipments of waste destined for disposal operations and waste not listed in Annex III, IIIA or IIIB destined for recovery operations, it is appropriate to ensure optimum supervision and control by requiring prior written consent to such shipments. Such a procedure should in turn entail prior notification, which enables the competent authorities to be duly informed so that they can take all necessary measures for the protection of human health and the environment. It should also enable those authorities to raise reasoned objections to such a shipment.

(15) In the case of shipments of waste listed in Annex III, IIIA or IIIB destined for recovery operations, it is appropriate to ensure a minimum level of supervision and control by requiring such shipments to be accompanied by certain information.

(16) In view of the need for uniform application of this Regulation and for the proper functioning of the internal market, it is necessary in the interests of efficiency to require that notifications be processed through the competent authority of dispatch.

(17) It is also important to clarify the system of financial guarantees or equivalent insurance.

(18) Considering the responsibility of waste producers for the environmentally sound management of waste, the notification and movement documents for waste shipments should, where practicable, be filled in by the waste producers.

(19) It is necessary to provide procedural safeguards for the notifier, both in the interests of legal certainty and to ensure uniform application of this Regulation and the proper functioning of the internal market.

(20) In the case of shipments of waste for disposal, Member States should take into

---

[11]   OJ L 273, 10.10.2002, p. 1. Regulation as last amended by Commission Regulation (EC) No 416/2005 (OJ L 66, 12.3.2005, p. 10).

account the principles of proximity, priority for recovery and self-sufficiency at Community and national levels, in accordance with Directive 2006/12/EC of the European Parliament and of the Council of 5 April 2006 on waste[12], by taking measures in accordance with the Treaty to prohibit generally or partially or to object systematically to such shipments. Account should also be taken of the requirement laid down in Directive 2006/12/EC, whereby Member States are to establish an integrated and adequate network of waste disposal installations, in order to enable the Community as a whole to become self-sufficient in waste disposal and the Member States to move towards that aim individually, taking into account geographical circumstances or the need for specialised installations for certain types of waste. Member States should also be able to ensure that the waste management facilities covered by Council Directive 96/61/EC of 24 September 1996 concerning integrated pollution prevention and control[13] apply best available techniques as defined in that Directive in compliance with the permit of the facility, and that the waste is treated in accordance with legally binding environmental protection standards in relation to disposal operations established in Community legislation.

(21) In the case of shipments of waste destined for recovery, Member States should be able to ensure that the waste management facilities covered by Directive 96/61/EC apply best available techniques as defined in that Directive in compliance with the permit of the facility. Member States should also be able to ensure that waste is treated in accordance with legally binding environmental protection standards in relation to recovery operations established in Community legislation and that, taking account of Article 7(4) of Directive 2006/12/EC, waste is treated in accordance with waste management plans established pursuant to that Directive with the purpose of ensuring the implementation of legally binding recovery or recycling obligations established in Community legislation.

(22) The development of mandatory requirements for waste facilities and the treatment of specific waste materials at Community level, in addition to the existing provisions of Community law, can contribute to the creation of a high level of environmental protection across the Community, assist in the creation of a level playing field for recycling and help to ensure that the development of an economically viable internal market for recycling is not hindered. Therefore there is a need to develop a Community level playing field for recycling through the application of common standards in certain areas, as appropriate and including in relation to secondary materials, in order to increase the quality of recycling. The Commission should submit, as appropriate, proposals for such standards for certain wastes and certain recycling facilities as soon as practicable based on further examination in the context of the waste strategy and taking into account existing Community legislation and legislation in the Member States. In the interim, it should be possible, under certain conditions, to object to planned shipments where the related recovery would not be in accordance with national legislation in the country of dispatch relating to the recovery of waste. In the interim, the Commission should also keep under review the situation regarding possible undesired shipments of waste to the new Member States and, if

---

[12] OJ L 114, 27.4.2006, p. 9.

[13] OJ L 257, 10.10.1996, p. 26. Directive as last amended by Regulation (EC) No 166/2006 of the European Parliament and of the Council (OJ L 33, 4.2.2006, p. 1).

necessary, submit appropriate proposals to deal with such situations.

(23) Member States should be required to ensure that, in accordance with the United Nations Economic Commission for Europe (UNECE) Convention on Access to Information, Public Participation in Decision-making and Access to Justice in Environmental Matters of 25 June 1998 (Aarhus Convention), the relevant competent authorities make publicly available by appropriate means information on notifications of shipments, where such information is not confidential under national or Community legislation.

(24) An obligation should be laid down to the effect that waste from a shipment that cannot be completed as intended is to be taken back to the country of dispatch or recovered or disposed of in an alternative way.

(25) It should also be made compulsory for the person whose action is the cause of an illegal shipment to take back the waste involved or make alternative arrangements for its recovery or disposal. Failing that, the competent authorities of dispatch or destination, as appropriate, should intervene themselves.

(26) It is necessary, in order to protect the environment of the countries concerned, to clarify the scope of the prohibition laid down in accordance with the Basel Convention of exports from the Community of any waste destined for disposal in a third country other than an EFTA (European Free Trade Association) country.

(27) Countries that are Parties to the Agreement on the European Economic Area may adopt the control procedures provided for shipments within the Community.

(28) It is also necessary, in order to protect the environment of the countries concerned, to clarify the scope of the prohibition of exports of hazardous waste destined for recovery in a country to which the OECD Decision does not apply, also laid down in accordance with the Basel Convention. In particular, it is necessary to clarify the list of waste to which that prohibition applies and to ensure that it also includes the waste listed in Annex II to the Basel Convention, namely waste collected from households and residues from the incineration of household waste.

(29) Specific arrangements should be maintained for exports of non-hazardous waste destined for recovery in countries to which the OECD Decision does not apply and provision should be made for them to be further streamlined at a later date.

(30) Imports into the Community of waste for disposal should be permitted where the exporting country is a Party to the Basel Convention. Imports into the Community of waste for recovery should be permitted where the exporting country is one to which the OECD Decision applies or is a Party to the Basel Convention. In other cases, however, imports should be allowed only if the exporting country is bound by a bilateral or multilateral agreement or arrangement compatible with Community legislation and in accordance with Article 11 of the Basel Convention, except when this is not possible during situations of crisis, peacemaking, peacekeeping or war.

(31) This Regulation should be applied in accordance with international maritime law.

(32) This Regulation should reflect the rules regarding exports and imports of waste to and from the overseas countries and territories laid down in Council Decision

2001/822/EC of 27 November 2001 on the association of the overseas countries and territories with the European Community (Overseas Association Decision)[14].

(33) The necessary steps should be taken to ensure that, in accordance with Directive 2006/12/EC and other Community legislation on waste, waste shipped within the Community and waste imported into the Community is managed, throughout the period of shipment and including recovery or disposal in the country of destination, without endangering human health and without using processes or methods which could harm the environment. As regards exports from the Community that are not prohibited, efforts should be made to ensure that the waste is managed in an environmentally sound manner throughout the period of shipment and including recovery or disposal in the third country of destination. The facility which receives the waste should be operated in accordance with human health and environmental protection standards that are broadly equivalent to those established in Community legislation. A list of non-binding guidelines should be established in which guidance may be sought on environmentally sound management.

(34) Member States should provide the Commission with information concerning the implementation of this Regulation, both through the reports submitted to the Secretariat of the Basel Convention and on the basis of a separate questionnaire.

(35) It is necessary to ensure the safe and environmentally sound management of ship dismantling in order to protect human health and the environment. Furthermore, it should be noted that a ship may become waste as defined in Article 2 of the Basel Convention and that at the same time it may be defined as a ship under other international rules. It is important to recall that work is ongoing, involving inter-agency cooperation between International Labour Organisation (ILO), International Maritime Organisation (IMO) and the Secretariat of the Basel Convention, to establish mandatory requirements at the global level ensuring an efficient and effective solution to the problem of ship dismantling.

(36) Efficient international cooperation regarding control of shipments of waste is instrumental in ensuring that shipments of hazardous waste are controlled. Information exchange, shared responsibility and cooperative efforts between the Community and its Member States and third countries should be promoted with a view to ensuring sound management of waste.

(37) Certain Annexes to this Regulation should be adopted by the Commission in accordance with the procedure referred to in Article 18(3) of Directive 2006/12/EC. This procedure should also apply to the amendment of the Annexes to take account of scientific and technical progress, of modifications in the relevant Community legislation or of events connected to the OECD Decision or to the Basel Convention and other related international conventions and agreements.

(38) In preparing the instructions for completing the notification and movement documents to be set out in Annex IC, the Commission, taking into account the OECD Decision and the Basel Convention, should specify, inter alia, that the notification and movement documents should, as far as possible, be on two pages and what the precise timing is for completion of the notification and movement documents in

---

[14]   OJ L 314, 30.11.2001, p. 1.

Annex IA and IB, taking into account Annex II. In addition, where terminology and requirements differ between the OECD Decision or the Basel Convention and this Regulation, the specific requirements should be clarified.

(39) In considering the mixtures of wastes to be added in Annex IIIA, the following information should be considered, inter alia: the properties of the waste, such as its possible hazardous characteristics, its potential for contamination and its physical state; the management aspects, such as the technological capacity to recover the waste, and the environmental benefits arising from the recovery operation, including whether the environmentally sound management of the waste may be impaired. The Commission should progress towards the completion of this Annex as far as possible before the date of entry into force of this Regulation and complete this task at the latest six months after that date.

(40) Additional measures related to the implementation of this Regulation should also be adopted by the Commission in accordance with the procedure referred to in Article 18(3) of Directive 2006/12/EC. These measures should include a method for calculating the financial guarantee or equivalent insurance to be completed by the Commission, if possible, before the date of application of this Regulation.

(41) The measures necessary for the implementation of this Regulation should be adopted in accordance with Council Decision 1999/468/EC of 28 June 1999 laying down the procedures for the exercise of implementing powers conferred on the Commission[15].

(42) Since the objective of this Regulation, namely to ensure protection of the environment when waste is subject to shipment, cannot be sufficiently achieved by the Member States and can therefore, by reason of the scale and effects thereof, be better achieved at Community level, the Community may adopt measures in accordance with the principle of subsidiarity as set out in Article 5 of the Treaty. In accordance with the principle of proportionality, as set out in that Article, this Regulation does not go beyond what is necessary in order to achieve that objective,

HAVE ADOPTED THIS REGULATION:

TITLE I

SCOPE AND DEFINITIONS

Article 1

Scope

1.   This Regulation establishes procedures and control regimes for the shipment of waste, depending on the origin, destination and route of the shipment, the type of waste shipped and the type of treatment to be applied to the waste at its destination.

2.   This Regulation shall apply to shipments of waste:

(a)   between Member States, within the Community or with transit through third

---

[15]   OJ L 184, 17.7.1999, p. 23.

countries;

(b)   imported into the Community from third countries;

(c)   exported from the Community to third countries;

(d)   in transit through the Community, on the way from and to third countries.

3.   The following shall be excluded from the scope of this Regulation:

(a)   the offloading to shore of waste, including waste water and residues, generated by the normal operation of ships and offshore platforms, provided that such waste is subject to the requirements of the International Convention for the Prevention of Pollution from Ships, 1973, as modified by the Protocol of 1978 relating thereto (Marpol 73/78), or other binding international instruments;

(b)   waste generated on board vehicles, trains, aeroplanes and ships, until such waste is offloaded in order to be recovered or disposed of;

(c)   shipments of radioactive waste as defined in Article 2 of Council Directive 92/3/Euratom of 3 February 1992 on the supervision and control of shipments of radioactive waste between Member States and into and out of the Community[16];

(d)   shipments which are subject to the approval requirements of Regulation (EC) No 1774/2002;

(e)   shipments of the waste referred to in point 1(b)(ii), (iv) and (v) of Article 2 of Directive 2006/12/EC, where such shipments are already covered by other Community legislation containing similar provisions;

(f)   shipments of waste from the Antarctic into the Community which are in accordance with the requirements of the Protocol on Environmental Protection to the Antarctic Treaty (1991);

(g)   imports into the Community of waste generated by armed forces or relief organisations in situations of crisis, peacemaking or peacekeeping operations where such waste is shipped, by the armed forces or relief organisations concerned or on their behalf, directly or indirectly to the country of destination. In such cases, any competent authority of transit and the competent authority of destination in the Community shall be informed in advance concerning the shipment and its destination;

(h)   shipments of CO2 for the purposes of geological storage in accordance with Directive 2009/31/EC of the European Parliament and of the Council of 23 April 2009 on the geological storage of carbon dioxide[17];

(i)   ships flying the flag of a Member State falling under the scope of Regulation

---

[16]   OJ L 35, 12.2.1992, p. 24.
[17]   OJ L 140, 5.6.2009, p. 114.

(EU) No 1257/2013 of the European Parliament and of the Council[18].

4.   Shipments of waste from the Antarctic to countries outside the Community, which transit through the Community, shall be subject to Articles 36 and 49.

5.   Shipments of waste exclusively within a Member State shall be subject only to Article 33.

Article 2

Definitions

For the purposes of this Regulation:

1.   'waste' is as defined in Article 1(1)(a) of Directive 2006/12/EC;

2.   'hazardous waste' is as defined in Article 1(4) of Council Directive 91/689/EEC of 12 December 1991 on hazardous waste[19];

3.   'mixture of wastes' means waste that results from an intentional or unintentional mixing of two or more different wastes and for which mixture no single entry exists in Annexes III, IIIB, IV and IVA. Waste shipped in a single shipment of wastes, consisting of two or more wastes, where each waste is separated, is not a mixture of wastes;

4.   'disposal' is as defined in Article 1(1)(e) of Directive 2006/12/EC;

5.   'interim disposal' means disposal operations D 13 to D 15 as defined in Annex II A to Directive 2006/12/EC;

6.   'recovery' is as defined in Article 1(1)(f) of Directive 2006/12/EC;

7.   'interim recovery' means recovery operations R 12 and R 13 as defined in Annex II B to Directive 2006/12/EC;

7a.   're-use' is as defined in Article 3(13) of Directive 2008/98/EC of the European Parliament and of the Council[20];

8.   'environmentally sound management' means taking all practicable steps to ensure that waste is managed in a manner that will protect human health and the environment against adverse effects which may result from such waste;

9.   'producer' is anyone whose activities produce waste (original producer) and/or anyone who carries out pre-processing, mixing or other operations resulting in a change in the nature or composition of this waste (new producer) (as defined in Article 1(1)(b) of Directive 2006/12/EC);

10.   'holder' is the producer of the waste or the natural or legal person who is in

---

[18]   Regulation (EU) No 1257/2013 of the European Parliament and of the Council of 20 November 2013 on ship recycling and amending Regulation (EC) No 1013/2006 and Directive 2009/16/EC (OJ L 330, 10.12.2013, p. 1.).

[19]   OJ L 377, 31.12.1991, p. 20. Directive as amended by Directive 94/31/EC (OJ L 168, 2.7.1994, p. 28).

[20]   Directive 2008/98/EC of the European Parliament and of the Council of 19 November 2008 on waste and repealing certain Directives (OJ L 312, 22.11.2008, p. 3).

possession of it (and as defined in Article 1(1)(c) of Directive 2006/12/EC);

11. 'collector' is anyone carrying out waste collection as defined in Article 1(1)(g) of Directive 2006/12/EC;

12. 'dealer' is anyone who acts in the role of principal to purchase and subsequently sell waste, including such dealers who do not take physical possession of the waste, and as referred to in Article 12 of Directive 2006/12/EC;

13. 'broker' is anyone arranging the recovery or disposal of waste on behalf of others, including such brokers who do not take physical possession of the waste, as referred to in Article 12 of Directive 2006/12/EC;

14. 'consignee' means the person or undertaking under the jurisdiction of the country of destination to whom or to which the waste is shipped for recovery or disposal;

15. 'notifier' means:

(a) in the case of a shipment originating from a Member State, any natural or legal person under the jurisdiction of that Member State who intends to carry out a shipment of waste or intends to have a shipment of waste carried out and to whom the duty to notify is assigned. The notifier is one of the persons or bodies listed below, selected in accordance with the ranking established in this listing:

(i) the original producer, or

(ii) the licensed new producer who carries out operations prior to shipment, or

(iii) a licensed collector who, from various small quantities of the same type of waste collected from a variety of sources, has assembled the shipment which is to start from a single notified location, or

(iv) a registered dealer who has been authorised in writing by the original producer, new producer or licensed collector specified in (i), (ii) and (iii) to act on his/her behalf as notifier,

(v) a registered broker who has been authorised in writing by the original producer, new producer or licensed collector specified in (i), (ii) and (iii) to act on his/her behalf as notifier,

(vi) where all of the persons specified in (i), (ii), (iii), (iv) and (v) if applicable, are unknown or insolvent, the holder.

Should a notifier specified in (iv) or (v) fail to fulfil any of the take-back obligations set out in Articles 22 to 25, the original producer, new producer or licensed collector specified in (i), (ii) or (iii) respectively who authorised that dealer or broker to act on his/her behalf shall be deemed to be the notifier for the purposes of the said take-back obligations. In circumstances of illegal shipment notified by a dealer or broker specified in (iv) or (v), the person specified in (i), (ii) or (iii) who authorised that dealer or broker to act on his/her behalf shall be deemed to be the notifier for the purposes of this Regulation;

(b) in the case of import into, or transit through, the Community of waste that does not originate in a Member State, any of the following natural or legal persons under the jurisdiction of the country of dispatch who intends to carry out a shipment of waste or intends to have, or who has had, a shipment of waste carried out, being either:

    (i) the person designated by the law of the country of dispatch; or, in the absence of any such designation,

    (ii) the holder at the time the export took place;

16.   'Basel Convention' means the Basel Convention of 22 March 1989 on the control of transboundary movements of hazardous wastes and their disposal;

17.   'OECD Decision' means Decision C(2001)107/Final of the OECD Council concerning the revision of Decision C(92)39/Final on control of transboundary movements of wastes destined for recovery operations;

18.   'competent authority' means:

(a) in the case of Member States, the body designated by the Member State concerned in accordance with Article 53; or

(b) in the case of a non-Member State that is a Party to the Basel Convention, the body designated by that country as the competent authority for the purposes of that Convention in accordance with Article 5 thereof; or

(c) in the case of any country not referred to in either (a) or (b), the body that has been designated as the competent authority by the country or region concerned or, in the absence of such designation, the regulatory authority for the country or region, as appropriate, which has jurisdiction over shipments of waste for recovery or disposal or transit, as the case may be;

19.   'competent authority of dispatch' means the competent authority for the area from which the shipment is planned to be initiated or is initiated;

20.   'competent authority of destination' means the competent authority for the area to which the shipment is planned or takes place, or in which waste is loaded prior to recovery or disposal in an area not under the national jurisdiction of any country;

21.   'competent authority of transit' means the competent authority for any country, other than that of the competent authority of dispatch or destination, through which the shipment is planned or takes place;

22.   'country of dispatch' means any country from which a shipment of waste is planned to be initiated or is initiated;

23.   'country of destination' means any country to which a shipment of waste is planned or takes place for recovery or disposal therein, or for the purpose of loading prior to recovery or disposal in an area not under the national jurisdiction of any country;

24. 'country of transit' means any country, other than the country of dispatch or destination, through which a shipment of waste is planned or takes place;

25. 'area under the national jurisdiction of a country' means any land or marine area within which a state exercises administrative and regulatory responsibility in accordance with international law as regards the protection of human health or the environment;

26. 'overseas countries and territories' means the overseas countries and territories as listed in Annex IA to Decision 2001/822/EC;

27. 'customs office of export from the Community' is the customs office as defined in Article 161(5) of Council Regulation (EEC) No 2913/92 of 12 October 1992 establishing the Community Customs Code[21];

28. customs office of exit from the Community' is the customs office as defined in Article 793(2) of Commission Regulation (EEC) No 2454/93 of 2 July 1993 laying down provisions for the implementation of Council Regulation (EEC) No 2913/92 establishing the Community Customs Code[22];

29. 'customs office of entry into the Community' is the customs office where waste brought into the customs territory of the Community shall be conveyed to in accordance with Article 38(1) of Regulation (EEC) No 2913/92;

30. 'import' means any entry of waste into the Community but excluding transit through the Community;

31. 'export' means the action of waste leaving the Community but excluding transit through the Community;

32. 'transit' means a shipment of waste or a planned shipment of waste through one or more countries other than the country of dispatch or destination;

33. 'transport' means the carriage of waste by road, rail, air, sea or inland waterways;

34. 'shipment' means the transport of waste destined for recovery or disposal which is planned or takes place:

(a) between a country and another country; or

(b) between a country and overseas countries and territories or other areas, under that country's protection; or

(c) between a country and any land area which is not part of any country under international law; or

(d) between a country and the Antarctic; or

(e) from one country through any of the areas referred to above; or

---

[21] OJ L 302, 19.10.1992, p. 1. Regulation as last amended by Regulation (EC) No 648/2005 of the European Parliament and of the Council (OJ L 117, 4.5.2005, p. 13).

[22] OJ L 253, 11.10.1993, p. 1. Regulation as last amended by Regulation (EC) No 215/2006 (OJ L 38, 9.2.2006, p. 11).

(f) within a country through any of the areas referred to above and which origi-nates in and ends in the same country; or

(g) from a geographic area not under the jurisdiction of any country, to a coun-try;

35. 'illegal shipment' means any shipment of waste effected:

(a) without notification to all competent authorities concerned pursuant to this Regulation; or

(b) without the consent of the competent authorities concerned pursuant to this Regulation; or

(c) with consent obtained from the competent authorities concerned through falsification, misrepresentation or fraud; or

(d) in a way which is not specified materially in the notification or movement documents; or

(e) in a way which results in recovery or disposal in contravention of Commu-nity or international rules; or

(f) contrary to Articles 34, 36, 39, 40, 41 and 43; or

(g) which, in relation to shipments of waste as referred to in Article 3(2) and (4), has resulted from:

(i)   the waste being discovered not to be listed in Annexes III, IIIA or IIIB, or

(ii)   non-compliance with Article 3(4),

(iii)   the shipment being effected in a way which is not specified materially in the document set out in Annex VII;

35a. 'inspection' means actions undertaken by the authorities involved to ascertain whether an establishment, an undertaking, a broker, a dealer, a shipment of waste or the related recovery or disposal complies with the relevant require-ments set out in this Regulation.

TITLE II

SHIPMENTS WITHIN THE COMMUNITY WITH OR WITHOUT TRANSIT THROUGH THIRD COUNTRIES

Article 3

Overall procedural framework

1.   Shipments of the following wastes shall be subject to the procedure of prior written notification and consent as laid down in the provisions of this Title:

(a)   if destined for disposal operations:

all wastes;

(b)   if destined for recovery operations:

    (i)   wastes listed in Annex IV, which include, inter alia, wastes listed in Annexes II and VIII to the Basel Convention,

    (ii)   wastes listed in Annex IVA,

    (iii)   wastes not classified under one single entry in either Annex III, IIIB, IV or IVA,

    (iv)   mixtures of wastes not classified under one single entry in either Annex III, IIIB, IV or IVA unless listed in Annex IIIA.

2.   Shipments of the following wastes destined for recovery shall be subject to the general information requirements laid down in Article 18, if the amount of waste shipped exceeds 20 kg:

(a)   waste listed in Annex III or IIIB;

(b)   mixtures, not classified under one single entry in Annex III, of two or more wastes listed in Annex III, provided that the composition of these mixtures does not impair their environmentally sound recovery and provided that such mixtures are listed in Annex IIIA, in accordance with Article 58.

3.   For wastes listed in Annex III, in exceptional cases, the relevant provisions shall apply as if they had been listed in Annex IV, if they display any of the hazardous characteristics listed in Annex III to Directive 91/689/EEC. These cases shall be treated in accordance with Article 58.

4.   Shipments of waste explicitly destined for laboratory analysis to assess either its physical or chemical characteristics or to determine its suitability for recovery or disposal operations shall not be subject to the procedure of prior written notification and consent as described in paragraph 1. Instead, the procedural requirements of Article 18 shall apply. The amount of such waste exempted when explicitly destined for laboratory analysis shall be determined by the minimum quantity reasonably needed to adequately perform the analysis in each particular case, and shall not exceed 25 kg.

5.   Shipments of mixed municipal waste (waste entry 20 03 01) collected from private households, including where such collection also covers such waste from other producers, to recovery or disposal facilities shall, in accordance with this Regulation, be subject to the same provisions as shipments of waste destined for disposal.

CHAPTER 1

Prior written notification and consent

Article 4

Notification

Where the notifier intends to ship waste as referred to in Article 3(1)(a) or (b), he/she shall submit a prior written notification to and through the competent authority of dispatch and, if submitting a general notification, comply with Article 13.

When a notification is submitted, the following requirements shall be fulfilled:

1. notification and movement documents:

   Notification shall be effected by means of the following documents:

   (a) the notification document set out in Annex IA; and

   (b) the movement document set out in Annex IB.

   In submitting a notification, the notifier shall fill in the notification document and, where relevant, the movement document.

   When the notifier is not the original producer in accordance with point 15(a)(i) of Article 2, the notifier shall ensure that this producer or one of the persons indicated in point 15(a)(ii) or (iii) of Article 2, where practicable, also signs the notification document set out in Annex IA.

   The notification document and the movement document shall be issued to the notifier by the competent authority of dispatch;

2. information and documentation in the notification and movement documents:

   The notifier shall supply on, or annex to, the notification document information and documentation as listed in Annex II, Part 1. The notifier shall supply on, or annex to, the movement document information and documentation referred to in Annex II, Part 2, to the extent possible at the time of notification.

   A notification shall be considered properly carried out when the competent authority of dispatch is satisfied that the notification document and movement document have been completed in accordance with the first subparagraph;

3. additional information and documentation:

   If requested by any of the competent authorities concerned, the notifier shall supply additional information and documentation. A list of additional information and documentation that may be requested is set out in Annex II, Part 3.

   A notification shall be considered properly completed when the competent authority of destination is satisfied that the notification document and the movement document have been completed and that the information and documentation as listed in Annex II, Parts 1 and 2, as well as any additional information and documentation requested in accordance with this paragraph and as listed in Annex II, Part 3, have been supplied by the notifier;

4. conclusion of a contract between the notifier and the consignee:

   The notifier shall conclude a contract as described in Article 5 with the consignee for the recovery or disposal of the notified waste.

   Evidence of this contract or a declaration certifying its existence in accordance with Annex IA shall be supplied to the competent authorities involved at the time of notification. A copy of the contract or such evidence to the satisfaction of the competent authority concerned shall be provided by the notifier or consignee upon request by the competent authority;

5. establishment of a financial guarantee or equivalent insurance:

    A financial guarantee or equivalent insurance shall be established as described in Article 6. A declaration to this effect shall be made by the notifier through completion of the appropriate part of the notification document set out in Annex IA.

    The financial guarantee or equivalent insurance (or if the competent authority so allows, evidence of that guarantee or insurance or a declaration certifying its existence) shall be supplied as part of the notification document at the time of notification or, if the competent authority so allows, pursuant to national legislation, at such time before the shipment starts;

6. coverage of the notification:

    A notification shall cover the shipment of waste from its initial place of dispatch and including its interim and non-interim recovery or disposal.

    If subsequent interim or non-interim operations take place in a country other than the first country of destination, the non-interim operation and its destination shall be indicated in the notification and Article 15(f) shall apply.

    Only one waste identification code shall be covered for each notification, except for:

    (a) wastes not classified under one single entry in either Annex III, IIIB, IV or IVA. In this case, only one type of waste shall be specified;

    (b) mixtures of wastes not classified under one single entry in either Annex III, IIIB, IV or IVA unless listed in Annex IIIA. In this case, the code for each fraction of the waste shall be specified in order of importance.

Article 5

Contract

1. All shipments of waste for which notification is required shall be subject to the requirement of the conclusion of a contract between the notifier and the consignee for the recovery or disposal of the notified waste.

2. The contract shall be concluded and effective at the time of notification and for the duration of the shipment until a certificate is issued in accordance with Article 15(e), Article 16(e) or, where appropriate, Article 15(d).

3. The contract shall include obligations:

(a) on the notifier to take the waste back if the shipment or the recovery or disposal has not been completed as intended or if it has been effected as an illegal shipment, in accordance with Article 22 and Article 24(2);

(b) on the consignee to recover or dispose of the waste if it has been effected as an illegal shipment, in accordance with Article 24(3); and

(c) on the facility to provide, in accordance with Article 16(e), a certificate that the waste has been recovered or disposed of, in accordance with the notification and the conditions specified therein and the requirements of this Regulation.

4.   If the waste shipped is destined for interim recovery or disposal operations, the contract shall include the following additional obligations:

(a)   the obligation on the facility of destination to provide, in accordance with Article 15(d) and, where appropriate, Article 15(e), the certificates that the waste has been recovered or disposed of in accordance with the notification and the conditions specified therein and the requirements of this Regulation; and

(b)   the obligation on the consignee to submit, where applicable, a notification to the initial competent authority of the initial country of dispatch in accordance with Article 15(f)(ii).

5.   If the waste is shipped between two establishments under the control of the same legal entity, the contract may be replaced by a declaration by the entity in question undertaking to recover or dispose of the notified waste.

Article 6

Financial guarantee

1.   All shipments of waste for which notification is required shall be subject to the requirement of a financial guarantee or equivalent insurance covering:

(a)   costs of transport;

(b)   costs of recovery or disposal, including any necessary interim operation; and

(c)   costs of storage for 90 days.

2.   The financial guarantee or equivalent insurance is intended to cover costs arising in the context of:

(a)   cases where a shipment or the recovery or disposal cannot be completed as intended, as referred to in Article 22; and

(b)   cases where a shipment or the recovery or disposal is illegal as referred to in Article 24.

3.   The financial guarantee or equivalent insurance shall be established by the notifier or by another natural or legal person on its behalf and shall be effective at the time of the notification or, if the competent authority which approves the financial guarantee or equivalent insurance so allows, at the latest when the shipment starts, and shall apply to the notified shipment at the latest when the shipment starts.

4.   The competent authority of dispatch shall approve the financial guarantee or equivalent insurance, including the form, wording and amount of the cover.

However, in cases of import into the Community, the competent authority of destination in the Community shall review the amount of cover and, if necessary, approve an additional financial guarantee or equivalent insurance.

5.   The financial guarantee or equivalent insurance shall be valid for and cover a

notified shipment and completion of recovery or disposal of the notified waste.

The financial guarantee or equivalent insurance shall be released when the competent authority concerned has received the certificate referred to in Article 16(e) or, where appropriate, in Article 15(e) as regards interim recovery or disposal operations.

6.   By way of derogation from paragraph 5, if the waste shipped is destined for interim recovery or disposal operations and a further recovery or disposal operation takes place in the country of destination, the financial guarantee or equivalent insurance may be released when the waste leaves the interim facility and the competent authority concerned has received the certificate referred to in Article 15(d). In this case, any further shipment to a recovery or disposal facility shall be covered by a new financial guarantee or equivalent insurance unless the competent authority of destination is satisfied that such a financial guarantee or equivalent insurance is not required. In these circumstances, the competent authority of destination shall be responsible for obligations arising in the case of an illegal shipment or for take-back where the shipment or the further recovery or disposal operation cannot be completed as intended.

7.   The competent authority within the Community which has approved the financial guarantee or equivalent insurance shall have access thereto and shall make use of the funding, including for the purpose of payments to other authorities concerned, in order to meet the obligations arising in accordance with Articles 23 and 25.

8.   In the case of a general notification pursuant to Article 13, a financial guarantee or equivalent insurance covering parts of the general notification may be established, instead of one covering the entire general notification. In such cases, the financial guarantee or equivalent insurance shall apply to the shipment at the latest when the notified shipment it covers starts.

The financial guarantee or equivalent insurance shall be released when the competent authority concerned has received the certificate referred to in Article 16(e) or, where appropriate, in Article 15(e) as regards interim recovery or disposal operations for the relevant waste. Paragraph 6 shall apply mutatis mutandis.

9.   Member States shall inform the Commission of provisions of national law adopted pursuant to this Article.

Article 7

Transmission of the notification by the competent authority of dispatch

1.   Once the notification has been properly carried out, as described in the second subparagraph, point 2 of Article 4, the competent authority of dispatch shall retain a copy of the notification and transmit the notification to the competent authority of destination with copies to any competent authority(ies) of transit, and shall inform the notifier of the transmission. This shall be done within three working days of receipt of the notification.

2.   If the notification is not properly carried out, the competent authority of dispatch

shall request information and documentation from the notifier in accordance with the second subparagraph, point 2 of Article 4.

This shall be done within three working days of receipt of the notification.

In such cases the competent authority of dispatch shall have three working days following the receipt of the information and/or documentation requested in which to comply with paragraph 1.

3.   Once the notification has been properly carried out, as described in the second subparagraph, point 2 of Article 4, the competent authority of dispatch may decide, within three working days, not to proceed with the notification, if it has objections to the shipment in accordance with Articles 11 and 12.

It shall immediately inform the notifier of its decision and of these objections.

4.   If, within 30 days of receipt of the notification, the competent authority of dispatch has not transmitted the notification as required under paragraph 1, it shall provide the notifier with a reasoned explanation upon his/her request. This shall not apply when the request for information, referred to in paragraph 2, has not been complied with.

Article 8

Requests for information and documentation by the competent authorities concerned and acknowledgement by the competent authority of destination

1.   Following the transmission of the notification by the competent authority of dispatch, if any of the competent authorities concerned considers that additional information and documentation is required as referred to in the second subparagraph, point 3 of Article 4, it shall request such information and documentation from the notifier and inform the other competent authorities of such request. This shall be done within three working days of receipt of the notification. In such cases the competent authorities concerned shall have three working days following the receipt of the information and documentation requested in which to inform the competent authority of destination.

2.   When the competent authority of destination considers that the notification has been properly completed, as described in the second subparagraph, point 3 of Article 4, it shall send an acknowledgement to the notifier and copies to the other competent authorities concerned. This shall be done within three working days of receipt of the properly completed notification.

3.   If, within 30 days of receipt of the notification, the competent authority of destination has not acknowledged the notification as required under paragraph 2, it shall provide the notifier, upon his/her request, with a reasoned explanation.

Article 9

Consents by the competent authorities of destination, dispatch and transit and time periods for transport, recovery or disposal

1.   The competent authorities of destination, dispatch and transit shall have 30 days following the date of transmission of the acknowledgement by the competent authority of destination in accordance with Article 8 in which to take one of the following duly reasoned decisions in writing as regards the notified shipment:

(a)   consent without conditions;

(b)   consent with conditions in accordance with Article 10; or

(c)   objections in accordance with Articles 11 and 12.

Tacit consent by the competent authority of transit may be assumed if no objection is lodged within the said 30-day time limit.

2.   The competent authorities of destination, dispatch and, where appropriate, transit shall transmit their decision and the reasons therefor to the notifier in writing within the 30-day time limit referred to in paragraph 1, with copies to the other competent authorities concerned.

3.   The competent authorities of destination, dispatch and, where appropriate, transit shall signify their written consent by appropriately stamping, signing and dating the notification document or their copies thereof.

4.   A written consent to a planned shipment shall expire one calendar year after it is issued or on such later date as is indicated in the notification document. However, this shall not apply if a shorter period is indicated by the competent authorities concerned.

5.   Tacit consent to a planned shipment shall expire one calendar year after the expiry of the 30-day time limit referred to in paragraph 1.

6.   The planned shipment may take place only after fulfilment of the requirements of Article 16(a) and (b) and during the period of validity of the tacit or written consents of all competent authorities.

7.   The recovery or disposal of waste in relation to a planned shipment shall be completed no later than one calendar year from the receipt of the waste by the facility, unless a shorter period is indicated by the competent authorities concerned.

8.   The competent authorities concerned shall withdraw their consent when they have knowledge that:

(a)   the composition of the waste is not as notified; or

(b)   the conditions imposed on the shipment are not respected; or

(c)   the waste is not recovered or disposed of in compliance with the permit of the facility that performs the said operation; or

(d)   the waste is to be, or has been, shipped, recovered or disposed of in a way that is not in accordance with the information supplied on, or annexed to, the notification and movement documents.

9.　　Any withdrawal of consent shall be transmitted by means of official notice to the notifier with copies to the other competent authorities concerned and to the consignee.

Article 10

Conditions for a shipment

1.　　The competent authorities of dispatch, destination and transit may, within 30 days following the date of transmission of the acknowledgement of the competent authority of destination in accordance with Article 8, lay down conditions in connection with their consent to a notified shipment. Such conditions may be based on one or more of the reasons specified in either Article 11 or Article 12.

2.　　The competent authorities of dispatch, destination and transit may also, within the 30-day time limit referred to in paragraph 1, lay down conditions in respect of the transport of waste within their jurisdiction. Such transport conditions shall not be more stringent than those laid down in respect of similar shipments occurring wholly within their jurisdiction and shall take due account of existing agreements, in particular relevant international agreements.

3.　　The competent authorities of dispatch, destination and transit may also, within the 30-day time limit referred to in paragraph 1, lay down a condition that their consent is to be considered withdrawn if the financial guarantee or equivalent insurance is not applicable at the latest when the notified shipment starts, as required by Article 6(3).

4.　　Conditions shall be transmitted to the notifier in writing by the competent authority that lays them down, with copies to the competent authorities concerned.

Conditions shall be supplied on, or annexed to, the notification document by the relevant competent authority.

5.　　The competent authority of destination may also, within the 30-day time limit referred to in paragraph 1, lay down a condition that the facility which receives the waste shall keep a regular record of inputs, outputs and/or balances for wastes and the related recovery or disposal operations as contained in the notification, and for the period of validity of the notification. Such records shall be signed by a person legally responsible for the facility and be sent to the competent authority of destination within one month of completion of the notified recovery or disposal operation.

Article 11

Objections to shipments of waste destined for disposal

1.　　Where a notification is submitted regarding a planned shipment of waste destined for disposal, the competent authorities of destination and dispatch may, within

30 days following the date of transmission of the acknowledgement of the competent authority of destination in accordance with Article 8, raise reasoned objections based on one or more of the following grounds and in accordance with the Treaty:

(a)   that the planned shipment or disposal would not be in accordance with measures taken to implement the principles of proximity, priority for recovery and self-sufficiency at Community and national levels in accordance with Directive 2006/12/EC, to prohibit generally or partially or to object systematically to shipments of waste; or

(b)   that the planned shipment or disposal would not be in accordance with national legislation relating to environmental protection, public order, public safety or health protection concerning actions taking place in the objecting country; or

(c)   that the notifier or the consignee has previously been convicted of illegal shipment or some other illegal act in relation to environmental protection. In this case, the competent authorities of dispatch and destination may refuse all shipments involving the person in question in accordance with national legislation; or

(d)   that the notifier or the facility has repeatedly failed to comply with Articles 15 and 16 in connection with past shipments; or

(e)   that the Member State wishes to exercise its right pursuant to Article 4(1) of the Basel Convention to prohibit the import of hazardous waste or of waste listed in Annex II to that Convention; or

(f)   that the planned shipment or disposal conflicts with obligations resulting from international conventions concluded by the Member State(s) concerned or the Community; or

(g)   that the planned shipment or disposal is not in accordance with Directive 2006/12/EC, in particular Articles 5 and 7 thereof, while taking into account geographical circumstances or the need for specialised installations for certain types of waste:

    (i) in order to implement the principle of self-sufficiency at Community and national levels, or

    (ii) in cases where the specialised installation has to dispose of waste from a nearer source and the competent authority has given priority to this waste, or

    (iii) in order to ensure that shipments are in accordance with waste management plans, or

(h)   that the waste will be treated in a facility which is covered by Directive 96/61/EC, but which does not apply best available techniques as defined in Article 9(4) of that Directive in compliance with the permit of the facility; or

(i)   that the waste is mixed municipal waste collected from private households (waste entry 20 03 01); or

(j)   that the waste concerned will not be treated in accordance with legally binding environmental protection standards in relation to disposal operations established in Community legislation (also in cases where temporary derogations are granted).

2.   The competent authority(ies) of transit may, within the 30-day time limit referred to in paragraph 1, raise reasoned objections based only on paragraph 1(b), (c), (d) and (f).

3.   In the case of hazardous waste produced in a Member State of dispatch in such a small quantity overall per year that the provision of new specialised disposal installations within that Member State would be uneconomic, paragraph 1(a) shall not apply.

The competent authority of destination shall cooperate with the competent authority of dispatch which considers that this paragraph and not paragraph 1(a) should apply, with a view to resolving the issue bilaterally.

If there is no satisfactory solution, either Member State may refer the matter to the Commission. The issue shall then be determined in accordance with the regulatory procedure referred to in Article 59a (2).

4.   If, within the 30-day time limit referred to in paragraph 1, the competent authorities consider that the problems which gave rise to their objections have been resolved, they shall immediately inform the notifier in writing, with copies to the consignee and to the other competent authorities concerned.

5.   If the problems giving rise to the objections have not been resolved within the 30-day time limit referred to in paragraph 1, the notification shall cease to be valid. In cases where the notifier still intends to carry out the shipment, a new notification shall be submitted, unless all the competent authorities concerned and the notifier agree otherwise.

6.   Measures taken by Member States in accordance with paragraph 1(a), to prohibit generally or partially or to object systematically to shipments of waste destined for disposal, or in accordance with paragraph 1(e), shall immediately be notified to the Commission which shall inform the other Member States.

Article 12

Objections to shipments of waste destined for recovery

1.   Where a notification is submitted regarding a planned shipment of waste destined for recovery, the competent authorities of destination and dispatch may, within 30 days following the date of transmission of the acknowledgement of the competent authority of destination in accordance with Article 8, raise reasoned objections based on one or more of the following grounds and in accordance with the Treaty:

(a)   that the planned shipment or recovery would not be in accordance with Directive 2006/12/EC, in particular Articles 3, 4, 7 and 10 thereof; or

(b)   that the planned shipment or recovery would not be in accordance with national legislation relating to environmental protection, public order, public safety or health protection concerning actions taking place in the objecting country; or

(c) that the planned shipment or recovery would not be in accordance with national legislation in the country of dispatch relating to the recovery of waste, including where the planned shipment would concern waste destined for recovery in a facility which has lower treatment standards for the particular waste than those of the country of dispatch, respecting the need to ensure the proper functioning of the internal market;

This shall not apply if:

(i) there is corresponding Community legislation, in particular related to waste, and if requirements that are at least as stringent as those laid down in the Community legislation have been introduced in national legislation transposing such Community legislation,

(ii) the recovery operation in the country of destination takes place under conditions that are broadly equivalent to those prescribed in the national legislation of the country of dispatch,

(iii) the national legislation in the country of dispatch, other than that covered by (i), has not been notified in accordance with Directive 98/34/EC of the European Parliament and of the Council of 22 June 1998 laying down a procedure for the provision of information in the field of technical standards and regulations and of rules on Information Society services[23], where required by that Directive, or

(d) that the notifier or the consignee has previously been convicted of illegal shipment or some other illegal act in relation to environmental protection. In this case, the competent authorities of dispatch and destination may refuse all shipments involving the person in question in accordance with national legislation; or

(e) that the notifier or the facility has repeatedly failed to comply with Articles 15 and 16 in connection with past shipments; or

(f) that the planned shipment or recovery conflicts with obligations resulting from international conventions concluded by the Member State(s) concerned or the Community; or

(g) that the ratio of the recoverable and non-recoverable waste, the estimated value of the materials to be finally recovered or the cost of the recovery and the cost of the disposal of the non-recoverable fraction do not justify the recovery, having regard to economic and/or environmental considerations; or

(h) that the waste shipped is destined for disposal and not for recovery; or

(i) that the waste will be treated in a facility which is covered by Directive 96/61/EC, but which does not apply best available techniques as defined in Article 9(4) of that Directive in compliance with the permit of the facility; or

(j) that the waste concerned will not be treated in accordance with legally binding environmental protection standards in relation to recovery operations, or legally

---

[23] OJ L 204, 21.7.1998, p. 37. Directive as last amended by the 2003 Act of Accession.

binding recovery or recycling obligations established in Community legislation (also in cases where temporary derogations are granted); or

(k)   that the waste concerned will not be treated in accordance with waste management plans drawn up pursuant to Article 7 of Directive 2006/12/EC with the purpose of ensuring the implementation of legally binding recovery or recycling obligations established in Community legislation.

2.   The competent authority(ies) of transit may, within the 30-day time limit referred to in paragraph 1, raise reasoned objections to the planned shipment based only on paragraph 1(b), (d), (e) and (f).

3.   If, within the 30-day time limit referred to in paragraph 1, the competent authorities consider that the problems which gave rise to their objections have been resolved, they shall immediately inform the notifier in writing, with copies to the consignee and to the other competent authorities concerned.

4.   If the problems giving rise to the objections are not resolved within the 30-day time limit referred to in paragraph 1, the notification shall cease to be valid. In cases where the notifier still intends to carry out the shipment, a new notification shall be submitted, unless all the competent authorities concerned and the notifier agree otherwise.

5.   Objections raised by competent authorities in accordance with paragraph 1(c) shall be reported by Member States to the Commission in accordance with Article 51.

6.   The Member State of dispatch shall inform the Commission and the other Member States of the national legislation on which objections raised by competent authorities in accordance with paragraph 1(c) may be based, and shall state to which waste and waste recovery operations those objections apply, before such legislation is invoked in order to raise reasoned objections.

Article 13

General notification

1.   The notifier may submit a general notification to cover several shipments if, in the case of each shipment:

(a)   the waste has essentially similar physical and chemical characteristics; and

(b)   the waste is shipped to the same consignee and the same facility; and

(c)   the route of the shipment as indicated in the notification document is the same.

2.   If, owing to unforeseen circumstances, the same route cannot be followed, the notifier shall inform the competent authorities concerned as soon as possible and, if possible, before the shipment starts if the need for modification is already known.

Where the route modification is known before the shipment starts and involves competent authorities other than those concerned by the general notification, the general notification may not be used and a new notification shall be submitted.

3.   The competent authorities concerned may make their agreement to the use of a general notification subject to the subsequent provision of additional information and documentation, in accordance with the second subparagraph, points 2 and 3 of Article 4.

Article 14

Pre-consented recovery facilities

1.   The competent authorities of destination which have jurisdiction over specific recovery facilities may decide to issue pre-consents to such facilities.

Such decisions shall be limited to a specific period and may be revoked at any time.

2.   In the case of a general notification submitted in accordance with Article 13, the period of validity of the consent referred to in Article 9(4) and (5) may be extended to up to three years by the competent authority of destination in agreement with the other competent authorities concerned.

3.   Competent authorities which decide to issue a pre-consent to a facility in accordance with paragraphs 1 and 2 shall inform the Commission and, where appropriate, the OECD Secretariat of:

(a)   the name, registration number and address of the recovery facility;

(b)   the description of technologies employed, including R-code(s);

(c)   the wastes as listed in Annexes IV and IVA or the wastes to which the decision applies;

(d)   the total pre-consented quantity;

(e)   the period of validity;

(f)   any change in the pre-consent;

(g)   any change in the information notified; and

(h)   any revocation of the pre-consent.

For this purpose the form set out in Annex VI shall be used.

4.   By way of derogation from Articles 9, 10 and 12, the consent given in accordance with Article 9, conditions imposed in accordance with Article 10 or objections raised in accordance with Article 12 by the competent authorities concerned shall be subject to a time limit of seven working days following the date of transmission of the acknowledgement of the competent authority of destination in accordance with Article 8.

5.   Notwithstanding paragraph 4, the competent authority of dispatch may decide that more time is needed in order to receive further information or documentation from the notifier.

In such cases, the competent authority shall, within seven working days, inform the notifier in writing with copies to the other competent authorities concerned.

The total time needed shall not exceed 30 days following the date of transmission of the acknowledgement of the competent authority of destination in accordance with Article 8.

Article 15

Additional provisions regarding interim recovery and disposal operations

Shipments of waste destined for interim recovery or disposal operations shall be subject to the following additional provisions:

(a) Where a shipment of waste is destined for an interim recovery or disposal operation, all the facilities where subsequent interim as well as non-interim recovery and disposal operations are envisaged shall also be indicated in the notification document in addition to the initial interim recovery or disposal operation.

(b) The competent authorities of dispatch and destination may give their consent to a shipment of waste destined for an interim recovery or disposal operation only if there are no grounds for objection, in accordance with Articles 11 or 12, to the shipment(s) of waste to the facilities performing any subsequent interim or non-interim recovery or disposal operations.

(c) Within three days of the receipt of the waste by the facility which carries out this interim recovery or disposal operation, that facility shall provide confirmation in writing that the waste has been received.

This confirmation shall be supplied on, or annexed to, the movement document. The said facility shall send signed copies of the movement document containing this confirmation to the notifier and to the competent authorities concerned.

(d) As soon as possible, but no later than 30 days after completion of the interim recovery or disposal operation, and no later than one calendar year, or a shorter period in accordance with Article 9(7), following the receipt of the waste, the facility carrying out this operation shall, under its responsibility, certify that the interim recovery or disposal has been completed.

This certificate shall be contained in, or annexed to, the movement document.

The said facility shall send signed copies of the movement document containing this certificate to the notifier and to the competent authorities concerned.

(e) When a recovery or disposal facility which carries out an interim recovery or disposal operation delivers the waste for any subsequent interim or non-interim recovery or disposal operation to a facility located in the country of destination, it shall obtain as soon as possible but no later than one calendar year following delivery of the waste, or a shorter period in accordance with Article 9(7), a certificate from that facility that the subsequent non-interim recovery or disposal operation has been completed.

The said facility that carries out an interim recovery or disposal operation shall promptly transmit the relevant certificate(s) to the notifier and the competent authorities concerned, identifying the shipment(s) to which the certificate(s) pertain.

(f) When a delivery as described in subparagraph (e) is made to a facility respectively located:

   (i) in the initial country of dispatch or in another Member State, a new notification shall be required in accordance with the provisions of this Title, or

   (ii) in a third country, a new notification shall be required in accordance with the provisions of this Regulation, with the addition that the provisions concerning the competent authorities concerned shall also apply to the initial competent authority of the initial country of dispatch.

Article 16

Requirements following consent to a shipment

After consent has been given to a notified shipment by the competent authorities involved, all undertakings involved shall complete the movement document, or, in the case of a general notification, the movement documents at the points indicated, sign it or them and retain a copy or copies. The following requirements shall be fulfilled:

(a) Completion of the movement document by the notifier: once the notifier has received consent from the competent authorities of dispatch, destination and transit or, in relation to the competent authority of transit, can assume tacit consent, he/she shall insert the actual date of shipment and otherwise complete the movement document to the extent possible.

(b) Prior information regarding actual start of shipment: the notifier shall send signed copies of the then completed movement document, as described in point (a), to the competent authorities concerned and to the consignee at least three working days before the shipment starts.

(c) Documents to accompany each transport: the notifier shall retain a copy of the movement document. The movement document and copies of the notification document containing the written consents and the conditions of the competent authorities concerned shall accompany each transport. The movement document shall be retained by the facility which receives the waste.

(d) Written confirmation of receipt of the waste by the facility: within three days of receipt of the waste, the facility shall provide confirmation in writing that the waste has been received.

This confirmation shall be contained in, or annexed to, the movement document.

The facility shall send signed copies of the movement document containing this confirmation to the notifier and to the competent authorities concerned.

(e) Certificate for non-interim recovery or disposal by the facility: as soon as possible, but no later than 30 days after completion of the non-interim recovery or disposal operation, and no later than one calendar year, or a shorter period in accordance with Article 9(7), following receipt of the waste, the facility carrying out the operation shall, under its responsibility, certify that the non-interim recovery or disposal has been completed.

This certificate shall be contained in, or annexed to, the movement document.

The facility shall send signed copies of the movement document containing this certificate to the notifier and to the competent authorities concerned.

Article 17

Changes in the shipment after consent

1.　If any essential change is made to the details and/or conditions of the consented shipment, including changes in the intended quantity, route, routing, date of shipment or carrier, the notifier shall inform the competent authorities concerned and the consignee immediately and, where possible, before the shipment starts.

2.　In such cases a new notification shall be submitted, unless all the competent authorities concerned consider that the proposed changes do not require a new notification.

3.　Where such changes involve competent authorities other than those concerned in the original notification, a new notification shall be submitted.

CHAPTER 2

General information requirements

Article 18

Waste to be accompanied by certain information

1.　Waste as referred to in Article 3(2) and (4) that is intended to be shipped shall be subject to the following procedural requirements:

(a)　In order to assist the tracking of shipments of such waste, the person under the jurisdiction of the country of dispatch who arranges the shipment shall ensure that the waste is accompanied by the document contained in Annex VII.

(b)　The document contained in Annex VII shall be signed by the person who arranges the shipment before the shipment takes place and shall be signed by the recovery facility or the laboratory and the consignee when the waste in question is received.

2.　The contract referred to in Annex VII between the person who arranges the shipment and the consignee for recovery of the waste shall be effective when the shipment starts and shall include an obligation, where the shipment of waste or its recovery cannot be completed as intended or where it has been effected as an illegal shipment, on the person who arranges the shipment or, where that person is not in a position to complete the shipment of waste or its recovery (for example, is insolvent), on the consignee, to:

(a)　take the waste back or ensure its recovery in an alternative way; and

(b)　provide, if necessary, for its storage in the meantime.

The person who arranges the shipment or the consignee shall provide a copy of the contract upon request by the competent authority concerned.

3. For inspection, enforcement, planning and statistical purposes, Member States may in accordance with national legislation require information as referred to in paragraph 1 on shipments covered by this Article.

4. The information referred to in paragraph 1 shall be treated as confidential where this is required by Community and national legislation.

CHAPTER 3

General requirements

Article 19

Prohibition on mixing waste during shipment

From the start of the shipment to the receipt in a recovery or disposal facility, waste, as specified on the notification document or as referred to in Article 18, shall not be mixed with other waste.

Article 20

Keeping of documents and information

1. All documents sent to or by the competent authorities in relation to a notified shipment shall be kept in the Community for at least three years from the date when the shipment starts, by the competent authorities, the notifier, the consignee and the facility which receives the waste.

2. Information given pursuant to Article 18(1) shall be kept in the Community for at least three years from the date when the shipment starts, by the person who arranges for the shipment, the consignee and the facility which receives the waste.

Article 21

Public access to notifications

The competent authorities of dispatch or destination may make publicly available by appropriate means, such as the Internet, information on notifications of shipments they have consented to, where such information is not confidential under national or Community legislation.

CHAPTER 4

Take-back obligations

Article 22

Take-back when a shipment cannot be completed as intended

1.   Where any of the competent authorities concerned becomes aware that a shipment of waste, including its recovery or disposal, cannot be completed as intended in accordance with the terms of the notification and movement documents and/or contract referred to in the second subparagraph, point 4 of Article 4 and in Article 5, it shall immediately inform the competent authority of dispatch. Where a recovery or disposal facility rejects a shipment received, it shall immediately inform the competent authority of destination.

2.   The competent authority of dispatch shall ensure that, except in cases referred to in paragraph 3, the waste in question is taken back to its area of jurisdiction or elsewhere within the country of dispatch by the notifier as identified in accordance with the ranking established in point 15 of Article 2, or, if impracticable, by that competent authority itself or by a natural or legal person on its behalf.

This shall take place within 90 days, or such other period as may be agreed between the competent authorities concerned, after the competent authority of dispatch becomes aware or has been advised in writing by the competent authorities of destination or transit that the consented shipment of waste or its recovery or disposal cannot be completed and has been informed of the reason(s) therefor. Such advice may result from information submitted to the competent authorities of destination or transit, inter alia, by other competent authorities.

3.   The take-back obligation in paragraph 2 shall not apply if the competent authorities of dispatch, transit and destination involved in disposing of or recovering the waste are satisfied that the waste can be recovered or disposed of in an alternative way in the country of destination or elsewhere by the notifier or, if impracticable, by the competent authority of dispatch or by a natural or legal person on its behalf.

The take-back obligation in paragraph 2 shall not apply if the waste shipped has, in the course of the operation at the facility concerned, been irreversibly mixed with other waste before a competent authority concerned has become aware of the fact that the notified shipment cannot be completed as referred to in paragraph 1. Such mixture shall be recovered or disposed of in an alternative way in accordance with the first subparagraph.

4.   In cases of take-back as referred to in paragraph 2, a new notification shall be submitted, unless the competent authorities concerned agree that a duly reasoned request by the initial competent authority of dispatch is sufficient.

A new notification, where appropriate, shall be submitted by the initial notifier or, if impracticable, by any other natural or legal persons identified in accordance with point 15 of Article 2, or, if impracticable, by the initial competent authority of dispatch or by a natural or legal person on its behalf.

No competent authority shall oppose or object to the return of waste from a shipment that cannot be completed or to the related recovery and disposal operation.

5.   In cases of alternative arrangements outside the initial country of destination as referred to in paragraph 3, a new notification, where appropriate, shall be submitted by the initial notifier or, if impracticable, by any other natural or legal persons identified in accordance with point 15 of Article 2, or, if impracticable, by the initial competent authority of dispatch or by a natural or legal person on its behalf.

When such a new notification is submitted by the notifier, this notification shall also be submitted to the competent authority of the initial country of dispatch.

6.   In cases of alternative arrangements in the initial country of destination as referred to in paragraph 3, a new notification shall not be required and a duly reasoned request shall suffice. Such a duly reasoned request, seeking agreement to the alternative arrangement, shall be transmitted to the competent authority of destination and dispatch by the initial notifier or, if impracticable, to the competent authority of destination by the initial competent authority of dispatch.

7.   If no new notification is to be submitted in accordance with paragraphs 4 or 6, a new movement document shall be completed in accordance with Article 15 or Article 16 by the initial notifier or, if impracticable, by any other natural or legal persons identified in accordance with point 15 of Article 2, or, if impracticable, by the initial competent authority of dispatch or by a natural or legal person on its behalf.

If a new notification is submitted by the initial competent authority of dispatch in accordance with paragraphs 4 or 5, a new financial guarantee or equivalent insurance shall not be required.

8.   The obligation of the notifier and the subsidiary obligation of the country of dispatch to take the waste back or arrange for alternative recovery or disposal shall end when the facility issues the certificate of non-interim recovery or disposal as referred to in Article 16(e) or, where appropriate, in Article 15(e). In the cases of interim recovery or disposal referred to in Article 6(6), the subsidiary obligation of the country of dispatch shall end when the facility issues the certificate referred to in Article 15(d).

If a facility issues a certificate of recovery or disposal in such a way as to result in an illegal shipment, with the consequence that the financial guarantee is released, Article 24(3) and Article 25(2) shall apply.

9.   Where waste from a shipment which cannot be completed, including its recovery or disposal, is discovered within a Member State, the competent authority with jurisdiction over the area where the waste was discovered shall be responsible for ensuring that arrangements are made for the safe storage of the waste pending its return or non-interim recovery or disposal in an alternative way.

Article 23

Costs for take-back when a shipment cannot be completed

1.   Costs arising from the return of waste from a shipment that cannot be completed, including costs of its transport, recovery or disposal pursuant to Article 22(2) or (3) and, from the date on which the competent authority of dispatch becomes aware that a shipment of waste or its recovery or disposal cannot be completed, storage costs pursuant to Article 22(9) shall be charged:

(a)   to the notifier as identified in accordance with the ranking established in point 15 of Article 2; or, if impracticable;

(b)   to other natural or legal persons as appropriate; or, if impracticable;

(c)   to the competent authority of dispatch; or, if impracticable;

(d)   as otherwise agreed between the competent authorities concerned.

2.   This Article shall be without prejudice to Community and national provisions concerning liability.

Article 24

Take-back when a shipment is illegal

1.   Where a competent authority discovers a shipment that it considers to be an illegal shipment, it shall immediately inform the other competent authorities concerned.

2.   If an illegal shipment is the responsibility of the notifier, the competent authority of dispatch shall ensure that the waste in question is:

(a)   taken back by the notifier de facto; or, if no notification has been submitted;

(b)   taken back by the notifier de jure; or, if impracticable;

(c)   taken back by the competent authority of dispatch itself or by a natural or legal person on its behalf; or, if impracticable;

(d)   alternatively recovered or disposed of in the country of destination or dispatch by the competent authority of dispatch itself or by a natural or legal person on its behalf; or, if impracticable;

(e)   alternatively recovered or disposed of in another country by the competent authority of dispatch itself or by a natural or legal person on its behalf if all the competent authorities concerned agree.

This take-back, recovery or disposal shall take place within 30 days, or such other period as may be agreed between the competent authorities concerned after the competent authority of dispatch becomes aware of or has been advised in writing by the competent authorities of destination or transit of the illegal shipment and informed of the reason(s) therefor. Such advice may result from information submitted to the competent authorities of destination or transit, inter alia, by other competent authorities.

In cases of take-back as referred to in (a), (b) and (c), a new notification shall be submitted, unless the competent authorities concerned agree that a duly reasoned request by the initial competent authority of dispatch is sufficient.

The new notification shall be submitted by the person or authority listed in (a), (b) or (c) and in accordance with that order.

No competent authority shall oppose or object to the return of waste of an illegal shipment. In the case of alternative arrangements as referred to in (d) and (e) by the competent authority of dispatch, a new notification shall be submitted by the initial competent authority of dispatch or by a natural or legal person on its behalf unless the competent authorities concerned agree that a duly reasoned request by that authority is sufficient.

3.   If an illegal shipment is the responsibility of the consignee the competent authority of destination shall ensure that the waste in question is recovered or disposed of in an environmentally sound manner:

(a)   by the consignee; or, if impracticable,

(b)   by the competent authority itself or by a natural or legal person on its behalf.

This recovery or disposal shall take place within 30 days, or such other period as may be agreed between the competent authorities concerned after the competent authority of destination becomes aware of or has been advised in writing by the competent authorities of dispatch or transit of the illegal shipment and informed of the reason(s) therefor. Such advice may result from information submitted to the competent authorities of dispatch and transit, inter alia, by other competent authorities.

To this end, the competent authorities concerned shall cooperate, as necessary, in the recovery or disposal of the waste.

4.   If no new notification is to be submitted, a new movement document shall be completed in accordance with Article 15 or 16 by the person responsible for take-back or, if impracticable, by the initial competent authority of dispatch.

If a new notification is submitted by the initial competent authority of dispatch, a new financial guarantee or equivalent insurance shall not be required.

5.   In particular in cases where responsibility for the illegal shipment cannot be imputed to either the notifier or the consignee, the competent authorities concerned shall cooperate to ensure that the waste in question is recovered or disposed of.

6.   In the cases of interim recovery or disposal referred to in Article 6(6) where an illegal shipment is discovered after completion of the interim recovery or disposal operation, the subsidiary obligation of the country of dispatch to take the waste back or arrange for alternative recovery or disposal shall end when the facility has issued the certificate referred to in Article 15(d).

If a facility issues a certificate of recovery or disposal in such a way as to result in an illegal shipment, with the consequence that the financial guarantee is released, paragraph 3 and Article 25(2) shall apply.

7.   Where the waste of an illegal shipment is discovered within a Member State, the competent authority with jurisdiction over the area where the waste was discovered shall be responsible for ensuring that arrangements are made for the safe storage of the waste pending its return or non-interim recovery or disposal in an alternative way.

8.   Articles 34 and 36 shall not apply in cases where illegal shipments are returned to the country of dispatch and that country of dispatch is a country covered by the prohibitions set out in those Articles.

9.   In the case of an illegal shipment as defined in point 35(g) of Article 2, the person who arranges the shipment shall be subject to the same obligations established in this Article as the notifier.

10.   This Article shall be without prejudice to Community and national provisions concerning liability.

Article 25

Costs for take-back when a shipment is illegal

1.   Costs arising from the take-back of waste of an illegal shipment, including costs of its transport, recovery or disposal pursuant to Article 24(2) and, from the date on which the competent authority of dispatch becomes aware that a shipment is illegal, storage costs pursuant to Article 24(7), shall be charged to:

(a)   the notifier de facto, as identified in accordance with the ranking established in point 15 of Article 2; or, if no notification has been submitted;

(b)   the notifier de jure or other natural or legal persons as appropriate; or, if impracticable;

(c)   the competent authority of dispatch.

2.   Costs arising from recovery or disposal pursuant to Article 24(3), including possible transport and storage costs pursuant to Article 24(7), shall be charged to:

(a)   the consignee; or, if impracticable;

(b)   the competent authority of destination.

3.   Costs arising from recovery or disposal pursuant to Article 24(5), including possible transport and storage costs pursuant to Article 24(7), shall be charged to:

(a)   the notifier, as identified in accordance with the ranking established in point 15 of Article 2, and/or the consignee, depending upon the decision by the competent authorities involved; or, if impracticable,

(b)   other natural or legal persons as appropriate; or, if impracticable,

(c)   the competent authorities of dispatch and destination.

4.   In the case of an illegal shipment as defined in point 35(g) of Article 2, the person who arranges the shipment shall be subject to the same obligations established in this Article as the notifier.

5.   This Article shall be without prejudice to Community and national provisions concerning liability.

CHAPTER 5

General administrative provisions

Article 26

Format of the communications

1.   The information and documents listed below may be submitted by post:

(a)   notification of a planned shipment pursuant to Articles 4 and 13;

(b)   request for information and documentation pursuant to Articles 4, 7 and 8;

(c)   submission of information and documentation pursuant to Articles 4, 7 and 8;

(d)   written consent to a notified shipment pursuant to Article 9;

(e)   conditions for a shipment pursuant to Article 10;

(f)   objections to a shipment pursuant to Articles 11 and 12;

(g)   information on decisions to issue pre-consents to specific recovery facilities pursuant to Article 14(3);

(h)   written confirmation of receipt of the waste pursuant to Articles 15 and 16;

(i)   certificate for recovery or disposal of the waste pursuant to Articles 15 and 16;

(j)   prior information regarding actual start of the shipment pursuant to Article 16;

(k)   information on changes in the shipment after consent pursuant to Article 17; and

(l)   written consents and movement documents to be sent pursuant to Titles IV, V and VI.

2.   Subject to the agreement of the competent authorities concerned and the notifier, the documents referred to in paragraph 1 may alternatively be submitted using any of the following methods of communication:

(a)   by fax; or

(b)   by fax followed by post; or

(c)   by e-mail with digital signature. In this case, any stamp or signature required shall be replaced by the digital signature; or

(d)   by e-mail without digital signature followed by post.

3.   The documents to accompany each transport in accordance with Article 16(c) and Article 18 may be in an electronic form with digital signatures if they can be made readable at any time during the transport and if this is acceptable to the competent authorities concerned.

4.   Subject to the agreement of the competent authorities concerned and of the notifier, the information and documents listed in paragraph 1 may be submitted and exchanged by means of electronic data interchange with electronic signature or elec-

tronic authentication in accordance with Directive 1999/93/EC of the European Parliament and of the Council[24], or a comparable electronic authentication system which provides the same level of security.

With a view to facilitating the implementation of the first subparagraph, the Commission shall, where feasible, adopt implementing acts establishing the technical and organisational requirements for the practical implementation of electronic data interchange for the submission of documents and information. The Commission shall take into consideration any relevant international standards, and shall ensure that those requirements are in conformity with Directive 1999/93/EC or provide at least the same level of security as provided for under that Directive. Those implementing acts shall be adopted in accordance with the examination procedure referred to in Article 59a (2).

Article 27

Language

1.   Any notification, information, documentation or other communication submitted pursuant to the provisions of this Title shall be supplied in a language acceptable to the competent authorities concerned.

2.   The notifier shall provide the competent authorities concerned with authorised translation(s) into a language which is acceptable to them, should they so request.

Article 28

Disagreement on classification issues

1.   If the competent authorities of dispatch and of destination cannot agree on the classification as regards the distinction between waste and non-waste, the subject matter shall be treated as if it were waste. This shall be without prejudice to the right of the country of destination to deal with the shipped material in accordance with its national legislation, following arrival of the shipped material and where such legislation is in accordance with Community or international law.

2.   If the competent authorities of dispatch and of destination cannot agree on the classification of the notified waste as being listed in Annex III, IIIA, IIIB or IV, the waste shall be regarded as listed in Annex IV.

3.   If the competent authorities of dispatch and destination cannot agree on the classification of the waste treatment operation notified as being recovery or disposal, the provisions regarding disposal shall apply.

4.   Paragraphs 1 to 3 shall apply only for the purposes of this Regulation, and shall be without prejudice to rights of interested parties to resolve any dispute related to these questions before a court of law or tribunal.

---

[24]   Directive 1999/93/EC of the European Parliament and of the Council of 13 December 1999 on a Community framework for electronic signatures (OJ L 13, 19.1.2000, p. 12).

Article 29

Administrative costs

Appropriate and proportionate administrative costs of implementing the notification and supervision procedures and usual costs of appropriate analyses and inspections may be charged to the notifier.

Article 30

Border-area agreements

1.   In exceptional cases, and if the specific geographical or demographical situation warrants such a step, Member States may conclude bilateral agreements making the notification procedure for shipments of specific flows of waste less stringent in respect of cross-border shipments to the nearest suitable facility located in the border area between the two Member States concerned.

2.   Such bilateral agreements may also be concluded where waste is shipped from and treated in the country of dispatch but transits another Member State.

3.   Member States may also conclude such agreements with countries that are Parties to the Agreement on the European Economic Area.

4.   Such agreements shall be notified to the Commission before they take effect.

CHAPTER 6

Shipments within the Community with transit via third countries

Article 31

Shipments of waste destined for disposal

Where a shipment of waste takes place within the Community with transit via one or more third countries, and the waste is destined for disposal, the competent authority of dispatch shall, in addition to the provisions of this Title, ask the competent authority in the third countries whether it wishes to send its written consent to the planned shipment:

(a)   in the case of Parties to the Basel Convention, within 60 days, unless it has waived this right in accordance with the terms of that Convention; or

(b)   in the case of countries not Parties to the Basel Convention, within a period agreed between the competent authorities.

Article 32

Shipments of waste destined for recovery

1.   When a shipment of waste takes place within the Community with transit via one or more third countries to which the OECD Decision does not apply, and the waste is destined for recovery, Article 31 shall apply.

2.   When a shipment of waste takes place within the Community, including shipments between localities in the same Member State, with transit via one or more third countries to which the OECD Decision applies, and the waste is destined for recovery, the consent referred to in Article 9 may be provided tacitly, and if no objection has been lodged or no conditions have been specified, the shipment may start 30 days after the date of transmission of the acknowledgement by the competent authority of destination in accordance with Article 8.

TITLE III

SHIPMENTS EXCLUSIVELY WITHIN MEMBER STATES

Article 33

Application of this Regulation to shipments exclusively within Member States

1.   Member States shall establish an appropriate system for the supervision and control of shipments of waste exclusively within their jurisdiction. This system shall take account of the need for coherence with the Community system established by Titles II and VII.

2.   Member States shall inform the Commission of their system for supervision and control of shipments of waste. The Commission shall inform the other Member States thereof.

3.   Member States may apply the system provided for in Titles II and VII within their jurisdiction.

TITLE IV

EXPORTS FROM THE COMMUNITY TO THIRD COUNTRIES

CHAPTER 1

Exports of waste for disposal

Article 34

Export prohibited except to EFTA countries

1.   All exports of waste from the Community destined for disposal shall be prohibited.

2.   The prohibition in paragraph 1 shall not apply to exports of waste destined for disposal in EFTA countries which are also Parties to the Basel Convention.

3.   However, exports of waste for disposal to an EFTA country Party to the Basel Convention shall also be prohibited:

(a)   where the EFTA country prohibits imports of such waste; or

(b)   if the competent authority of dispatch has reason to believe that the waste will not be managed in an environmentally sound manner, as referred to in Article 49, in the country of destination concerned.

4.   This provision shall be without prejudice to the take-back obligations as laid down in Articles 22 and 24.

Article 35

Procedures when exporting to EFTA countries

1.   Where waste is exported from the Community and destined for disposal in EFTA countries Parties to the Basel Convention, the provisions of Title II shall apply mutatis mutandis, with the adaptations and additions listed in paragraphs 2 and 3.

2.   The following adaptations shall apply:

(a)   the competent authority of transit outside the Community shall have 60 days following the date of transmission of its acknowledgement of receipt of the notification in which to request additional information on the notified shipment, to provide, if the country concerned has decided not to require prior written consent and has informed the other Parties thereof in accordance with Article 6(4) of the Basel Convention, tacit consent or to give a written consent with or without conditions; and

(b)   the competent authority of dispatch in the Community shall take the decision to consent to the shipment as referred to in Article 9 only after having received written consent from the competent authority of destination and, where appropriate, the tacit or written consent of the competent authority of transit outside the Community, and not earlier than 61 days following the date of transmission of the acknowledgement by the competent authority of transit. The competent authority of dispatch may take the decision before the conclusion of the 61-day time limit if it has the written consent of the other competent authorities concerned.

3.   The following additional provisions shall apply:

(a)   the competent authority of transit in the Community shall acknowledge the receipt of the notification to the notifier;

(b)   the competent authorities of dispatch and, where appropriate, transit in the Community shall send a stamped copy of their decisions to consent to the shipment to the customs office of export and to the customs office of exit from the Community;

(c)   a copy of the movement document shall be delivered by the carrier to the customs office of export and the customs office of exit from the Community;

(d)   as soon as the waste has left the Community, the customs office of exit from the Community shall send a stamped copy of the movement document to the competent authority of dispatch in the Community stating that the waste has left the Community;

(e) if, 42 days after the waste has left the Community, the competent authority of dispatch in the Community has received no information from the facility about receipt of the waste, it shall without delay inform the competent authority of destination; and

(f) the contract referred to in the second subparagraph, point 4 of Article 4 and in Article 5 shall stipulate that:

(i) if a facility issues an incorrect certificate of disposal with the consequence that the financial guarantee is released, the consignee shall bear the costs arising from the duty to return the waste to the area of jurisdiction of the competent authority of dispatch and from its recovery or disposal in an alternative and environmentally sound manner,

(ii) within three days of receipt of the waste for disposal, the facility shall send signed copies of the completed movement document, except for the certificate of disposal referred to in subpoint iii, to the notifier and the competent authorities concerned, and

(iii) as soon as possible but no later than 30 days after completion of disposal, and no later than one calendar year following the receipt of the waste the facility shall, under its responsibility, certify that the disposal has been completed and shall send signed copies of the movement document containing this certification to the notifier and to the competent authorities concerned.

4. The shipment may take place only if:

(a) the notifier has received written consent from the competent authorities of dispatch, destination and, where appropriate, transit outside the Community and if the conditions laid down are met;

(b) a contract between the notifier and consignee has been concluded and is effective, as required in the second subparagraph, point 4 of Article 4 and in Article 5;

(c) a financial guarantee or equivalent insurance has been established and is effective, as required in the second subparagraph, point 5 of Article 4 and in Article 6; and

(d) environmentally sound management, as referred to in Article 49, is ensured.

5. Where waste is exported, it shall be destined for disposal operations within a facility which, under applicable national law, is operating or is authorised to operate in the country of destination.

6. If a customs office of export or a customs office of exit from the Community discovers an illegal shipment, it shall without delay inform the competent authority in the country of the customs office which shall:

(a) without delay inform the competent authority of dispatch in the Community; and

(b)   ensure detention of the waste until the competent authority of dispatch has decided otherwise and has communicated that decision in writing to the
competent authority in the country of the customs office in which the waste is
detained.

CHAPTER 2

Exports of waste for recovery

Section 1

Exports to non-OECD Decision countries

Article 36

Exports prohibition

1.   Exports from the Community of the following wastes destined for recovery in
countries to which the OECD Decision does not apply are prohibited:

(a)   wastes listed as hazardous in Annex V;

(b)   wastes listed in Annex V, Part 3;

(c)   hazardous wastes not classified under one single entry in Annex V;

(d)   mixtures of hazardous wastes and mixtures of hazardous wastes with non-hazardous wastes not classified under one single entry in Annex V;

(e)   wastes that the country of destination has notified to be hazardous under Article 3 of the Basel Convention;

(f)   wastes the import of which has been prohibited by the country of destination;
or

(g)   wastes which the competent authority of dispatch has reason to believe will not
be managed in an environmentally sound manner, as referred to in Article 49,
in the country of destination concerned.

2.   This provision shall be without prejudice to the take-back obligations as set out
in Articles 22 and 24.

3.   Member States may, in exceptional cases, adopt provisions to determine, on
the basis of documentary evidence provided in an appropriate way by the notifier,
that a specific hazardous waste listed in Annex V is excluded from the export prohibition if it does not display any of the properties listed in Annex III to Directive
91/689/EEC, taking into account, as regards the properties H3 to H8, H10 and H11
defined in that Annex, the limit values laid down in Commission Decision 2000/532/EC
of 3 May 2000 replacing Decision 94/3/EC establishing a list of wastes pursuant to
Article 1(a) of Council Directive 75/442/EEC on waste and Council Decision
94/904/EC establishing a list of hazardous waste pursuant to Article 1(4) of Council

Directive 91/689/EEC on hazardous waste[25].

4.   The fact that waste is not listed as hazardous in Annex V, or that it is listed in Annex V, Part 1, List B, shall not preclude, in exceptional cases, characterisation of such waste as hazardous and therefore subject to the export prohibition if it displays any of the properties listed in Annex III to Directive 91/689/EEC, taking into account, as regards the properties H3 to H8, H10 and H11 defined in that Annex, the limit values laid down in Commission Decision 2000/532/EC, as provided for in Article 1(4), second indent, of Directive 91/689/EEC and in the introductory paragraph of Annex III to this Regulation.

5.   In the cases referred to in paragraphs 3 and 4, the Member State concerned shall inform the envisaged country of destination prior to taking a decision. Member States shall notify such cases to the Commission before the end of each calendar year. The Commission shall forward the information to all Member States and to the Secretariat of the Basel Convention. On the basis of the information provided, the Commission may make comments and, where appropriate, adapt Annex V in accordance with Article 58.

Article 37

Procedures when exporting waste listed in Annex III or IIIA

1.   In the case of waste which is listed in Annex III or IIIA and the export of which is not prohibited under Article 36, the Commission shall, within 20 days of the entry into force of this Regulation, send a written request to each country to which the OECD Decision does not apply, seeking:

(i)   confirmation in writing that the waste may be exported from the Community for recovery in that country, and

(ii)   an indication as to which control procedure, if any, would be followed in the country of destination.

Each country to which the OECD Decision does not apply shall be given the following options:

(a)   a prohibition; or

(b)   a procedure of prior written notification and consent as described in Article 35; or

(c)   no control in the country of destination.

2.   Before the date of application of this Regulation, the Commission shall adopt a Regulation taking into account all replies received pursuant to paragraph 1 and shall inform the Committee established pursuant to Article 18 of Directive 2006/12/EC.

If a country has not issued a confirmation as referred to in paragraph 1 or if a country

---

[25]   OJ L 226, 6.9.2000, p. 3. Decision as last amended by Council Decision 2001/573/EC (OJ L 203, 28.7.2001, p. 18).

for any reason has not been contacted, paragraph 1(b) shall apply.

The Commission shall periodically update the Regulation adopted.

3.   If a country indicates in its reply that certain shipments of waste are not subject to any control, Article 18 shall apply mutatis mutandis to such shipments.

4.   Where waste is exported, it shall be destined for recovery operations within a facility which, under applicable national law, is operating or is authorised to operate in the country of destination.

5.   In the case of a shipment of waste not classified under one single entry in Annex III or a shipment of mixtures of wastes not classified under one single entry in Annex III or IIIA or a shipment of waste classified in Annex IIIB, and provided that the export is not prohibited pursuant to Article 36, paragraph 1(b) of this Article shall apply.

Section 2

Exports to OECD-Decision countries

Article 38

Exports of waste listed in Annexes III, IIIA, IIIB, IV and IVA

1.   Where waste listed in Annexes III, IIIA, IIIB, IV and IVA, waste not classified or mixtures of wastes not classified under one single entry in either Annex III, IV or IVA are exported from the Community and destined for recovery in countries to which the OECD Decision applies, with or without transit through countries to which the OECD Decision applies, the provisions of Title II shall apply mutatis mutandis, with the adaptations and additions listed in paragraphs 2, 3 and 5.

2.   The following adaptations shall apply:

(a)   mixtures of wastes listed in Annex IIIA destined for an interim operation shall be subject to the procedure of prior written notification and consent if any subsequent interim or non-interim recovery or disposal operation is to take place in a country to which the OECD Decision does not apply;

(b)   waste listed in Annex IIIB shall be subject to the procedure of prior written notification and consent;

(c)   the consent as required in accordance with Article 9 may be provided in the form of tacit consent from the competent authority of destination outside the Community.

3.   As regards exports of waste listed in Annexes IV and IVA, the following additional provisions shall apply:

(a)   the competent authorities of dispatch and, where appropriate, transit in the Community shall send a stamped copy of their decisions to consent to the shipment to the customs office of export and to the customs office of exit from the Community;

(b)   a copy of the movement document shall be delivered by the carrier to the customs office of export and customs office of exit from the Community;

(c)   as soon as the waste has left the Community, the customs office of exit from the Community shall send a stamped copy of the movement document to the competent authority of dispatch in the Community stating that the waste has left the Community;

(d)   if, 42 days after the waste has left the Community, the competent authority of dispatch in the Community has received no information from the facility about receipt of the waste, it shall without delay inform the competent authority of destination; and

(e)   the contract referred to in the second subparagraph, point 4 of Article 4 and in Article 5 shall stipulate that:

   (i)   if a facility issues an incorrect certificate of recovery with the consequence that the financial guarantee is released, the consignee shall bear the costs arising from the duty to return the waste to the area of jurisdiction of the competent authority of dispatch and from its recovery or disposal in an alternative and environmentally sound manner,

   (ii)   within three days of receipt of the waste for recovery, the facility shall send signed copies of the completed movement document, except for the certificate of recovery referred to in subpoint iii, to the notifier and the competent authorities concerned, and

   (iii) as soon as possible but no later than 30 days after completion of recovery, and no later than one calendar year following the receipt of the waste the facility shall, under its responsibility, certify that the recovery has been completed and shall send signed copies of the movement document containing this certification to the notifier and to the competent authorities concerned.

4.   The shipment may take place only if:

(a)   the notifier has received written consent from the competent authorities of dispatch, destination and, where appropriate, transit or, if tacit consent from the competent authorities of destination and transit outside the Community is provided or can be assumed and if the conditions laid down are met;

(b)   Article 35(4)(b), (c) and (d) is complied with.

5.   If an export as described in paragraph 1 of waste listed in Annexes IV and IVA is in transit through a country to which the OECD Decision does not apply, the following adaptations shall apply:

(a)   the competent authority of transit to which the OECD Decision does not apply shall have 60 days following the date of transmission of its acknowledgement of receipt of the notification in which to request additional information on the notified shipment, to provide, if the country concerned has decided not to require prior written consent and has informed the other Parties thereof in accordance with Article 6(4) of the Basel Convention, tacit consent or to give a written consent with or without conditions; and

(b) the competent authority of dispatch in the Community shall take the decision to consent to the shipment as referred to in Article 9 only after having received tacit or written consent from that competent authority of transit to which the OECD Decision does not apply, and not earlier than 61 days following the date of transmission of the acknowledgement of the competent authority of transit. The competent authority of dispatch may take the decision before the conclusion of the 61-day time limit if it has the written consent of the other competent authorities concerned.

6. Where waste is exported, it shall be destined for recovery operations within a facility which, under applicable national law, is operating or is authorised to operate in the country of destination.

7. If a customs office of export or a customs office of exit from the Community discovers an illegal shipment, it shall without delay inform the competent authority in the country of the customs office which shall:

(a) without delay inform the competent authority of dispatch in the Community; and

(b) ensure detention of the waste until the competent authority of dispatch has decided otherwise and has communicated that decision in writing to the competent authority in the country of the customs office in which the waste is detained.

CHAPTER 3

General provisions

Article 39

Exports to the Antarctic

Exports of waste from the Community to the Antarctic shall be prohibited.

Article 40

Exports to overseas countries or territories

1. Exports from the Community of waste destined for disposal in overseas countries or territories shall be prohibited.

2. As regards exports of waste destined for recovery in overseas countries or territories, the prohibition set out in Article 36 shall apply mutatis mutandis.

3. As regards exports of waste destined for recovery in overseas countries or territories not covered by the prohibition set out in paragraph 2, the provisions of Title II shall apply mutatis mutandis.

TITLE V

IMPORTS INTO THE COMMUNITY FROM THIRD COUNTRIES

CHAPTER 1

Imports of waste for disposal

Article 41

Imports prohibited except from a country Party to the Basel Convention or with an agreement in place or from other areas during situations of crisis or war

1.	Imports into the Community of waste destined for disposal shall be prohibited except those from:

(a)	countries which are Parties to the Basel Convention; or

(b)	other countries with which the Community, or the Community and its Member States, have concluded bilateral or multilateral agreements or arrangements compatible with Community legislation and in accordance with Article 11 of the Basel Convention; or

(c)	other countries with which individual Member States have concluded bilateral agreements or arrangements in accordance with paragraph 2; or

(d)	other areas in cases where, on exceptional grounds during situations of crisis, peacemaking, peacekeeping or war, no bilateral agreements or arrangements pursuant to points (b) or (c) can be concluded or where a competent authority in the country of dispatch has either not been designated or is unable to act.

2.	In exceptional cases, individual Member States may conclude bilateral agreements and arrangements for the disposal of specific waste in those Member States, where such waste will not be managed in an environmentally sound manner, as referred to in Article 49, in the country of dispatch.

These agreements and arrangements shall be compatible with Community legislation and in accordance with Article 11 of the Basel Convention.

These agreements and arrangements shall guarantee that the disposal operations will be carried out in an authorised facility and will comply with the requirements for environmentally sound management.

These agreements and arrangements shall also guarantee that the waste is produced in the country of dispatch and that disposal will be carried out exclusively in the Member State which has concluded the agreement or arrangement.

These agreements or arrangements shall be notified to the Commission prior to their conclusion. However, in emergency situations they may be notified up to one month after conclusion.

3.	Bilateral or multilateral agreements or arrangements entered into in accordance with paragraph 1(b) and (c) shall be based upon the procedural requirements of Article 42.

4.   The countries referred to in paragraph 1(a), (b) and (c) shall be required to present a prior duly reasoned request to the competent authority of the Member State of destination on the basis that they do not have and cannot reasonably acquire the technical capacity and the necessary facilities in order to dispose of the waste in an environmentally sound manner.

Article 42

Procedural requirements for imports from a country Party to the Basel Convention or from other areas during situations of crisis or war

1.   Where waste is imported into the Community and destined for disposal from countries Parties to the Basel Convention, the provisions of Title II shall apply mutatis mutandis, with the adaptations and additions listed in paragraphs 2 and 3.

2.   The following adaptations shall apply:

(a)   the competent authority of transit outside the Community shall have 60 days following the date of transmission of its acknowledgement of receipt of the notification in which to request additional information on the notified shipment, to provide, if the country concerned has decided not to require prior written consent and has informed the other Parties thereof in accordance with Article 6(4) of the Basel Convention, tacit consent or to give a written consent with or without conditions; and

(b)   in the cases referred to in Article 41(1)(d) involving situations of crisis, peace-making, peacekeeping or war, the consent of the competent authorities of dispatch shall not be required.

3.   The following additional provisions shall apply:

(a)   the competent authority of transit in the Community shall acknowledge the receipt of the notification to the notifier, with copies to the competent authorities concerned;

(b)   the competent authorities of destination and, where appropriate, transit in the Community shall send a stamped copy of their decisions to consent to the shipment to the customs office of entry into the Community;

(c)   a copy of the movement document shall be delivered by the carrier to the customs office of entry into the Community; and

(d)   having carried out the necessary customs formalities, the customs office of entry into the Community shall send a stamped copy of the movement document to the competent authorities of destination and transit in the Community, stating that the waste has entered the Community.

4.   The shipment may take place only if:

(a)   the notifier has received written consent from the competent authorities of dispatch, destination and, where appropriate, transit and if the conditions laid down are met;

(b) a contract between the notifier and consignee has been concluded and is effective, as required in the second subparagraph, point 4 of Article 4 and in Article 5;

(c) a financial guarantee or equivalent insurance has been established and is effective, as required in the second subparagraph, point 5 of Article 4 and in Article 6; and

(d) environmentally sound management, as referred to in Article 49, is ensured.

5.   If a customs office of entry into the Community discovers an illegal shipment, it shall without delay inform the competent authority in the country of the customs office which shall:

(a) without delay inform the competent authority of destination in the Community which shall inform the competent authority of dispatch outside the Community; and

(b) ensure detention of the waste until the competent authority of dispatch outside the Community has decided otherwise and has communicated that decision in writing to the competent authority in the country of the customs office in which the waste is detained.

CHAPTER 2

Imports of waste for recovery

Article 43

Imports prohibited except from an OECD Decision country or a country Party to the Basel Convention or with an agreement in place or from other areas during situations of crisis or war

1.   All imports into the Community of waste destined for recovery shall be prohibited except those from:

(a) countries to which the OECD Decision applies; or

(b) other countries which are Parties to the Basel Convention; or

(c) other countries with which the Community, or the Community and its Member States, have concluded bilateral or multilateral agreements or arrangements compatible with Community legislation and in accordance with Article 11 of the Basel Convention; or

(d) other countries with which individual Member States have concluded bilateral agreements or arrangements in accordance with paragraph 2; or

(e) other areas in cases where, on exceptional grounds during situations of crisis, peacemaking, peacekeeping or war, no bilateral agreements or arrangements pursuant to points (c) or (d) can be concluded or where a competent authority in the country of dispatch has either not been designated or is unable to act.

2.   In exceptional cases, individual Member States may conclude bilateral agreements and arrangements for the recovery of specific waste in those Member States, where such waste will not be managed in an environmentally sound manner, as referred to in Article 49, in the country of dispatch.

In such cases Article 41(2) shall apply.

3.   Bilateral or multilateral agreements or arrangements entered into in accordance with paragraph 1(c) and (d) shall be based upon the procedural requirements of Article 42 in so far as may be relevant.

Article 44

Procedural requirements for imports from an OECD Decision country or from other areas during situations of crisis or war

1.   Where waste destined for recovery is imported into the Community from countries and through countries to which the OECD Decision applies, the provisions of Title II shall apply mutatis mutandis, with the adaptations and additions listed in paragraphs 2 and 3.

2.   The following adaptations shall apply:

(a)   the consent as required in accordance with Article 9 may be provided in the form of tacit consent from the competent authority of dispatch outside the Community;

(b)   prior written notification in accordance with Article 4 may be submitted by the notifier; and

(c)   in the cases referred to in Article 43(1)(e) involving situations of crisis, peace-making, peacekeeping or war, the consent of the competent authorities of dispatch shall not be required.

3.   In addition, Article 42(3)(b), (c) and (d) shall be complied with.

4.   The shipment may take place only if:

(a)   the notifier has received written consent from the competent authorities of dispatch, destination and, where appropriate, transit or if tacit consent from the competent authority of dispatch outside the Community is provided or can be assumed and if the conditions laid down are met;

(b)   a contract between the notifier and consignee has been concluded and is effective, as required in the second subparagraph, point 4 of Article 4 and in Article 5;

(c)   a financial guarantee or equivalent insurance has been established and is effective, as required in the second subparagraph, point 5 of Article 4 and in Article 6; and

(d)   environmentally sound management, as referred to in Article 49, is ensured.

5.   If a customs office of entry into the Community discovers an illegal shipment, it

shall without delay inform the competent authority in the country of the customs office which shall:

(a) without delay inform the competent authority of destination in the Community which shall inform the competent authority of dispatch outside the Community; and

(b) ensure detention of the waste until the competent authority of dispatch outside the Community has decided otherwise and has communicated that decision in writing to the competent authority in the country of the customs office in which the waste is detained.

Article 45

Procedural requirements for imports from a non-OECD Decision country Party to the Basel Convention or from other areas during situations of crisis or war

Where waste destined for recovery is imported into the Community:

(a) from a country to which the OECD Decision does not apply; or

(b) through any country to which the OECD Decision does not apply and which is also Party to the Basel Convention,

Article 42 shall apply mutatis mutandis.

CHAPTER 3

General provisions

Article 46

Imports from overseas countries or territories

1. Where waste is imported into the Community from overseas countries or territories, Title II shall apply mutatis mutandis.

2. One or more overseas countries and territories and the Member State to which they are linked may apply national procedures to shipments from the overseas country or territory to that Member State.

3. Member States which apply paragraph 2 shall notify the Commission of the national procedures applied.

TITLE VI

TRANSIT THROUGH THE COMMUNITY FROM AND TO THIRD COUNTRIES

CHAPTER 1

Transit of waste for disposal

Article 47

Transit through the Community of waste destined for disposal

Where waste destined for disposal is shipped through Member States from and to third countries, Article 42 shall apply mutatis mutandis, with the adaptations and additions listed below:

(a) the first and last competent authority of transit in the Community shall, where appropriate, send a stamped copy of the decisions to consent to the shipment or, if they have provided tacit consent, a copy of the acknowledgement in accordance with Article 42(3)(a) to the customs offices of entry into and exit from the Community respectively; and

(b) as soon as the waste has left the Community, the customs office of exit from the Community shall send a stamped copy of the movement document to the competent authority(ies) of transit in the Community, stating that the waste has left the Community.

CHAPTER 2

Transit of waste for recovery

Article 48

Transit through the Community of waste destined for recovery

1. Where waste destined for recovery is shipped through Member States from and to a country to which the OECD Decision does not apply, Article 47 shall apply mutatis mutandis.

2. Where waste destined for recovery is shipped through Member States from and to a country to which the OECD Decision applies, Article 44 shall apply mutatis mutandis, with the adaptations and additions listed below:

(a) the first and last competent authority of transit in the Community shall, where appropriate, send a stamped copy of the decisions to consent to the shipment or, if they have provided tacit consent, a copy of the acknowledgement in accordance with Article 42(3)(a) to the customs offices of entry into and exit from the Community respectively; and

(b) as soon as the waste has left the Community, the customs office of exit from the Community shall send a stamped copy of the movement document to the competent authority(ies) of transit in the Community, stating that the waste has left the Community.

3. Where waste destined for recovery is shipped through Member States from a

country to which the OECD Decision does not apply to a country to which the OECD Decision applies or vice versa, paragraph 1 shall apply as regards the country to which the OECD Decision does not apply and paragraph 2 shall apply as regards the country to which the OECD Decision applies.

TITLE VII

OTHER PROVISIONS

CHAPTER 1

Additional obligations

Article 49

Protection of the environment

1.   The producer, the notifier and other undertakings involved in a shipment of waste and/or its recovery or disposal shall take the necessary steps to ensure that any waste they ship is managed without endangering human health and in an environmentally sound manner throughout the period of shipment and during its recovery and disposal. In particular, when the shipment takes place in the Community, the requirements of Article 4 of Directive 2006/12/EC and other Community legislation on waste shall be respected.

2.   In the case of exports from the Community, the competent authority of dispatch in the Community shall:

(a)   require and endeavour to secure that any waste exported is managed in an environmentally sound manner throughout the period of shipment, including recovery as referred to in Articles 36 and 38 or disposal as referred to in Article 34, in the third country of destination;

(b)   prohibit an export of waste to third countries if it has reason to believe that the waste will not be managed in accordance with the requirements of point (a).

Environmentally sound management may, inter alia, be assumed as regards the waste recovery or disposal operation concerned, if the notifier or the competent authority in the country of destination can demonstrate that the facility which receives the waste will be operated in accordance with human health and environmental protection standards that are broadly equivalent to standards established in Community legislation.

This assumption shall, however, be without prejudice to the overall assessment of environmentally sound management throughout the period of shipment and including recovery or disposal in the third country of destination.

For the purposes of seeking guidance on environmentally sound management, the guidelines listed in Annex VIII may be considered.

3.   In the case of imports into the Community, the competent authority of destination in the Community shall:

(a)   require and take the necessary steps to ensure that any waste shipped into its area of jurisdiction is managed without endangering human health and without using processes or methods which could harm the environment, and in accordance with Article 4 of Directive 2006/12/EC and other Community legislation on waste throughout the period of shipment, including recovery or disposal in the country of destination;

(b)   prohibit an import of waste from third countries if it has reason to believe that the waste will not be managed in accordance with the requirements of point (a).

Article 50

Enforcement in Member States

1.   Member States shall lay down the rules on penalties applicable for infringement of the provisions of this Regulation and shall take all measures necessary to ensure that they are implemented. The penalties provided for must be effective, proportionate and dissuasive. Member States shall notify the Commission of their national legislation relating to prevention and detection of illegal shipments and penalties for such shipments.

2.   Member States shall, by way of measures for the enforcement of this Regulation, provide, inter alia, for inspections of establishments, undertakings, brokers and dealers in accordance with Article 34 of Directive 2008/98/EC, and for inspections of shipments of waste and of the related recovery or disposal.

2a.   By 1 January 2017, Member States shall ensure that, in respect of their entire geographical territory, one or more plans are established, either separately or as a clearly defined part of other plans, for inspections carried out pursuant to paragraph 2 ('inspection plan'). Inspection plans shall be based on a risk assessment covering specific waste streams and sources of illegal shipments and considering, if available and where appropriate, intelligence-based data such as data on investigations by police and customs authorities and analyses of criminal activities. That risk assessment shall aim, inter alia, to identify the minimum number of inspections required, including physical checks on establishments, undertakings, brokers, dealers and shipments of waste or on the related recovery or disposal. An inspection plan shall include the following elements:

(a)   the objectives and priorities of the inspections, including a description of how those priorities have been identified;

(b)   the geographical area covered by that inspection plan;

(c)   information on planned inspections, including on physical checks;

(d)   the tasks assigned to each authority involved in inspections;

(e)   arrangements for cooperation between authorities involved in inspections;

(f)   information on the training of inspectors on matters relating to inspections; and

(g)   information on the human, financial and other resources for the implementation of that inspection plan.

An inspection plan shall be reviewed at least every three years and, where appropriate, updated. That review shall evaluate to which extent the objectives and other elements of that inspection plan have been implemented.

3.   Inspections of shipments may take place in particular:

(a)   at the point of origin, carried out with the producer, holder or notifier;

(b)   at the point of destination, including interim and non-interim recovery or disposal, carried out with the consignee or the facility;

(c)   at the frontiers of the Union; and/or

(d)   during the shipment within the Union.

4.   Inspections of shipments shall include the verification of documents, the confirmation of identity and, where appropriate, physical checking of the waste.

4a.   In order to ascertain that a substance or object being carried by road, rail, air, sea or inland waterway is not waste, the authorities involved in inspections may, without prejudice to Directive 2012/19/EU of the European Parliament and of the Council[26], require the natural or legal person who is in possession of the substance or object concerned, or who arranges the carriage thereof, to submit documentary evidence:

(a)   as to the origin and destination of the substance or object concerned; and

(b)   that it is not waste, including, where appropriate, evidence of functionality.

For the purpose of the first subparagraph, the protection of the substance or object concerned against damage during transportation, loading and unloading, such as adequate packaging and appropriate stacking, shall also be ascertained.

4b.   The authorities involved in inspections may conclude that the substance or object concerned is waste where:

—   the evidence referred to in paragraph 4a or required under other Union legislation to ascertain that a substance or object is not waste, has not been submitted within the period specified by them, or

—   they consider the evidence and information available to them to be insufficient to reach a conclusion, or they consider the protection provided against damage referred to in the second subparagraph of paragraph 4a to be insufficient.

In such circumstances, the carriage of the substance or object concerned or the shipment of waste concerned shall be considered as an illegal shipment. Consequently, it shall be dealt with in accordance with Articles 24 and 25 and the authorities involved in inspections shall, without delay, inform the competent authority of the country where the inspection concerned took place accordingly.

4c.   In order to ascertain whether a shipment of waste complies with this Regulation, the authorities involved in inspections may require the notifier, the person who

---

[26]   Directive 2012/19/EU of the European Parliament and of the Council of 4 July 2012 on waste electrical and electronic equipment (OJ L 197, 24.7.2012, p. 38).

arranges the shipment, the holder, the carrier, the consignee and the facility that receives the waste to submit relevant documentary evidence to them within a period specified by them.

In order to ascertain whether a shipment of waste falling under the general information requirements of Article 18 is destined for recovery operations which are in accordance with Article 49, the authorities involved in inspections may require the person who arranges the shipment to submit relevant documentary evidence, provided by the interim and non-interim recovery facility and, if necessary, approved by the competent authority of destination.

4d.   Where the evidence referred to in paragraph 4c has not been submitted to the authorities involved in inspections within the period specified by them, or they consider the evidence and information available to them to be insufficient to reach a conclusion, the shipment concerned shall be considered as an illegal shipment. Consequently, it shall be dealt with in accordance with Articles 24 and 25 and the authorities involved in inspections shall, without delay, inform the competent authority of the country where the inspection concerned took place accordingly.

4e.   By 18 July 2015, the Commission shall adopt, by means of implementing acts, a preliminary correlation table between the codes of the combined nomenclature, provided for in Council Regulation (EEC) No 2658/87[27] and the entries of waste listed in Annexes III, IIIA, IIIB, IV, IVA and V to this Regulation. The Commission shall maintain that correlation table up-to-date in order to reflect changes to that nomenclature and to the entries listed in those Annexes, as well as to include any new waste-related codes of the Harmonised System Nomenclature that the World Customs Organisation may adopt.

Those implementing acts shall be adopted in accordance with the examination procedure referred to in Article 59a (2).

5.   Member States shall cooperate, bilaterally and multilaterally, with one another in order to facilitate the prevention and detection of illegal shipments. They shall exchange relevant information on shipments of waste, flows of waste, operators and facilities and share experience and knowledge on enforcement measures, including the risk assessment carried out pursuant to paragraph 2a of this Article, within established structures, in particular, through the network of correspondents designated in accordance with Article 54.

6.   Member States shall identify those members of their permanent staff responsible for the cooperation referred to in paragraph 5 and identify the focal point(s) for the physical checks referred to in paragraph 4. The information shall be sent to the Commission which shall distribute a compiled list to the correspondents referred to in Article 54.

7.   At the request of another Member State, a Member State may take enforcement action against persons suspected of being engaged in the illegal shipment of waste who are present in that Member State.

---

[27]   Council Regulation (EEC) No 2658/87 of 23 July 1987 on the tariff and statistical nomenclature and on the Common Customs Tariff (OJ L 256, 7.9.1987, p. 1).

Article 51

Reports by Member States

1.   Before the end of each calendar year, each Member State shall send the Commission a copy of the report for the previous calendar year which, in accordance with Article 13(3) of the Basel Convention, it has drawn up and submitted to the Secretariat of that Convention.

2.   Before the end of each calendar year, Member States shall also draw up a report for the previous year based on the additional reporting questionnaire in Annex IX, and shall send it to the Commission.

3.   The reports drawn up by Member States in accordance with paragraphs 1 and 2 shall be submitted to the Commission in an electronic version.

4.   The Commission shall establish every three years a report, based on these reports, on the implementation of this Regulation by the Community and its Member States.

Article 52

International cooperation

Member States, where appropriate and necessary in liaison with the Commission, shall cooperate with other Parties to the Basel Convention and inter-State organisations, inter alia, via the exchange and/or sharing of information, the promotion of environmentally sound technologies and the development of appropriate codes of good practice.

Article 53

Designation of competent authorities

Member States shall designate the competent authority or authorities responsible for the implementation of this Regulation. Each Member State shall designate only one single competent authority of transit.

Article 54

Designation of correspondents

Member States and the Commission shall each designate one or more correspondents responsible for informing or advising persons or undertakings making enquiries. The Commission correspondent shall forward to the correspondents of the Member States any questions put to him/her which concern the latter, and vice versa.

Article 55

Designation of customs offices of entry into and exit from the Community

Member States may designate specific customs offices of entry into and exit from the Community for shipments of waste entering and leaving the Community. If Member States decide to designate such customs offices, no shipment of waste shall be allowed to use any other frontier crossing points within a Member State for the purposes of entering or leaving the Community.

Article 56

Notification of, and information regarding, designations

1.   Member States shall notify the Commission of designations of:

(a)   competent authorities, pursuant to Article 53;

(b)   correspondents, pursuant to Article 54; and,

(c)   where appropriate, customs offices of entry into and exit from the Community, pursuant to Article 55.

2.   In relation to those designations, Member States shall notify the Commission of the following information:

(a)   name(s);

(b)   postal address(es);

(c)   e-mail address(es);

(d)   telephone number(s);

(e)   fax number(s); and

(f)   languages acceptable to the competent authorities.

3.   Member States shall immediately notify the Commission of any changes in this information.

4.   This information as well as any changes in the information shall be submitted to the Commission in an electronic as well as a paper version if so required.

5.   The Commission shall publish on its web-site lists of the designated competent authorities, correspondents and customs offices of entry into and exit from the Community, and shall update these lists as appropriate.

CHAPTER 2

Other provisions

Article 57

Meeting of the correspondents

The Commission shall, if requested by Member States or if otherwise appropriate, periodically hold a meeting of the correspondents to examine the questions raised by the implementation of this Regulation. Relevant stakeholders shall be invited to such meetings, or parts of meetings, where all Member States and the Commission are in agreement that this is appropriate.

Article 58

Amendment of the Annexes

1.   The Commission shall be empowered to adopt delegated acts in accordance with Article 58a to amend the following:

(a)   Annexes IA, IB, IC, II, III, IIIA, IIIB, IV, V, VI and VII to take account of changes agreed under the Basel Convention and the OECD Decision;

(b)   Annex V to reflect agreed changes to the list of waste adopted in accordance with Article 7 of Directive 2008/98/EC;

(c)   Annex VIII to reflect decisions taken under relevant international conventions and agreements.

Article 58a

Exercise of the delegation

1.   The power to adopt delegated acts is conferred on the Commission subject to the conditions laid down in this Article.

2.   The delegation of power referred to in Article 58 shall be conferred on the Commission for a period of five years from 17 July 2014. The Commission shall draw up a report in respect of the delegation of power not later than nine months before the end of the five-year period. The delegation of power shall be tacitly extended for periods of an identical duration, unless the European Parliament or the Council opposes such extension not later than three months before the end of each period.

3.   The delegation of power referred to in Article 58 may be revoked at any time by the European Parliament or by the Council. A decision to revoke shall put an end to the delegation of the power specified in that decision. It shall take effect the day following the publication of the decision in the Official Journal of the European Union or at a later date specified therein. It shall not affect the validity of any delegated acts already in force.

4.   As soon as it adopts a delegated act, the Commission shall notify it simultaneously to the European Parliament and to the Council.

5.   A delegated act adopted pursuant to Article 58 shall enter into force only if no objection has been expressed either by the European Parliament or the Council within a period of two months of notification of that act to the European Parliament and the Council or if, before the expiry of that period, the European Parliament and the Council have both informed the Commission that they will not object. That period shall be extended by two months at the initiative of the European Parliament or of the Council.

Article 59a

Committee procedure

1.   The Commission shall be assisted by the committee established by Article 39 of Directive 2008/98/EC. That committee is a committee within the meaning of Regulation (EU) No 182/2011.

2.   Where reference is made to this paragraph, Article 5 of Regulation (EU) No 182/2011 shall apply.

Where the committee delivers no opinion, the Commission shall not adopt the draft implementing act and the third subparagraph of Article 5(4) of Regulation (EU) No 182/2011 shall apply.

Article 60

Review

1.   By 15 July 2006, the Commission shall complete its review of the relationship between existing sectoral legislation on animal and public health, including shipments of waste covered by Regulation (EC) No 1774/2002, and the provisions of this Regulation. If necessary, this review shall be accompanied by appropriate proposals with a view to achieving an equivalent level of procedures and control regime for the shipment of such waste.

2.   Within five years from 12 July 2007, the Commission shall review the implementation of Article 12(1)(c), including its effect on environment protection and the functioning of the internal market. If necessary, this review shall be accompanied by appropriate proposals to amend this provision.

2a.   By 31 December 2020, the Commission shall, taking into account, inter alia, the reports drawn up in accordance with Article 51, carry out a review of this Regulation and submit a report on the results thereof to the European Parliament and to the Council, accompanied, if appropriate, by a legislative proposal. In that review, the Commission shall consider, in particular, the effectiveness of Article 50(2a) in combating illegal shipments, taking into account environmental, social and economic aspects.

Article 61

Repeals

1.   Regulation (EEC) No 259/93 and Decision 94/774/EC are hereby repealed with effect from 12 July 2007.

2.   References made to the repealed Regulation (EEC) No 259/93 shall be construed as being made to this Regulation.

3.   Decision 1999/412/EC is hereby repealed with effect from 1 January 2008.

Article 62

Transition rules

1.   Any shipment that has been notified and for which the competent authority of destination has given acknowledgement before 12 July 2007 shall be subject to the provisions of Regulation (EEC) No 259/93.

2.   Any shipment for which the competent authorities concerned have given their consent pursuant to Regulation (EEC) No 259/93 shall be completed not later than one year from 12 July 2007.

3.   Reporting pursuant to Article 41(2) of Regulation (EEC) No 259/93 and Article 51 of this Regulation for the year 2007 shall be based on the questionnaire contained in Decision 1999/412/EC.

Article 63

Transitional arrangements for certain Member States

As the transition periods for Latvia, Poland, Slovakia, Bulgaria and Romania have expired, this article is not reproduced here.

Article 64

Entry into force and application

1.   This Regulation shall enter into force on the third day following that of its publication in the Official Journal of the European Union.

It shall apply from 12 July 2007.

2.   Should the date of accession of Bulgaria or Romania be later than the date of application specified in paragraph 1, Article 63(4) and (5) shall, by way of derogation from paragraph 1 of this Article, apply from the date of accession.

3.   Subject to the agreement of the Member States concerned, Article 26(4) may be applied before 12 July 2007.

This Regulation shall be binding in its entirety and directly applicable in all Member States.

ANNEX IA

Notification document for transboundary movements/shipments of waste

[Not published herein]

Cf. sect. I.4.2.2. fig. 4 and 5 of explanatory notes

ANNEX IB

Movement document for transboundary movements/shipments of waste

[Not published herein]

Cf. sect. I 4.5.1 fig. 7 and 8 of explanatory notes

ANNEX IC

SPECIFIC INSTRUCTIONS FOR COMPLETING THE NOTIFICATION AND MOVEMENT DOCUMENTS

[Not published herein]

Cf. sect. I.4.2.2.2 and I.4.5.1 of explanatory notes

ANNEX II

INFORMATION AND DOCUMENTATION RELATED TO NOTIFICATION

Part 1:   Information to be supplied on, or annexed to, the notification document:

1. Serial number or other accepted identifier of the notification document and intended total number of shipments.

2. Notifier's name, address, telephone number, fax number, e-mail address, registration number and contact person.

3. If the notifier is not the producer: producer's (producers') name, address, telephone number, fax number, e-mail address and contact person.

4. Dealer's (dealers') or broker's (brokers') name, address, telephone number, fax number, e-mail address and contact person, where the notifier has authorised him in accordance with point 15 of Article 2.

5. Recovery or disposal facility's name, address, telephone number, fax number, e-mail address, registration number, contact person, technologies employed and possible status as pre-consented in accordance with Article 14.

    If the waste is destined for an interim recovery or disposal operation, similar information regarding all facilities where subsequent interim and non-interim recovery or disposal operations are envisaged shall be indicated.

    If the recovery or disposal facility is listed in Annex I, Category 5 of Directive 96/61/EC, evidence (e.g. a declaration certifying its existence) of a valid permit issued in accordance with Articles 4 and 5 of that Directive shall be provided.

6. Consignee's name, address, telephone number, fax number, e-mail address, registration number and contact person.

7. Intended carrier's (carriers') and/or their agent's (agents') name, address, telephone number, fax number, e-mail address, registration number and contact person.

8. Country of dispatch and relevant competent authority.

9. Countries of transit and relevant competent authorities.

10. Country of destination and relevant competent authority.

11. Single notification or general notification. If general notification, period of validity requested.

12. Date(s) envisaged for start of the shipment(s).

13. Means of transport envisaged.

14. Intended routing (point of exit from and entry into each country concerned, including customs offices of entry into and/or exit from and/or export from the Community) and intended route (route between points of exit and entry), including possible alternatives, also in case of unforeseen circumstances.

15. Evidence of registration of the carrier(s) regarding waste transports (e.g. a declaration certifying its existence).

16. Designation of the waste on the appropriate list, the source(s), description, composition and any hazardous characteristics. In the case of waste from various sources, also a detailed inventory of the waste.

17. Estimated maximum and minimum quantities.

18. Type of packaging envisaged.

19. Specification of the recovery or disposal operation(s) as referred to in Annexes II A and II B to Directive 2006/12/EC.

20. If the waste is destined for recovery:

    (a) the planned method of disposal for the non-recoverable fraction after recovery;

    (b) the amount of recovered material in relation to non-recoverable waste;

    (c) the estimated value of the recovered material;

    (d) the cost of recovery and the cost of disposal of the non-recoverable fraction.

21. Evidence of insurance against liability for damage to third parties (e.g. a declaration certifying its existence).

22. Evidence of a contract (or a declaration certifying its existence) between the notifier and consignee for the recovery or disposal of the waste that has been concluded and is effective at the time of the notification, as required in the second subparagraph, point 4 of Article 4 and in Article 5.

23. A copy of the contract or evidence of the contract (or a declaration certifying its existence) between the producer, new producer or collector and the broker or dealer, in the event that the broker or dealer acts as notifier.

24. Evidence of a financial guarantee or equivalent insurance (or a declaration certifying its existence if the competent authority so allows) that has been established and is effective at the time of the notification or, if the competent authority which approves the financial guarantee or equivalent insurance so allows, at the latest when the shipment starts, as required in the second subparagraph, point 5 of Article 4 and in Article 6.

25. Certification by the notifier that the information is complete and correct to the best of his/her knowledge.

26. When the notifier is not the producer in accordance with point 15(a)(i) of Article 2, the notifier shall ensure that the producer or one of the persons indicated in point 15(a)(ii) or (iii) of Article 2, where practicable, also signs the notification document provided for in Annex IA.

Part 2: Information to be supplied on, or annexed to, the movement document:

Supply all information listed in Part 1, updated in accordance with the points set out below, and the other additional information specified:

1. Serial and total number of shipments.

2. Date shipment started.

3. Means of transport.

4. Carrier's (carriers') name, address, telephone number, fax number and e-mail address.

5. Routing (point of exit from and entry into each country concerned, including customs offices of entry into and/or exit from and/or export from the Community) and route (route between points of exit and entry), including possible alternatives, also in case of unforeseen circumstances.

6. Quantities.

7. Type of packaging.

8. Any special precautions to be taken by the carrier(s).

9. Declaration by the notifier that all necessary consents have been received from the competent authorities of the countries concerned. This declaration must be signed by the notifier.

10. Appropriate signatures for each custody transfer.

Part 3: Additional information and documentation that may be requested by the competent authorities:

1. The type and duration of the authorisation pursuant to which the recovery or disposal facility operates.

2. Copy of the permit issued in accordance with Articles 4 and 5 of Directive 96/61/EC.

3. Information concerning the measures to be taken to ensure transport safety.

4. The transport distance(s) between the notifier and the facility, including possible alternative routes, also in case of unforeseen circumstances and, in the event of intermodal transport, the place where the transfer will take place.

5. Information about costs of transport between the notifier and the facility.

6. Copy of the registration of the carrier(s) regarding the waste transport.

7. Chemical analysis of the composition of the waste.

8. Description of the production process of the waste.

9. Description of the treatment process of the facility which receives the waste.

10. The financial guarantee or equivalent insurance or a copy thereof.

11. Information concerning the calculation of the financial guarantee or equivalent insurance as required in the second subparagraph, point 5 of Article 4 and in Article 6.

12. Copy of the contracts referred to in Part 1, points 22 and 23.

13. Copy of the policy of insurance against liability for damage to third parties.

14. Any other information which is pertinent to the assessment of the notification in accordance with this Regulation and national legislation.

ANNEX III

**LIST OF WASTES SUBJECT TO THE GENERAL INFORMATION REQUIRE-
MENTS LAID DOWN IN ARTICLE 18 ('GREEN' LISTED WASTE)[28]**

[Not published herein]

Cf. appendix 1 of explanatory notes

---

[28] This list originates from the OECD Decision, Appendix 3.

ANNEX IIIA

MIXTURES OF TWO OR MORE WASTES LISTED IN ANNEX III AND NOT CLASSIFIED UNDER ONE SINGLE ENTRY AS REFERRED TO IN ARTICLE 3(2)

1.　Regardless of whether or not mixtures are included on this list, they may not be subject to the general information requirements laid down in Article 18 if they are contaminated by other materials to an extent which:

(a)　increases the risks associated with the wastes sufficiently to render them appropriate for submission to the procedure of prior written notification and consent, when taking into account the hazardous characteristics listed in Annex III to Directive 91/689/EEC; or

(b)　prevents the recovery of the wastes in an environmentally sound manner.

2.　The following mixtures of wastes are included in this Annex:

(a)　mixtures of wastes classified under Basel entries B1010 and B1050;

(b)　mixtures of wastes classified under Basel entries B1010 and B1070;

(c)　mixtures of wastes classified under Basel entries B3040 and B3080;

(d)　mixtures of wastes classified under (OECD) entry GB040 and under Basel entry B1100 restricted to hard zinc spelter, zinc-containing drosses, aluminium skimmings (or skims) excluding salt slag and wastes of refractory linings, including crucibles, originating from copper smelting;

(e)　mixtures of wastes classified under (OECD) entry GB040, under Basel entry B1070 and under Basel entry B1100 restricted to wastes of refractory linings, including crucibles, originating from copper smelting.

The entries referred to in points (d) and (e) shall not apply for exports to countries to which the OECD Decision does not apply.

3.　The following mixtures of wastes classified under separate indents or sub-indents of one single entry are included in this Annex:

(a)　mixtures of wastes classified under Basel entry B1010;

(b)　mixtures of wastes classified under Basel entry B2010;

(c)　mixtures of wastes classified under Basel entry B2030;

(g)　mixtures of wastes classified under Basel entry B3020 restricted to unbleached paper or paperboard or of corrugated paper or paperboard, other paper or paperboard, made mainly of bleached chemical pulp, not coloured in the mass, paper or paperboard made mainly of mechanical pulp (for example, newspapers, journals and similar printed matter);

(h)　mixtures of wastes classified under Basel entry B3030;

(i)   mixtures of wastes classified under Basel entry B3040;

(j)   mixtures of wastes classified under Basel entry B3050.

4.   The following mixtures of wastes classified under separate indents or sub-indents of one single entry are included in this Annex only for the purposes of shipments within the Union:

(a)   mixtures of wastes classified under entry EU3011 and listed under the indent referring to non-halogenated polymers;

(b)   mixtures of wastes classified under entry EU3011 and listed under the indent referring to cured resins or condensation products;

(c)   mixtures of wastes classified under entry EU3011 and listed under 'perfluoroalkoxy alkanes'.

ANNEX IIIB

ADDITIONAL GREEN LISTED WASTE AWAITING INCLUSION IN THE RELEVANT ANNEXES TO THE BASEL CONVENTION OR THE OECD DECISION AS REFERRED TO IN ARTICLE 58(1)(b)

1.   Regardless of whether or not wastes are included on this list, they may not be subject to the general information requirements laid down in Article 18 if they are contaminated by other materials to an extent which:

(a)   increases the risks associated with the wastes sufficiently to render them appropriate for submission to the procedure of prior written notification and consent, when taking into account the hazardous characteristics listed in Annex III to Directive 2008/98/EC of the European Parliament and of the Council[29]; or

(b)   prevents the recovery of the wastes in an environmentally sound manner.

2.   The following wastes are included in this Annex:

BEU04 Composite packaging consisting of mainly paper and some plastic, not containing residues and not covered by Basel entry B3020

BEU05 Clean biodegradable waste from agriculture, horticulture, forestry, gardens, parks and cemeteries

3.   The shipments of waste listed in this Annex are without prejudice to the provisions of Council Directive 2000/29/EC[30], including measures adopted pursuant to Article 16(3) thereof.

---

[29]   OJ L 312, 22.11.2008, p. 3.
[30]   OJ L 169, 10.7.2000, p. 1.

ANNEX IV

LIST OF WASTES SUBJECT TO THE PROCEDURE OF PRIOR WRITTEN NOTIFICATION AND CONSENT ('AMBER' LISTED WASTE)[31]

[Not published herein]

Cf. appendix 1 of explanatory notes

ANNEX IVA

WASTE LISTED IN ANNEX III BUT SUBJECT TO THE PROCEDURE OF PRIOR WRITTEN NOTIFICATION AND CONSENT (ARTICLE 3(3))

empty

ANNEX V

WASTE SUBJECT TO THE EXPORT PROHIBITION IN ARTICLE 36

[Not published herein]

ANNEX VI

FORM FOR PRE-CONSENTED FACILITIES (ARTICLE 14)

[Not published herein]

ANNEX VII

INFORMATION ACCOMPANYING SHIPMENTS OF WASTE AS REFERRED TO IN ARTICLE 3(2) AND (4)

[Not published herein]

Cf. sect. I.4.5.1 of explanatory notes

---

[31] This list originates from the OECD Decision, Appendix 4.

ANNEX VIII

GUIDELINES ON ENVIRONMENTALLY SOUND MANAGEMENT (ARTICLE 49)

I. **Guidelines and guidance documents adopted under the Basel Convention:**

1. Technical Guidelines on the Environmentally Sound Management of Biomedical and Health Care Wastes (Y1; Y3)[32]

2. Technical Guidelines on the Environmentally Sound Management of Waste Lead Acid Batteries[33]

3. Technical Guidelines on the Environmentally Sound Management of the Full and Partial Dismantling of Ships[34]

4. Technical Guidelines on the Environmentally Sound Recycling/Reclamation of Metals and Metal Compounds (R4)[35]

5. General technical guidelines for the environmentally sound management of wastes consisting of, containing or contaminated with persistent organic pollutants[36]

6. Technical guidelines on the environmentally sound management of wastes consisting of, containing or contaminated with 1,1,1 trichloro 2,2 bis (4 chlorophenyl) ethane (DDT)[37]

7. Technical guidelines on the environmentally sound management of wastes consisting of, containing or contaminated with hexabromocyclododecane (HBCD)[38]

8. Technical guidelines on the environmentally sound management of wastes consisting of, containing or contaminated with perfluorooctane sulfonic acid (PFOS), its salts and perfluorooctane sulfonyl fluoride (PFOSF)[39]

9. Technical guidelines on the environmentally sound management of wastes consisting of, containing or contaminated with pentachlorophenol and its salts

---

[32] Adopted by the sixth meeting of the Conference of the Parties to the Basel Convention on the Control of Transboundary Movements of Hazardous Waste and Their Disposal, 9-13 December 2002.

[33] Adopted by the sixth meeting of the Conference of the Parties to the Basel Convention on the Control of Transboundary Movements of Hazardous Waste and Their Disposal, 9-13 December 2002.

[34] Adopted by the sixth meeting of the Conference of the Parties to the Basel Convention on the Control of Transboundary Movements of Hazardous Waste and Their Disposal, 9-13 December 2002.

[35] Adopted by the seventh meeting of the Conference of the Parties to the Basel Convention on the Control of Transboundary Movements of Hazardous Wastes and Their Disposal, 25-29 October 2004.

[36] Adopted by the fourteenth meeting of the Conference of the Parties to the Basel Convention on the Control of Transboundary Movements of Hazardous Wastes and Their Disposal, May 2019.

[37] Adopted by the eighth meeting of the Conference of the Parties to the Basel Convention on the Control of Transboundary Movements of Hazardous Wastes and Their Disposal, December 2006.

[38] Adopted by the twelfth meeting of the Conference of the Parties to the Basel Convention on the Control of Transboundary Movements of Hazardous Wastes and Their Disposal, May 2015.

[39] Adopted by the twelfth meeting of the Conference of the Parties to the Basel Convention on the Control of Transboundary Movements of Hazardous Wastes and Their Disposal, May 2015.

and esters (PCP)[40]

10. Technical guidelines on the environmentally sound management of wastes consisting of, containing or contaminated with the pesticides aldrin, alpha hexachlorocyclohexane, beta hexachlorocyclohexane, chlordane, chlordecone, dieldrin, endrin, heptachlor, hexachlorobenzene, hexachlorobutadiene, lindane, mirex, pentachlorobenzene, pentachlorophenol and its salts, perfluorooctane sulfonic acid, technical endosulfan and its related isomers or toxaphene or with hexachlorobenzene as an industrial chemical (POP Pesticides)[41]

11. Technical guidelines on the environmentally sound management of wastes consisting of, containing or contaminated with polychlorinated biphenyls, polychlorinated terphenyls, polychlorinated naphthalenes or polybrominated biphenyls including hexabromobiphenyl (PCBs, PCTs, PCNs or PBBs, including HBB)[42]

12. Technical guidelines on the environmentally sound management of wastes consisting of, containing or contaminated with hexabromodiphenyl ether and heptabromodiphenyl ether, or tetrabromodiphenyl ether and pentabromodiphenyl ether or decabromodiphenyl ether (POP-BDEs)[43]

13. Technical Guidelines for the Environmentally Sound Management of Wastes Containing or Contaminated with unintentionally produced polychlorinated dibenzo-p-dioxins, polychlorinated dibenzofurans, hexachlorobenzene, polychlorinated biphenyls, pentachlorobenzene, polychlorinated naphthalenes or hexachlorobutadiene[44]

14. Technical guidelines on the environmentally sound management of wastes consisting of, containing or contaminated with hexachlorobutadiene[45]

15. Technical guidelines on the environmentally sound management of wastes consisting of, containing or contaminated with short-chain chlorinated paraffins[46]

16. Technical guidelines for the environmentally sound management of used and waste pneumatic tyres[47]

17. Technical guidelines for the environmentally sound management of wastes

---

[40] Adopted by the thirteenth meeting of the Conference of the Parties to the Basel Convention on the Control of Transboundary Movements of Hazardous Wastes and Their Disposal, May 2017.

[41] Adopted by the thirteenth meeting of the Conference of the Parties to the Basel Convention on the Control of Transboundary Movements of Hazardous Wastes and Their Disposal, May 2017.

[42] Adopted by the thirteenth meeting of the Conference of the Parties to the Basel Convention on the Control of Transboundary Movements of Hazardous Wastes and Their Disposal, May 2017.

[43] Adopted by the fourteenth meeting of the Conference of the Parties to the Basel Convention on the Control of Transboundary Movements of Hazardous Wastes and Their Disposal, May 2019.

[44] Adopted by the fourteenth meeting of the Conference of the Parties to the Basel Convention on the Control of Transboundary Movements of Hazardous Wastes and Their Disposal, May 2019.

[45] Adopted by the fourteenth meeting of the Conference of the Parties to the Basel Convention on the Control of Transboundary Movements of Hazardous Wastes and Their Disposal, May 2019.

[46] Adopted by the fourteenth meeting of the Conference of the Parties to the Basel Convention on the Control of Transboundary Movements of Hazardous Wastes and Their Disposal, May 2019.

[47] Adopted by the tenth meeting of the Conference of the Parties to the Basel Convention on the Control of Transboundary Movements of Hazardous Wastes and Their Disposal, October 2013.

consisting of, containing or contaminated with mercury or mercury compounds[48]

18. Technical guidelines for the environmentally sound co-processing of hazardous wastes in cement kilns[49]

19. Guidance document on the environmentally sound management of used and end-of-life computing equipment[50]

20. Guidance document on environmentally sound management of used and end-of-life mobile phones[51]

21. Framework for the environmentally sound management of hazardous wastes and other wastes[52]

22. Practical manuals for the promotion of the environmentally sound management of wastes[53]

II.   Guidelines adopted by the OECD:

Technical guidance for the environmentally sound management of specific waste streams:

Used and scrap personal computers[54]

III.   Guidelines adopted by the International Maritime Organization (IMO):

Guidelines on ship recycling[55]

V.   Guidelines adopted by the International Labour Organization (ILO):

Safety and health in shipbreaking: guidelines for Asian countries and Turkey[56]

---

[48] Adopted by the twelfth meeting of the Conference of the Parties to the Basel Convention on the Control of Transboundary Movements of Hazardous Wastes and Their Disposal, May 2015.

[49] Adopted by the tenth meeting of the Conference of the Parties to the Basel Convention on the Control of Transboundary Movements of Hazardous Wastes and Their Disposal, October 2013.

[50] Adopted by the thirteenth meeting of the Conference of the Parties to the Basel Convention on the Control of Transboundary Movements of Hazardous Wastes and Their Disposal, May 2017.

[51] Adopted by the tenth meeting of the Conference of the Parties to the Basel Convention on the Control of Transboundary Movements of Hazardous Wastes and Their Disposal, October 2013.

[52] Adopted by the eleventh meeting of the Conference of the Parties to the Basel Convention on the Control of Transboundary Movements of Hazardous Wastes and Their Disposal, October 2013.

[53] Adopted by the thirteenth and fourteenth meeting of the Conference of the Parties to the Basel Convention on the Control of Transboundary Movements of Hazardous Wastes and Their Disposal, May 2017 and May 2019.

[54] Adopted by the Environment Policy Committee of the OECD in February 2003 (document ENV/EPOC/WGWPR(2001)3/FINAL).

[55] Resolution A.962 adopted by the Assembly of the IMO at its 23rd Regular session, 24 November to 5 December 2003.

[56] Approved for publication by the Governing Body of the ILO at its 289th session, 11-26 March 2004.

ANNEX IX

ADDITIONAL QUESTIONNAIRE FOR REPORTS BY MEMBER STATES PURSUANT TO ARTICLE 51(2)

[Not published herein]

II.4 Waste Framework Directive

DIRECTIVE 2008/98/EC OF THE EUROPEAN PARLIAMENT AND OF THE COUNCIL

of 19 November 2008[1]

on waste and repealing certain Directives

(WASTE FRAMEWORK DIRECTIVE – WFD)

(Text with EEA relevance)

CHAPTER I

SUBJECT MATTER, SCOPE AND DEFINITIONS

Article 1

Subject matter and scope

This Directive lays down measures to protect the environment and human health by preventing or reducing the generation of waste, the adverse impacts of the generation and management of waste and by reducing overall impacts of resource use and improving the efficiency of such use, which are crucial for the transition to a circular economy and for guaranteeing the Union's long-term competitiveness.

Article 2

Exclusions from the scope

1.   The following shall be excluded from the scope of this Directive:

(a)   gaseous effluents emitted into the atmosphere;

(b)   land (in situ) including unexcavated contaminated soil and buildings permanently connected with land;

(c)   uncontaminated soil and other naturally occurring material excavated in the course of construction activities where it is certain that the material will be used for the purposes of construction in its natural state on the site from which it was excavated;

(d)   radioactive waste;

(e)   decommissioned explosives;

(f)   faecal matter, if not covered by paragraph 2(b), straw and other natural non-

---

[1]   OJ L 312, p. 3, as amended on 30 May 2018 OJ L 150, p. 109

hazardous agricultural or forestry material used in farming, forestry or for the production of energy from such biomass through processes or methods which do not harm the environment or endanger human health.

2.   The following shall be excluded from the scope of this Directive to the extent that they are covered by other Community legislation:

(a)   waste waters;

(b)   animal by-products including processed products covered by Regulation (EC) No 1774/2002, except those which are destined for incineration, landfilling or use in a biogas or composting plant;

(c)   carcasses of animals that have died other than by being slaughtered, including animals killed to eradicate epizootic diseases, and that are disposed of in accordance with Regulation (EC) No 1774/2002;

(d)   waste resulting from prospecting, extraction, treatment and storage of mineral resources and the working of quarries covered by Directive 2006/21/EC of the European Parliament and of the Council of 15 March 2006 on the management of waste from extractive industries[2];

(e)   substances that are destined for use as feed materials as defined in point (g) of Article 3(2) of Regulation (EC) No 767/2009 of the European Parliament and of the Council[3] and that do not consist of or contain animal by-products.

3.   Without prejudice to obligations under other relevant Community legislation, sediments relocated inside surface waters for the purpose of managing waters and waterways or of preventing floods or mitigating the effects of floods and droughts or land reclamation shall be excluded from the scope of this Directive if it is proved that the sediments are non-hazardous.

4.   Specific rules for particular instances, or supplementing those of this Directive, on the management of particular categories of waste, may be laid down by means of individual Directives.

Article 3

Definitions

For the purposes of this Directive, the following definitions shall apply:

1.   'waste' means any substance or object which the holder discards or intends or is required to discard;

2.   'hazardous waste' means waste which displays one or more of the hazardous

---

[2]   OJ L 102, 11.4.2006, p. 15.

[3]   Regulation (EC) No 767/2009 of the European Parliament and of the Council of 13 July 2009 on the placing on the market and use of feed, amending European Parliament and Council Regulation (EC) No 1831/2003 and repealing Council Directive 79/373/EEC, Commission Directive 80/511/EEC, Council Directives 82/471/EEC, 83/228/EEC, 93/74/EEC, 93/113/EC and 96/25/EC and Commission Decision 2004/217/EC (OJ L 229, 1.9.2009, p. 1).

properties listed in Annex III;

2a.   'non-hazardous waste' means waste which is not covered by point 2;

2b.   'municipal waste' means:

(a)   mixed waste and separately collected waste from households, including paper and cardboard, glass, metals, plastics, bio-waste, wood, textiles, packaging, waste electrical and electronic equipment, waste batteries and accumulators, and bulky waste, including mattresses and furniture;

(b)   mixed waste and separately collected waste from other sources, where such waste is similar in nature and composition to waste from households;

Municipal waste does not include waste from production, agriculture, forestry, fishing, septic tanks and sewage network and treatment, including sewage sludge, end-of-life vehicles or construction and demolition waste.

This definition is without prejudice to the allocation of responsibilities for waste management between public and private actors;

2c.   'construction and demolition waste' means waste generated by construction and demolition activities;

3.   'waste oils' means any mineral or synthetic lubrication or industrial oils which have become unfit for the use for which they were originally intended, such as used combustion engine oils and gearbox oils, lubricating oils, oils for turbines and hydraulic oils;

4.   'bio-waste' means biodegradable garden and park waste, food and kitchen waste from households, offices, restaurants, wholesale, canteens, caterers and retail premises and comparable waste from food processing plants;

4a.   'food waste' means all food as defined in Article 2 of Regulation (EC) No 178/2002 of the European Parliament and of the Council4 that has become waste;

5.   'waste producer' means anyone whose activities produce waste (original waste producer) or anyone who carries out pre- processing, mixing or other operations resulting in a change in the nature or composition of this waste;

6.   'waste holder' means the waste producer or the natural or legal person who is in possession of the waste;

7.   'dealer' means any undertaking which acts in the role of principal to purchase and subsequently sell waste, including such dealers who do not take physical possession of the waste;

8.   'broker' means any undertaking arranging the recovery or disposal of waste on behalf of others, including such brokers who do not take physical possession of the waste;

---

[4]   Regulation (EC) No 178/2002 of the European Parliament and of the Council of 28 January 2002 laying down the general principles and requirements of food law, establishing the European Food Safety Authority and laying down procedures in matters of food safety (OJ L 31, 1.2.2002, p. 1).

9.   'waste management' means the collection, transport, recovery (including sorting), and disposal of waste, including the super- vision of such operations and the after-care of disposal sites, and including actions taken as a dealer or broker;

10.   'collection' means the gathering of waste, including the preliminary sorting and preliminary storage of waste for the purposes of transport to a waste treatment facility;

11.   'separate collection' means the collection where a waste stream is kept separately by type and nature so as to facilitate a specific treatment;

12.   'prevention' means measures taken before a substance, material or product has become waste, that reduce:

(a)   the quantity of waste, including through the re-use of products or the extension of the life span of products;

(b)   the adverse impacts of the generated waste on the environment and human health; or

(c)   the content of hazardous substances in materials and products;

13.   're-use' means any operation by which products or components that are not waste are used again for the same purpose for which they were conceived;

14.   'treatment' means recovery or disposal operations, including preparation prior to recovery or disposal;

15.   'recovery' means any operation the principal result of which is waste serving a useful purpose by replacing other materials which would otherwise have been used to fulfil a particular function, or waste being prepared to fulfil that function, in the plant or in the wider economy. Annex II sets out a non-exhaustive list of recovery operations;

15a. 'material recovery' means any recovery operation, other than energy recovery and the reprocessing into materials that are to be used as fuels or other means to generate energy. It includes, inter alia, preparing for re-use, recycling and backfilling;

16.   'preparing for re-use' means checking, cleaning or repairing recovery operations, by which products or components of products that have become waste are prepared so that they can be re-used without any other pre-processing;

17.   'recycling' means any recovery operation by which waste materials are reprocessed into products, materials or substances whether for the original or other purposes. It includes the reprocessing of organic material but does not include energy recovery and the reprocessing into materials that are to be used as fuels or for backfilling operations;

17a. 'backfilling' means any recovery operation where suitable non-hazardous waste is used for purposes of reclamation in excavated areas or for engineering purposes in landscaping. Waste used for backfilling must substitute non-waste materials, be suitable for the aforementioned purposes, and be limited to the amount strictly necessary to achieve those purposes;

18.   'regeneration of waste oils' means any recycling operation whereby base oils

can be produced by refining waste oils, in particular by removing the contaminants, the oxidation products and the additives contained in such oils;

19. 'disposal' means any operation which is not recovery even where the operation has as a secondary consequence the reclamation of substances or energy. Annex I sets out a non-exhaustive list of disposal operations;

20. 'best available techniques' means best available techniques as defined in Article 2(11) of Directive 96/61/EC;

21. 'extended producer responsibility scheme' means a set of measures taken by Member States to ensure that producers of products bear financial responsibility or financial and organisational responsibility for the management of the waste stage of a product's life cycle.

Article 4

Waste hierarchy

1. The following waste hierarchy shall apply as a priority order in waste prevention and management legislation and policy:

(a)   prevention;

(b)   preparing for re-use;

(c)   recycling;

(d)   other recovery, e.g. energy recovery; and

(e)   disposal.

2. When applying the waste hierarchy referred to in paragraph 1, Member States shall take measures to encourage the options that deliver the best overall environmental outcome. This may require specific waste streams departing from the hierarchy where this is justified by life-cycle thinking on the overall impacts of the generation and management of such waste.

Member States shall ensure that the development of waste legislation and policy is a fully transparent process, observing existing national rules about the consultation and involvement of citizens and stakeholders.

Member States shall take into account the general environmental protection principles of precaution and sustainability, technical feasibility and economic viability, protection of resources as well as the overall environmental, human health, economic and social impacts, in accordance with Articles 1 and 13.

3. Member States shall make use of economic instruments and other measures to provide incentives for the application of the waste hierarchy, such as those indicated in Annex IVa or other appropriate instruments and measures.

Article 5

By-products

1.   Member States shall take appropriate measures to ensure that a substance or object resulting from a production process the primary aim of which is not the production of that substance or object is considered not to be waste, but to be a by-product if the following conditions are met:

(a)   further use of the substance or object is certain;

(b)   the substance or object can be used directly without any further processing other than normal industrial practice;

(c)   the substance or object is produced as an integral part of a production process; and

(d)   further use is lawful, i.e. the substance or object fulfils all relevant product, environmental and health protection requirements for the specific use and will not lead to overall adverse environmental or human health impacts.

2.   The Commission may adopt implementing acts in order to establish detailed criteria on the uniform application of the conditions laid down in paragraph 1 to specific substances or objects.

Those detailed criteria shall ensure a high level of protection of the environment and human health and facilitate the prudent and rational utilisation of natural resources.

Those implementing acts shall be adopted in accordance with the examination procedure referred to in Article 39(2). When adopting those implementing acts, the Commission shall take as a starting point the most stringent and environmentally protective of any criteria adopted by Member States in accordance with paragraph 3 of this Article and shall prioritise replicable practices of industrial symbiosis in the development of the detailed criteria.

3.   Where criteria have not been set at Union level under paragraph 2, Member States may establish detailed criteria on the application of the conditions laid down in paragraph 1 to specific substances or objects.

Member States shall notify the Commission of those detailed criteria in accordance with Directive (EU) 2015/1535 of the European Parliament and of the Council[5] where so required by that Directive.

Article 6

End-of-waste status

1.   Member States shall take appropriate measures to ensure that waste which has undergone a recycling or other recovery operation is considered to have ceased

---

[5]   Directive (EU) 2015/1535 of the European Parliament and of the Council of 9 September 2015 laying down a procedure for the provision of information in the field of technical regulations and of rules on Information Society services (OJ L 241, 17.9.2015, p. 1).

to be waste if it complies with the following conditions:

(a)   the substance or object is to be used for specific purposes;

(b)   a market or demand exists for such a substance or object;

(c)   the substance or object fulfils the technical requirements for the specific purposes and meets the existing legislation and standards applicable to products; and

(d)   the use of the substance or object will not lead to overall adverse environmental or human health impacts.

2.   The Commission shall monitor the development of national end- of-waste criteria in Member States, and assess the need to develop Union-wide criteria on this basis. To that end, and where appropriate, the Commission shall adopt implementing acts in order to establish detailed criteria on the uniform application of the conditions laid down in paragraph 1 to certain types of waste.

Those detailed criteria shall ensure a high level of protection of the environment and human health and facilitate the prudent and rational utilisation of natural resources. They shall include:

(a)   permissible waste input material for the recovery operation;

(b)   allowed treatment processes and techniques;

(c)   quality criteria for end-of-waste materials resulting from the recovery operation in line with the applicable product standards, including limit values for pollutants where necessary;

(d)   requirements for management systems to demonstrate compliance with the end-of-waste criteria, including for quality control and self-monitoring, and accreditation, where appropriate; and

(e)   a requirement for a statement of conformity.

Those implementing acts shall be adopted in accordance with the examination procedure referred to in Article 39(2).

When adopting those implementing acts, the Commission shall take account of the relevant criteria established by Member States in accordance with paragraph 3 and shall take as a starting point the most stringent and environmentally protective of those criteria.

3.   Where criteria have not been set at Union level under paragraph 2, Member States may establish detailed criteria on the application of the conditions laid down in paragraph 1 to certain types of waste. Those detailed criteria shall take into account any possible adverse environmental and human health impacts of the substance or object and shall satisfy the requirements laid down in points (a) to (e) of paragraph 2.

Member States shall notify the Commission of those criteria in accordance with Directive (EU) 2015/1535 where so required by that Directive.

4.   Where criteria have not been set at either Union or national level under paragraph 2 or 3, respectively, a Member State may decide on a case-by-case basis, or

take appropriate measures to verify, that certain waste has ceased to be waste on the basis of the conditions laid down in paragraph 1 and, where necessary, reflecting the requirements laid down in points (a) to (e) of paragraph 2, and taking into account limit values for pollutants and any possible adverse environmental and human health impacts. Such case-by-case decisions are not required to be notified to the Commission in accordance with Directive (EU) 2015/1535.

Member States may make information about case-by-case decisions and about the results of verification by competent authorities publicly available by electronic means.

5.   The natural or legal person who:

(a)   uses, for the first time, a material that has ceased to be waste and that has not been placed on the market; or

(b)   places a material on the market for the first time after it has ceased to be waste,

shall ensure that the material meets relevant requirements under the applicable chemical and product related legislation. The conditions laid down in paragraph 1 have to be met before the legislation on chemicals and products applies to the material that has ceased to be waste.

Article 7

List of waste

1.   The Commission is empowered to adopt delegated acts in accordance with Article 38a in order to supplement this Directive by establishing, and reviewing in accordance with paragraphs 2 and 3 of this Article, a list of waste. The list of waste shall include hazardous waste and shall take into account the origin and composition of the waste and, where necessary, the limit values of concentration of hazardous substances. The list of waste shall be binding as regards determination of the waste which is to be considered as hazardous waste. The inclusion of a substance or object in the list shall not mean that it is waste in all circumstances. A substance or object shall be considered to be waste only where the definition in point (1) of Article 3 is met.

2.   A Member State may consider waste as hazardous waste where, even though it does not appear as such on the list of waste, it displays one or more of the properties listed in Annex III. The Member State shall notify the Commission of any such cases without delay and provide the Commission with all relevant information. In the light of notifications received, the list shall be reviewed in order to decide on its adaptation.

3.   Where a Member State has evidence to show that specific waste that appears on the list as hazardous waste does not display any of the properties listed in Annex III, it may consider that waste as non- hazardous waste. The Member State shall notify the Commission of any such cases without delay and shall provide the Commission with the necessary evidence. In the light of notifications received, the list shall be reviewed in order to decide on its adaptation.

4.   The reclassification of hazardous waste as non-hazardous waste may not be achieved by diluting or mixing the waste with the aim of lowering the initial concentrations of hazardous substances to a level below the thresholds for defining waste as hazardous.

5.   *deleted*

6.   Member States may consider waste as non-hazardous waste in accordance with the list of waste referred to in paragraph 1.

7.   The Commission shall ensure that the list of waste and any review of this list adhere, as appropriate, to principles of clarity, comprehensibility and accessibility for users, particularly small and medium-sized enterprises (SMEs).

CHAPTER II

GENERAL REQUIREMENTS

Article 8

Extended producer responsibility

1.   In order to strengthen the re-use and the prevention, recycling and other recovery of waste, Member States may take legislative or non- legislative measures to ensure that any natural or legal person who professionally develops, manufactures, processes, treats, sells or imports products (producer of the product) has extended producer responsibility.

Such measures may include an acceptance of returned products and of the waste that remains after those products have been used, as well as the subsequent management of the waste and financial responsibility for such activities. These measures may include the obligation to provide publicly available information as to the extent to which the product is re-usable and recyclable.

Where such measures include the establishment of extended producer responsibility schemes, the general minimum requirements laid down in Article 8a shall apply.

Member States may decide that producers of products that undertake financial or financial and organisational responsibilities for the management of the waste stage of a product's life cycle of their own accord should apply some or all of the general minimum requirements laid down in Article 8a.

2.   Member States may take appropriate measures to encourage the design of products and components of products in order to reduce their environmental impact and the generation of waste in the course of the production and subsequent use of products, and in order to ensure that the recovery and disposal of products that have become waste take place in accordance with Articles 4 and 13.

Such measures may encourage, inter alia, the development, production and marketing of products and components of products that are suitable for multiple use, that contain recycled materials, that are technically durable and easily reparable and that are, after having become waste, suitable for preparing for re-use and recycling in order to facilitate proper implementation of the waste hierarchy. The measures shall take into account the impact of products throughout their life cycle, the waste hierarchy and, where appropriate, the potential for multiple recycling.

3.   When applying extended producer responsibility, Member States shall take into account the technical feasibility and economic viability and the overall environmental, human health and social impacts, respecting the need to ensure the proper functioning of the internal market.

4.   The extended producer responsibility shall be applied without prejudice to the responsibility for waste management as provided for in Article 15(1) and without prejudice to existing waste stream specific and product specific legislation.

5.   The Commission shall organise an exchange of information between Member States and the actors involved in extended producer responsibility schemes on the practical implementation of the general minimum requirements laid down in Article 8a. This includes, inter alia, exchange of information on best practices to ensure adequate governance, cross-border cooperation concerning extended producer responsibility schemes and a smooth functioning of the internal market, on the organisational features and the monitoring of organisations implementing extended producer responsibility obligations on behalf of producers of products, on the modulation of financial contributions, on the selection of waste management operators and on the prevention of littering. The Commission shall publish the results of the exchange of information and may provide guidelines on these and other relevant aspects.

The Commission shall publish guidelines, in consultation with Member States, on cross-border cooperation concerning extended producer responsibility schemes and on the modulation of financial contributions referred to in point (b) of Article 8a(4).

Where necessary to avoid distortion of the internal market, the Commission may adopt implementing acts in order to lay down criteria with a view to the uniform application of point (b) of Article 8a(4), but excluding any precise determination of the level of the contributions. Those implementing acts shall be adopted in accordance with the examination procedure referred to in Article 39(2).

Article 8a

General minimum requirements for extended producer responsibility schemes

1.   Where extended producer responsibility schemes are established in accordance with Article 8(1), including pursuant to other legislative acts of the Union, Member States shall:

(a)   define in a clear way the roles and responsibilities of all relevant actors involved, including producers of products placing products on the market of the Member State, organisations implementing extended producer responsibility obligations on their behalf, private or public waste operators, local authorities and, where appropriate, re-use and preparing for re-use operators and social economy enterprises;

(b)   in line with the waste hierarchy, set waste management targets, aiming to attain at least the quantitative targets relevant for the extended producer responsibility scheme as laid down in this Directive, Directive 94/62/EC, Directive 2000/53/EC,

Directive 2006/66/EC and Directive 2012/19/EU of the European Parliament and of the Council[6], and set other quantitative targets and/or qualitative objectives that are considered relevant for the extended producer responsibility scheme;

(c)   ensure that a reporting system is in place to gather data on the products placed on the market of the Member State by the producers of products subject to extended producer responsibility and data on the collection and treatment of waste resulting from those products specifying, where appropriate, the waste material flows, as well as other data relevant for the purposes of point (b);

(d)   ensure equal treatment of producers of products regardless of their origin or size, without placing a disproportionate regulatory burden on producers, including small and medium-sized enterprises, of small quantities of products.

2.   Member States shall take the necessary measures to ensure that the waste holders targeted by the extended producer responsibility schemes established in accordance with Article 8(1), are informed about waste prevention measures, centres for re-use and preparing for re-use, take- back and collection systems, and the prevention of littering. Member States shall also take measures to create incentives for the waste holders to assume their responsibility to deliver their waste into the separate collection systems in place, notably, where appropriate, through economic incentives or regulations.

3.   Member States shall take the necessary measures to ensure that any producer of products or organisation implementing extended producer responsibility obligations on behalf of producers of products:

(a)   has a clearly defined geographical, product and material coverage without limiting those areas to those where the collection and management of waste are the most profitable;

(b)   provides an appropriate availability of waste collection systems within the areas referred to in point (a);

(c)   has the necessary financial means or financial and organisational means to meet its extended producer responsibility obligations;

(d)   puts in place an adequate self-control mechanism, supported, where relevant, by regular independent audits, to appraise:

  (i)   its financial management, including compliance with the requirements laid down in points (a) and (b) of paragraph 4;

  (ii)   the quality of data collected and reported in accordance with point (c) of paragraph 1 of this Article and with the requirements of Regulation (EC) No 1013/2006;

(e)   makes publicly available information about the attainment of the waste management targets referred to in point (b) of paragraph 1, and, in the case of collective fulfilment of extended producer responsibility obligations, also information about:

---

[6]   Directive 2012/19/EU of the European Parliament and of the Council of 4 July 2012 on waste electrical and electronic equipment (WEEE) (OJ L 197, 24.7.2012, p. 38).

(i) its ownership and membership;

(ii) the financial contributions paid by producers of products per unit sold or per tonne of product placed on the market; and

(iii) the selection procedure for waste management operators.

4.   Member States shall take the necessary measures to ensure that the financial contributions paid by the producer of the product to comply with its extended producer responsibility obligations:

(a)   cover the following costs for the products that the producer puts on the market in the Member State concerned:

—   costs of separate collection of waste and its subsequent transport and treatment, including treatment necessary to meet the Union waste management targets, and costs necessary to meet other targets and objectives as referred to in point (b) of paragraph 1, taking into account the revenues from re-use, from sales of secondary raw material from its products and from unclaimed deposit fees,

—   costs of providing adequate information to waste holders in accordance with paragraph 2,

—   costs of data gathering and reporting in accordance with point (c) of paragraph 1.

This point shall not apply to extended producer responsibility schemes established pursuant to Directive 2000/53/EC, 2006/66/EC or 2012/19/EU;

(b)   in the case of collective fulfilment of extended producer responsibility obligations, are modulated, where possible, for individual products or groups of similar products, notably by taking into account their durability, reparability, re-usability and recyclability and the presence of hazardous substances, thereby taking a life- cycle approach and aligned with the requirements set by relevant Union law, and where available, based on harmonised criteria in order to ensure a smooth functioning of the internal market; and

(c)   do not exceed the costs that are necessary to provide waste management services in a cost-efficient way. Such costs shall be established in a transparent way between the actors concerned.

Where justified by the need to ensure proper waste management and the economic viability of the extended producer responsibility scheme, Member States may depart from the division of financial responsibility as laid down in point (a), provided that:

(i) in the case of extended producer responsibility schemes established to attain waste management targets and objectives established under legislative acts of the Union, the producers of products bear at least 80 % of the necessary costs;

(ii) in the case of extended producer responsibility schemes established on or after 4 July 2018 to attain waste management targets and objectives solely established in Member State legislation, the producers of products bear at least 80 % of the necessary costs;

(iii) in the case of extended producer responsibility schemes established before 4 July 2018 to attain waste management targets and objectives solely established in Member State legislation, the producers of products bear at least 50 % of the necessary costs, and provided that the remaining costs are borne by original waste producers or distributors.

This derogation may not be used to lower the proportion of costs borne by producers of products under extended producer responsibility schemes established before 4 July 2018.

5.   Member States shall establish an adequate monitoring and enforcement framework with a view to ensuring that producers of products and organisations implementing extended producer responsibility obligations on their behalf implement their extended producer responsibility obligations, including in the case of distance sales, that the financial means are properly used and that all actors involved in the implementation of the extended producer responsibility schemes report reliable data.

Where, in the territory of a Member State, multiple organisations implement extended producer responsibility obligations on behalf of producers of products, the Member State concerned shall appoint at least one body independent of private interests or entrust a public authority to oversee the implementation of extended producer responsibility obligations.

Each Member State shall allow the producers of products established in another Member State and placing products on its territory to appoint a legal or natural person established on its territory as an authorised representative for the purposes of fulfilling the obligations of a producer related to extended producer responsibility schemes on its territory.

For the purposes of monitoring and verifying compliance with the obligations of the producer of the product in relation to extended producer responsibility schemes, Member States may lay down requirements, such as registration, information and reporting requirements, to be met by a legal or natural person to be appointed as an authorised representative on their territory.

6.   Member States shall ensure a regular dialogue between relevant stakeholders involved in the implementation of extended producer responsibility schemes, including producers and distributors, private or public waste operators, local authorities, civil society organisations and, where applicable, social economy actors, re-use and repair networks and preparing for re-use operators.

7.   Member States shall take measures to ensure that extended producer responsibility schemes that have been established before 4 July 2018, comply with this Article by 5 January 2023.

8.   The provision of information to the public under this Article shall be without prejudice to preserving the confidentiality of commercially sensitive information in conformity with the relevant Union and national law.

Article 9

Prevention of waste

1. Member States shall take measures to prevent waste generation. Those measures shall, at least:

(a) promote and support sustainable production and consumption models;

(b) encourage the design, manufacturing and use of products that are resource-efficient, durable (including in terms of life span and absence of planned obsolescence), reparable, re-usable and upgradable;

(c) target products containing critical raw materials to prevent that those materials become waste;

(d) encourage the re-use of products and the setting up of systems promoting repair and re-use activities, including in particular for electrical and electronic equipment, textiles and furniture, as well as packaging and construction materials and products;

(e) encourage, as appropriate and without prejudice to intellectual property rights, the availability of spare parts, instruction manuals, technical information, or other instruments, equipment or software enabling the repair and re-use of products without compromising their quality and safety;

(f) reduce waste generation in processes related to industrial production, extraction of minerals, manufacturing, construction and demolition, taking into account best available techniques;

(g) reduce the generation of food waste in primary production, in processing and manufacturing, in retail and other distribution of food, in restaurants and food services as well as in households as a contribution to the United Nations Sustainable Development Goal to reduce by 50 % the per capita global food waste at the retail and consumer levels and to reduce food losses along production and supply chains by 2030;

(h) encourage food donation and other redistribution for human consumption, prioritising human use over animal feed and the reprocessing into non-food products;

(i) promote the reduction of the content of hazardous substances in materials and products, without prejudice to harmonised legal requirements concerning those materials and products laid down at Union level, and ensure that any supplier of an article as defined in point 33 of Article 3 of Regulation (EC) No 1907/2006 of the European Parliament and of the Council[7] provides the information pursuant to Article 33(1) of that Regulation to the European Chemicals Agency as from 5 January 2021;

---

[7] Regulation (EC) No 1907/2006 of the European Parliament and of the Council of 18 December 2006 concerning the Registration, Evaluation, Authorisation and Restriction of Chemicals (REACH), establishing a European Chemicals Agency, amending Directive 1999/45/EC and repealing Council Regulation (EEC) No 793/93 and Commission Regulation (EC) No 1488/94 as well as Council Directive 76/769/EEC and Commission Directives 91/155/EEC, 93/67/EEC, 93/105/EC and 2000/21/EC (OJ L 396, 30.12.2006, p. 1).

(j)   reduce the generation of waste, in particular waste that is not suitable for preparing for re-use or recycling;

(k)   identify products that are the main sources of littering, notably in natural and marine environments, and take appropriate measures to prevent and reduce litter from such products; where Member States decide to implement this obligation through market restrictions, they shall ensure that such restrictions are proportionate and non-discriminatory;

(l)   aim to halt the generation of marine litter as a contribution towards the United Nations Sustainable Development Goal to prevent and significantly reduce marine pollution of all kinds; and

(m)   develop and support information campaigns to raise awareness about waste prevention and littering.

2.   The European Chemicals Agency shall establish a database for the data to be submitted to it pursuant to point (i) of paragraph 1 by 5 January 2020 and maintain it. The European Chemicals Agency shall provide access to that database to waste treatment operators. It shall also provide access to that database to consumers upon request.

3.   Member States shall monitor and assess the implementation of the waste prevention measures. For that purpose, they shall use appropriate qualitative or quantitative indicators and targets, notably on the quantity of waste that is generated.

4.   Member States shall monitor and assess the implementation of their measures on re-use by measuring re-use on the basis of the common methodology established by the implementing act referred to in paragraph 7, as from the first full calendar year after the adoption of that implementing act.

5.   Member States shall monitor and assess the implementation of their food waste prevention measures by measuring the levels of food waste on the basis of the methodology established by the delegated act referred to in paragraph 8, as from the first full calendar year after the adoption of that delegated act.

6.   By 31 December 2023, the Commission shall examine the data on food waste provided by Member States in accordance with Article 37(3) with a view to considering the feasibility of establishing a Union-wide food waste reduction target to be met by 2030 on the basis of the data reported by Member States in accordance with the common methodology established pursuant to paragraph 8 of this Article. To that end, the Commission shall submit a report to the European Parliament and to the Council, accompanied, if appropriate, by a legislative proposal.

7.   The Commission shall adopt implementing acts to establish indicators to measure the overall progress in the implementation of waste prevention measures and shall, by 31 March 2019, adopt an implementing act to establish a common methodology to report on re-use of products. Those implementing acts shall be adopted in accordance with the examination procedure referred to in Article 39(2).

8.   By 31 March 2019, the Commission shall adopt, on the basis of the outcome of the work of the EU Platform on Food Losses and Food Waste, a delegated act in accordance with Article 38a to supplement this Directive by establishing a common

methodology and minimum quality requirements for the uniform measurement of levels of food waste.

9.   By 31 December 2024, the Commission shall examine data on re- use provided by Member States in accordance with Article 37(3) with a view to considering the feasibility of measures to encourage the re-use of products, including the setting of quantitative targets. The Commission shall also examine the feasibility of setting other waste prevention measures, including waste reduction targets. To that end, the Commission shall submit a report to the European Parliament and to the Council, accompanied, if appropriate, by a legislative proposal.

Article 10

Recovery

1.   Member States shall take the necessary measures to ensure that waste undergoes preparing for re-use, recycling or other recovery operations, in accordance with Articles 4 and 13.

2.   Where necessary to comply with paragraph 1 and to facilitate or improve preparing for re-use, recycling and other recovery operations, waste shall be subject to separate collection and shall not be mixed with other waste or other materials with different properties.

3.   Member States may allow derogations from paragraph 2 provided that at least one of the following conditions is met:

(a)   collecting certain types of waste together does not affect their potential to undergo preparing for re-use, recycling or other recovery operations in accordance with Article 4 and results in output from those operations which is of comparable quality to that achieved through separate collection;

(b)   separate collection does not deliver the best environmental outcome when considering the overall environmental impacts of the management of the relevant waste streams;

(c)   separate collection is not technically feasible taking into consideration good practices in waste collection;

(d)   separate collection would entail disproportionate economic costs taking into account the costs of adverse environmental and health impacts of mixed waste collection and treatment, the potential for efficiency improvements in waste collection and treatment, revenues from sales of secondary raw materials as well as the application of the polluter-pays principle and extended producer responsibility.

Member States shall regularly review derogations under this paragraph taking into account good practices in separate collection of waste and other developments in waste management.

4.   Member States shall take measures to ensure that waste that has been separately collected for preparing for re-use and recycling pursuant to Article 11(1) and Article 22 is not incinerated, with the exception of waste resulting from subsequent treatment operations of the separately collected waste for which incineration delivers

the best environmental outcome in accordance with Article 4.

5.   Where necessary to comply with paragraph 1 of this Article and to facilitate or improve recovery, Member States shall take the necessary measures, before or during recovery, to remove hazardous substances, mixtures and components from hazardous waste with a view to their treatment in accordance with Articles 4 and 13.

6.   By 31 December 2021, Member States shall submit a report to the Commission on the implementation of this Article as regards municipal waste and bio-waste, including on the material and territorial coverage of separate collection and any derogations under paragraph 3.

Article 11

Preparing for re-use and recycling

1.   Member States shall take measures to promote preparing for re-use activities, notably by encouraging the establishment of and support for preparing for re-use and repair networks, by facilitating, where compatible with proper waste management, their access to waste held by collection schemes or facilities that can be prepared for re-use but is not destined for preparing for re-use by those schemes or facilities, and by promoting the use of economic instruments, procurement criteria, quantitative objectives or other measures.

Member States shall take measures to promote high-quality recycling and, to this end, subject to Article 10(2) and (3), shall set up separate collection of waste.

Subject to Article 10(2) and (3), Member States shall set up separate collection at least for paper, metal, plastic and glass, and, by 1 January 2025, for textiles.

Member States shall take measures to promote selective demolition in order to enable removal and safe handling of hazardous substances and facilitate re-use and high-quality recycling by selective removal of materials, and to ensure the establishment of sorting systems for construction and demolition waste at least for wood, mineral fractions (concrete, bricks, tiles and ceramics, stones), metal, glass, plastic and plaster.

2.   In order to comply with the objectives of this Directive, and move to a European circular economy with a high level of resource efficiency, Member States shall take the necessary measures designed to achieve the following targets:

(a)   by 2020, the preparing for re-use and the recycling of waste materials such as at least paper, metal, plastic and glass from households and possibly from other origins as far as these waste streams are similar to waste from households, shall be increased to a minimum of overall 50 % by weight;

(b)   by 2020, the preparing for re-use, recycling and other material recovery, including backfilling operations using waste to substitute other materials, of non-hazardous construction and demolition waste excluding naturally occurring material defined in category 17 05 04 in the list of waste shall be increased to a minimum of 70 % by weight;

(c)   by 2025, the preparing for re-use and the recycling of municipal waste shall be increased to a minimum of 55 % by weight;

(d)   by 2030, the preparing for re-use and the recycling of municipal waste shall be increased to a minimum of 60 % by weight;

(e)   by 2035, the preparing for re-use and the recycling of municipal waste shall be increased to a minimum of 65 % by weight.

3.   A Member State may postpone the deadlines for attaining the targets referred to in points (c), (d) and (e) of paragraph 2 by up to five years provided that that Member State:

(a)   prepared for re-use and recycled less than 20 % or landfilled more than 60 % of its municipal waste generated in 2013 as reported under the Joint Questionnaire of the OECD and Eurostat; and

(b)   at the latest 24 months before the respective deadline laid down in point (c), (d) or (e) of paragraph 2, notifies the Commission of its intention to postpone the respective deadline and submits an implementation plan in accordance with Annex IVb.

4.   Within three months of receipt of the implementation plan submitted pursuant to point (b) of paragraph 3, the Commission may request a Member State to revise that plan if the Commission considers that the plan does not comply with the requirements set out in Annex IVb. The Member State concerned shall submit a revised plan within three months of receipt of the Commission's request.

5.   In the event of postponing the attainment of the targets in accordance with paragraph 3, the Member State concerned shall take the necessary measures to increase the preparing for re-use and the recycling of municipal waste:

(a)   to a minimum of 50 % by 2025 in the event of postponing the deadline for attaining the target referred to in point (c) of paragraph 2;

(b)   to a minimum of 55 % by 2030 in the event of postponing the deadline for attaining the target referred to in point (d) of paragraph 2;

(c)   to a minimum of 60 % by 2035 in the event of postponing the deadline for attaining the target referred to in point (e) of paragraph 2.

6.   By 31 December 2024, the Commission shall consider the setting of preparing for re-use and recycling targets for construction and demolition waste and its material-specific fractions, textile waste, commercial waste, non-hazardous industrial waste and other waste streams, as well as preparing for re-use targets for municipal waste and recycling targets for municipal bio-waste. To that end, the Commission shall submit a report to the European Parliament and to the Council, accompanied, if appropriate, by a legislative proposal.

7.   By 31 December 2028, the Commission shall review the target laid down in point (e) of paragraph 2. To that end, the Commission shall submit a report to the European Parliament and to the Council, accompanied, if appropriate, by a legislative proposal.

The Commission shall assess co-processing technology that allows the incorporation of minerals in the co-incineration process of municipal waste. Where a reliable methodology can be found, as part of this review, the Commission shall consider whether such minerals may be counted towards recycling targets.

Article 11a

Rules on the calculation of the attainment of the targets

1.   For the purpose of calculating whether the targets laid down in points (c), (d) and (e) of Article 11(2) and in Article 11(3) have been attained:

(a)   Member States shall calculate the weight of the municipal waste generated and prepared for re-use or recycled in a given calendar year;

(b)   the weight of the municipal waste prepared for re-use shall be calculated as the weight of products or components of products that have become municipal waste and have undergone all necessary checking, cleaning or repairing operations to enable re-use without further sorting or pre-processing;

(c)   the weight of the municipal waste recycled shall be calculated as the weight of waste which, having undergone all necessary checking, sorting and other preliminary operations to remove waste materials that are not targeted by the subsequent reprocessing and to ensure high-quality recycling, enters the recycling operation whereby waste materials are actually reprocessed into products, materials or substances.

2.   For the purposes of point (c) of paragraph 1, the weight of the municipal waste recycled shall be measured when the waste enters the recycling operation.

By way of derogation from the first subparagraph, the weight of municipal waste recycled may be measured at the output of any sorting operation provided that:

(a)   such output waste is subsequently recycled;

(b)   the weight of materials or substances that are removed by further operations preceding the recycling operation and are not subsequently recycled is not included in the weight of waste reported as recycled.

3.   Member States shall establish an effective system of quality control and traceability of municipal waste to ensure that the conditions laid down in point (c) of paragraph 1 of this Article and in paragraph 2 of this Article are met. To ensure the reliability and accuracy of the data gathered on recycled waste, the system may consist of electronic registries set up pursuant to Article 35(4), technical specifications for the quality requirements of sorted waste, or average loss rates for sorted waste for various waste types and waste management practices respectively. Average loss rates shall only be used in cases where reliable data cannot be obtained otherwise and shall be calculated on the basis of the calculation rules established in the delegated act adopted pursuant to paragraph 10 of this Article.

4.   For the purpose of calculating whether the targets laid down in points (c), (d) and (e) of Article 11(2) and in Article 11(3) have been attained, the amount of municipal biodegradable waste that enters aerobic or anaerobic treatment may be counted as recycled where that treatment generates compost, digestate, or other output with a similar quantity of recycled content in relation to input, which is to be used as a

recycled product, material or substance. Where the output is used on land, Member States may count it as recycled only if this use results in benefits to agriculture or ecological improvement.

As from 1 January 2027, Member States may count municipal bio-waste entering aerobic or anaerobic treatment as recycled only if, in accordance with Article 22, it has been separately collected or separated at source.

5.   For the purposes of calculating whether the targets laid down in points (c), (d) and (e) of Article 11(2) and in Article 11(3) have been attained, the amount of waste materials that have ceased to be waste as a result of a preparatory operation before being reprocessed may be counted as recycled provided that such materials are destined for subsequent reprocessing into products, materials or substances to be used for the original or other purposes. However, end-of-waste materials to be used as fuels or other means to generate energy, or to be incinerated, backfilled or landfilled, shall not be counted towards the attainment of the recycling targets.

6.   For the purposes of calculating whether the targets laid down in points (c), (d) and (e) of Article 11(2) and in Article 11(3) have been attained, Member States may take into account the recycling of metals separated after incineration of municipal waste provided that the recycled metals meet certain quality criteria laid down in the implementing act adopted pursuant to paragraph 9 of this Article.

7.   Waste sent to another Member State for the purposes of preparing for re-use, recycling or backfilling in that other Member State may only be counted towards the attainment of the targets laid down in Article 11(2) and (3) by the Member State in which that waste was collected.

8.   Waste exported from the Union for preparing for re-use or recycling shall count towards the attainment of the targets laid down in Article 11(2) and (3) of this Directive by the Member State in which it was collected only if the requirements of paragraph 3 of this Article are met and if, in accordance with Regulation (EC) No 1013/2006, the exporter can prove that the shipment of waste complies with the requirements of that Regulation and that the treatment of waste outside the Union took place in conditions that are broadly equivalent to the requirements of the relevant Union environmental law.

9.   In order to ensure uniform conditions for the application of this Article, the Commission shall adopt by 31 March 2019 implementing acts establishing rules for the calculation, verification and reporting of data, in particular as regards:

(a)   a common methodology for the calculation of the weight of metals that have been recycled in accordance with paragraph 6, including quality criteria for the recycled metals, and

(b)   bio-waste separated and recycled at source.

Those implementing acts shall be adopted in accordance with the examination procedure referred to in Article 39(2).

10.   By 31 March 2019, the Commission shall adopt a delegated act in accordance with Article 38a in order to supplement this Directive by establishing rules for the calculation, verification and reporting of the weight of materials or substances which

are removed after a sorting operation and which are not subsequently recycled, based on average loss rates for sorted waste.

Article 11b

Early warning report

1.   The Commission shall, in cooperation with the European Environment Agency, draw up reports on the progress towards the attainment of the targets laid down in points (c), (d) and (e) of Article 11(2) and in Article 11(3) at the latest three years before each deadline laid down therein.

2.   The reports referred to in paragraph 1 shall include the following:

(a)   an estimation of the attainment of the targets by each Member State;

(b)   a list of Member States at risk of not attaining the targets within the respective deadlines, accompanied by appropriate recommendations for the Member States concerned;

(c)   examples of best practices that are used throughout the Union which could provide guidance for progressing towards attaining the targets.

Article 12

Disposal

1.   Member States shall ensure that, where recovery in accordance with Article 10(1) is not undertaken, waste undergoes safe disposal operations which meet the provisions of Article 13 on the protection of human health and the environment.

2.   By 31 December 2024, the Commission shall carry out an assessment of the disposal operations listed in Annex I, in particular in light of Article 13, and shall submit a report to the European Parliament and to the Council, accompanied, if appropriate, by a legislative proposal, with a view to regulating disposal operations, including through possible restrictions, and to consider a disposal reduction target, to ensure environmentally sound waste management.

Article 13

Protection of human health and the environment

Member States shall take the necessary measures to ensure that waste management is carried out without endangering human health, without harming the environment and, in particular:

(a)   without risk to water, air, soil, plants or animals;

(b)   without causing a nuisance through noise or odours; and

(c)   without adversely affecting the countryside or places of special interest.

Article 14

Costs

1.   In accordance with the polluter-pays principle, the costs of waste management, including for the necessary infrastructure and its operation, shall be borne by the original waste producer or by the current or previous waste holders.

2.   Without prejudice to Articles 8 and 8a, Member States may decide that the costs of waste management are to be borne partly or wholly by the producer of the product from which the waste came and that the distributors of such product may share these costs.

CHAPTER III

WASTE MANAGEMENT

Article 15

Responsibility for waste management

1.   Member States shall take the necessary measures to ensure that any original waste producer or other holder carries out the treatment of waste himself or has the treatment handled by a dealer or an establishment or undertaking which carries out waste treatment operations or arranged by a private or public waste collector in accordance with Articles 4 and 13.

2.   When the waste is transferred from the original producer or holder to one of the natural or legal persons referred to in paragraph 1 for preliminary treatment, the responsibility for carrying out a complete recovery or disposal operation shall not be discharged as a general rule.

Without prejudice to Regulation (EC) No 1013/2006, Member States may specify the conditions of responsibility and decide in which cases the original producer is to retain responsibility for the whole treatment chain or in which cases the responsibility of the producer and the holder can be shared or delegated among the actors of the treatment chain.

3.   Member States may decide, in accordance with Article 8, that the responsibility for arranging waste management is to be borne partly or wholly by the producer of the product from which the waste came and that distributors of such product may share this responsibility.

4.   Member States shall take the necessary measures to ensure that, within their territory, the establishments or undertakings which collect or transport waste on a professional basis deliver the waste collected and transported to appropriate treatment installations respecting the provisions of Article 13.

Article 16

Principles of self-sufficiency and proximity

1.   Member States shall take appropriate measures, in cooperation with other Member States where this is necessary or advisable, to establish an integrated and adequate network of waste disposal installations and of installations for the recovery of mixed municipal waste collected from private households, including where such collection also covers such waste from other producers, taking into account best available techniques.

By way of derogation from Regulation (EC) No 1013/2006, Member States may, in order to protect their network, limit incoming shipments of waste destined to incinerators that are classified as recovery, where it has been established that such shipments would result in national waste having to be disposed of or waste having to be treated in a way that is not consistent with their waste management plans. Member States shall notify the Commission of any such decision. Member States may also limit outgoing shipments of waste on environmental grounds as set out in Regulation (EC) No 1013/2006.

2.   The network shall be designed to enable the Community as a whole to become self-sufficient in waste disposal as well as in the recovery of waste referred to in paragraph 1, and to enable Member States to move towards that aim individually, taking into account geographical circumstances or the need for specialised installations for certain types of waste.

3.   The network shall enable waste to be disposed of or waste referred to in paragraph 1 to be recovered in one of the nearest appropriate installations, by means of the most appropriate methods and technologies, in order to ensure a high level of protection for the environment and public health.

4.   The principles of proximity and self-sufficiency shall not mean that each Member State has to possess the full range of final recovery facilities within that Member State.

Article 17

Control of hazardous waste

Member States shall take the necessary action to ensure that the production, collection and transportation of hazardous waste, as well as its storage and treatment, are carried out in conditions providing protection for the environment and human health in order to meet the provisions of Article 13, including action to ensure traceability from production to final destination and control of hazardous waste in order to meet the requirements of Articles 35 and 36.

Article 18

Ban on the mixing of hazardous waste

1.   Member States shall take the necessary measures to ensure that hazardous

waste is not mixed, either with other categories of hazardous waste or with other waste, substances or materials. Mixing shall include the dilution of hazardous substances.

2.   By way of derogation from paragraph 1, Member States may allow mixing provided that:

(a)   the mixing operation is carried out by an establishment or undertaking which has obtained a permit in accordance with Article 23;

(b)   the provisions of Article 13 are complied with and the adverse impact of the waste management on human health and the environment is not increased; and

(c)   the mixing operation conforms to best available techniques.

3.   Where hazardous waste has been unlawfully mixed in breach of this Article, Member States shall ensure, without prejudice to Article 36, that separation is carried out where technically feasible and necessary to comply with Article 13.

Where separation is not required pursuant to the first subparagraph of this paragraph, Member States shall ensure that the mixed waste is treated in a facility that has obtained a permit in accordance with Article 23 to treat such a mixture.

Article 19

Labelling of hazardous waste

1.   Member States shall take the necessary measures to ensure that, in the course of collection, transport and temporary storage, hazardous waste is packaged and labelled in accordance with the international and Community standards in force.

2.   Whenever hazardous waste is transferred within a Member State, it shall be accompanied by an identification document, which may be in electronic format, containing the appropriate data specified in Annex IB to Regulation (EC) No 1013/2006.

Article 20

Hazardous waste produced by households

1.   By 1 January 2025, Member States shall set up separate collection for hazardous waste fractions produced by households to ensure that they are treated in accordance with Articles 4 and 13 and do not contaminate other municipal waste streams.

2.   Articles 17, 18, 19 and 35 shall not apply to mixed waste produced by households.

3.   Articles 19 and 35 shall not apply to separate fractions of hazardous waste produced by households until they are accepted for collection, disposal or recovery by an establishment or an undertaking which has obtained a permit or has been registered in accordance with Article 23 or 26.

4.   By 5 January 2020, the Commission shall draw up guidelines to assist and

facilitate Member States in the separate collection of hazardous waste fractions produced by households.

Article 21

Waste oils

1.	Without prejudice to the obligations related to the management of hazardous waste laid down in Articles 18 and 19, Member States shall take the necessary measures to ensure that:

(a)	waste oils are collected separately, unless separate collection is not technically feasible taking into account good practices;

(b)	waste oils are treated, giving priority to regeneration or alternatively to other recycling operations delivering an equivalent or a better overall environmental outcome than regeneration, in accordance with Articles 4 and 13;

(c)	waste oils of different characteristics are not mixed and waste oils are not mixed with other kinds of waste or substances, if such mixing impedes their regeneration or another recycling operation delivering an equivalent or a better overall environmental outcome than regeneration.

2.	For the purposes of separate collection of waste oils and their proper treatment, Member States may, according to their national conditions, apply additional measures such as technical requirements, producer responsibility, economic instruments or voluntary agreements.

3.	If waste oils, according to national legislation, are subject to requirements of regeneration, Member States may prescribe that such waste oils shall be regenerated if technically feasible and, where Articles 11 or 12 of Regulation (EC) No 1013/2006 apply, restrict the transboundary shipment of waste oils from their territory to incineration or co-incineration facilities in order to give priority to the regeneration of waste oils.

4.	By 31 December 2022, the Commission shall examine data on waste oils provided by Member States in accordance with Article 37(4) with a view to considering the feasibility of adopting measures for the treatment of waste oils, including quantitative targets on the regeneration of waste oils and any further measures to promote the regeneration of waste oils. To that end, the Commission shall submit a report to the European Parliament and to the Council, accompanied, if appropriate, by a legislative proposal.

Article 22

Bio-waste

1.	Member States shall ensure that, by 31 December 2023 and subject to Article 10(2) and (3), bio-waste is either separated and recycled at source, or is collected separately and is not mixed with other types of waste.

Member States may allow waste with similar biodegradability and compostability properties which complies with relevant European standards or any equivalent national standards for packaging recoverable through composting and biodegradation, to be collected together with bio-waste.

2. Member States shall take measures in accordance with Articles 4 and 13, to:

(a) encourage the recycling, including composting and digestion, of bio-waste in a way that fulfils a high level of environment protection and results in output which meets relevant high-quality standards;

(b) encourage home composting; and

(c) promote the use of materials produced from bio-waste.

3. By 31 December 2018, the Commission shall request the European standardisation organisations to develop European standards for bio-waste entering organic recycling processes, for compost and for digestate, based on best available practices.

CHAPTER IV

PERMITS AND REGISTRATIONS

Article 23

Issue of permits

1. Member States shall require any establishment or undertaking intending to carry out waste treatment to obtain a permit from the competent authority.

Such permits shall specify at least the following:

(a) the types and quantities of waste that may be treated;

(b) for each type of operation permitted, the technical and any other requirements relevant to the site concerned;

(c) the safety and precautionary measures to be taken;

(d) the method to be used for each type of operation;

(e) such monitoring and control operations as may be necessary;

(f) such closure and after-care provisions as may be necessary.

2. Permits may be granted for a specified period and may be renewable.

3. Where the competent authority considers that the intended method of treatment is unacceptable from the point of view of environmental protection, in particular when the method is not in accordance with Article 13, it shall refuse to issue the permit.

4. It shall be a condition of any permit covering incineration or co- incineration with energy recovery that the recovery of energy take place with a high level of energy efficiency.

5. Provided that the requirements of this Article are complied with, any permit produced pursuant to other national or Community legislation may be combined with the permit required under paragraph 1 to form a single permit, where such a format obviates the unnecessary duplication of information and the repetition of work by the operator or the competent authority.

Article 24

Exemptions from permit requirements

Member States may exempt from the requirement laid down in Article 23(1) establishments or undertakings for the following operations:

(a) disposal of their own non-hazardous waste at the place of production; or

(b) recovery of waste.

Article 25

Conditions for exemptions

1. Where a Member State wishes to allow exemptions, as provided for in Article 24, it shall lay down, in respect of each type of activity, general rules specifying the types and quantities of waste that may be covered by an exemption, and the method of treatment to be used.

Those rules shall be designed to ensure that waste is treated in accordance with Article 13. In the case of disposal operations referred to in point (a) of Article 24 those rules should consider best available techniques.

2. In addition to the general rules provided for in paragraph 1, Member States shall lay down specific conditions for exemptions relating to hazardous waste, including types of activity, as well as any other necessary requirement for carrying out different forms of recovery and, where relevant, the limit values for the content of hazardous substances in the waste as well as the emission limit values.

3. Member States shall inform the Commission of the general rules laid down pursuant to paragraphs 1 and 2.

Article 26

Registration

Where the following are not subject to permit requirements, Member States shall ensure that the competent authority keeps a register of:

(a) establishments or undertakings which collect or transport waste on a professional basis;

(b) dealers or brokers; and

(c) establishments or undertakings which are subject to exemptions from the permit requirements pursuant to Article 24.

Where possible, existing records held by the competent authority shall be used to obtain the relevant information for this registration process in order to reduce the administrative burden.

Article 27

Minimum standards

1.   The Commission shall adopt delegated acts in accordance with Article 38a in order to supplement this Directive by setting out technical minimum standards for treatment activities, including for sorting and recycling of waste, which require a permit pursuant to Article 23 where there is evidence that a benefit in terms of the protection of human health and the environment would be gained from such minimum standards.

2.   Such minimum standards shall cover only those waste treatment activities that are not covered by Directive 96/61/EC or are not appropriate for coverage by that Directive.

3.   Such minimum standards shall:

(a)   be directed to the main environmental impacts of the waste treatment activity;

(b)   ensure that the waste is treated in accordance with Article 13;

(c)   take into account best available techniques; and

(d)   as appropriate, include elements regarding the quality of treatment and the process requirements.

4.   The Commission shall adopt delegated acts in accordance with Article 38a in order to supplement this Directive by setting out the minimum standards for activities that require registration pursuant to points (a) and (b) of Article 26 where there is evidence that a benefit in terms of the protection of human health and the environment or in avoiding disruption to the internal market would be gained from such minimum standards.

CHAPTER V

PLANS AND PROGRAMMES

Article 28

Waste management plans

1.   Member States shall ensure that their competent authorities establish, in accordance with Articles 1, 4, 13 and 16, one or more waste management plans.

Those plans shall, alone or in combination, cover the entire geographical territory of the Member State concerned.

2.   The waste management plans shall set out an analysis of the current waste

management situation in the geographical entity concerned, as well as the measures to be taken to improve environ- mentally sound preparing for re-use, recycling, recovery and disposal of waste and an evaluation of how the plan will support the implementation of the objectives and provisions of this Directive.

3.   The waste management plans shall contain, as appropriate and taking into account the geographical level and coverage of the planning area, at least the following:

(a)   the type, quantity and source of waste generated within the territory, the waste likely to be shipped from or to the national territory, and an evaluation of the development of waste streams in the future;

(b)   existing major disposal and recovery installations, including any special arrangements for waste oils, hazardous waste, waste containing significant amounts of critical raw materials, or waste streams addressed by specific Union legislation;

(c)   an assessment of the need for closure of existing waste installations, and for additional waste installation infrastructure in accordance with Article 16.

Member States shall ensure that an assessment of the investments and other financial means, including for local authorities, required to meet those needs is carried out. This assessment shall be included in the relevant waste management plans or in other strategic documents covering the entire territory of the Member State concerned;

(ca)   information on the measures to attain the objective laid down in Article 5(3a) of Directive 1999/31/EC or in other strategic documents covering the entire territory of the Member State concerned;

(cb)   an assessment of existing waste collection schemes, including the material and territorial coverage of separate collection and measures to improve its operation, of any derogations granted in accordance with Article 10(3), and of the need for new collection schemes;

(d)   sufficient information on the location criteria for site identification and on the capacity of future disposal or major recovery installations, if necessary;

(e)   general waste management policies, including planned waste management technologies and methods, or policies for waste posing specific management problems;

(f)   measures to combat and prevent all forms of littering and to clean up all types of litter;

(g)   appropriate qualitative or quantitative indicators and targets, including on the quantity of generated waste and its treatment and on municipal waste that is disposed of or subject to energy recovery.

4.   The waste management plan may contain, taking into account the geographical level and coverage of the planning area, the following:

(a)   organisational aspects related to waste management including a description of the allocation of responsibilities between public and private actors carrying out the waste management;

(b)   an evaluation of the usefulness and suitability of the use of economic and other instruments in tackling various waste problems, taking into account the need to maintain the smooth functioning of the internal market;

(c)   the use of awareness campaigns and information provision directed at the general public or at a specific set of consumers;

(d)   historical contaminated waste disposal sites and measures for their rehabilitation.

5.   Waste management plans shall conform to the waste planning requirements laid down in Article 14 of Directive 94/62/EC, to the targets laid down in Article 11(2) and (3) of this Directive and to the requirements laid down in Article 5 of Directive 1999/31/EC, and for the purposes of litter prevention, to the requirements laid down in Article 13 of Directive 2008/56/EC of the European Parliament and of the Council[8] and Article 11 of Directive 2000/60/EC of the European Parliament and of the Council[9].

Article 29

Waste prevention programmes

1.   Member States shall establish waste prevention programmes setting out at least the waste prevention measures as laid down in Article 9(1) in accordance with Articles 1 and 4.

Such programmes shall be integrated either into the waste management plans required under Article 28 or into other environmental policy programmes, as appropriate, or shall function as separate programmes. If any such programme is integrated into the waste management plan or into those other programmes, the waste prevention objectives and measures shall be clearly identified.

2.   When establishing such programmes, Member States shall, where relevant, describe the contribution of instruments and measures listed in Annex IVa to waste prevention and shall evaluate the usefulness of the examples of measures indicated in Annex IV or other appropriate measures. The programmes shall also describe existing waste prevention measures and their contribution to waste prevention.

The aim of such objectives and measures shall be to break the link between economic growth and the environmental impacts associated with the generation of waste.

2a.   Member States shall adopt specific food waste prevention programmes within their waste prevention programmes.

3.   *deleted*

---

[8]   Directive 2008/56/EC of the European Parliament and of the Council of 17 June 2008 establishing a framework for Community action in the field of marine environmental policy (Marine Strategy Framework Directive) (OJ L 164, 25.6.2008, p. 19).

[9]   Directive 2000/60/EC of the European Parliament and of the Council of 23 October 2000 establishing a framework for Community action in the field of water policy (OJ L 327, 22.12.2000, p. 1).

4. deleted

5.   The Commission shall create a system for sharing information on best practice regarding waste prevention and shall develop guidelines in order to assist the Member States in the preparation of the Programmes.

Article 30

Evaluation and review of plans and programmes

1.   Member States shall ensure that the waste management plans and waste prevention programmes are evaluated at least every sixth year and revised as appropriate and, where relevant, in accordance with Articles 9 and 11.

2.   The European Environment Agency shall publish, every two years, a report containing a review of the progress made in the completion and implementation of waste prevention programmes, including an assessment of the evolution as regards the prevention of waste generation for each Member State and for the Union as a whole, and as regards the decoupling of waste generation from economic growth and the transition towards a circular economy.

Article 31

Public participation

Member States shall ensure that relevant stakeholders and authorities and the general public have the opportunity to participate in the elaboration of the waste management plans and waste prevention programmes, and have access to them once elaborated, in accordance with Directive 2003/35/EC or, if relevant, Directive 2001/42/EC of the European Parliament and of the Council of 27 June 2001 on the assessment of the effects of certain plans and programmes on the environment[10]. They shall place the plans and programmes on a publicly available website.

Article 32

Cooperation

Member States shall cooperate as appropriate with the other Member States concerned and the Commission to draw up the waste management plans and the waste prevention programmes in accordance with Articles 28 and 29.

Article 33

Information to be submitted to the Commission

1.   Member States shall inform the Commission of the waste management plans

---

[10]   OJ L 197, 21.7.2001, p. 30.

and waste prevention programmes referred to in Articles 28 and 29, once adopted, and of any substantial revisions to the plans and programmes.

2.   The Commission shall adopt implementing acts to establish the format for notifying the information on the adoption and substantial revisions of the waste management plans and the waste prevention programmes. Those implementing acts shall be adopted in accordance with the examination procedure referred to in Article 39(2).

CHAPTER VI

INSPECTIONS AND RECORDS

Article 34

Inspections

1.   Establishments or undertakings which carry out waste treatment operations, establishments or undertakings which collect or transport waste on a professional basis, brokers and dealers, and establishments or undertakings which produce hazardous waste shall be subject to appropriate periodic inspections by the competent authorities.

2.   Inspections concerning collection and transport operations shall cover the origin, nature, quantity and destination of the waste collected and transported.

3.   Member States may take account of registrations obtained under the Community Eco-Management and Audit Scheme (EMAS), in particular regarding the frequency and intensity of inspections.

Article 35

Record keeping

1.   The establishments and undertakings referred to in Article 23(1), the producers of hazardous waste, and the establishments and undertakings which collect or transport hazardous waste on a professional basis, or act as dealers and brokers of hazardous waste, shall keep a chronological record of:

(a)   the quantity, nature and origin of that waste and the quantity of products and materials resulting from preparing for re-use, recycling or other recovery operations; and

(b)   where relevant, the destination, frequency of collection, mode of transport and treatment method foreseen in respect of the waste.

They shall make that data available to the competent authorities through the electronic registry or registries to be established pursuant to paragraph 4 of this Article.

2.   For hazardous waste, the records shall be preserved for at least three years except in the case of establishments and undertakings transporting hazardous waste which must keep such records for at least 12 months.

Documentary evidence that the management operations have been carried out shall be supplied at the request of the competent authorities or of a previous holder.

3.   Member States may require the producers of non-hazardous waste to comply with paragraphs 1 and 2.

4.   Member States shall set up an electronic registry or coordinated registries to record the data on hazardous waste referred to in paragraph 1 covering the entire geographical territory of the Member State concerned. Member States may establish such registries for other waste streams, in particular for those waste streams for which targets are set in legislative acts of the Union. Member States shall use the data on waste reported by industrial operators in the European Pollutant Release and Transfer Register set up under Regulation (EC) No 166/2006 of the European Parliament and of the Council[11].

5.   The Commission may adopt implementing acts to establish minimum conditions for the operation of such registries. Those implementing acts shall be adopted in accordance with the examination procedure referred to in Article 39(2).

Article 36

Enforcement and penalties

1.   Member States shall take the necessary measures to prohibit the abandonment, dumping or uncontrolled management of waste, including littering.

2.   Members States shall lay down provisions on the penalties applicable to infringements of the provisions of this Directive and shall take all measures necessary to ensure that they are implemented. The penalties shall be effective, proportionate and dissuasive.

CHAPTER VII

FINAL PROVISIONS

Article 37

Reporting

1.   Member States shall report the data concerning the implementation of points (a) to (e) of Article 11(2) and Article 11(3) for each calendar year to the Commission.

They shall report the data electronically within 18 months of the end of the reporting year for which the data are collected. The data shall be reported in the format established by the Commission in accordance with paragraph 7 of this Article.

The first reporting period shall start in the first full calendar year after the adoption of the implementing act that establishes the format for reporting, in accordance with

---

[11]   Regulation (EC) No 166/2006 of the European Parliament and of the Council of 18 January 2006 concerning the establishment of a European Pollutant Release and Transfer Register and amending Council Directives 91/689/EEC and 96/61/EC (OJ L 33, 4.2.2006, p. 1).

paragraph 7 of this Article.

2.   For the purposes of verifying compliance with point (b) of Article 11(2), Member States shall report the amount of waste used for backfilling and other material recovery operations separately from the amount of waste prepared for re-use or recycled. Member States shall report the reprocessing of waste into materials that are to be used for backfilling operations as backfilling.

For the purposes of verifying compliance with points (c), (d) and (e) of Article 11(2) and Article 11(3), Member States shall report the amount of waste prepared for re-use separately from the amount of waste recycled.

3.   Member States shall report the data concerning the implementation of Article 9(4) and (5) to the Commission every year.

They shall report the data electronically within 18 months of the end of the reporting year for which the data are collected. The data shall be reported in the format established by the Commission in accordance with paragraph 7 of this Article.

The first reporting period shall start in the first full calendar year after the adoption of the implementing act that establishes the format for reporting, in accordance with paragraph 7 of this Article.

4.   Member States shall report the data on mineral or synthetic lubrication or industrial oils placed on the market and waste oils separately collected and treated for each calendar year to the Commission.

They shall report the data electronically within 18 months of the end of the reporting year for which the data are collected. The data shall be reported in the format established by the Commission in accordance with paragraph 7.

The first reporting period shall start in the first full calendar year after the adoption of the implementing act that establishes the format for reporting, in accordance with paragraph 7.

5.   The data reported by Member States in accordance with this Article shall be accompanied by a quality check report and a report on the measures taken pursuant to Article 11a(3) and (8), including detailed information about the average loss rates where applicable. That information shall be reported in the format for reporting established by the Commission in accordance with paragraph 7 of this Article.

6.   The Commission shall review the data reported in accordance with this Article and publish a report on the results of its review. The report shall assess the organisation of the data collection, the sources of data and the methodology used in Member States as well as the completeness, reliability, timeliness and consistency of that data. The assessment may include specific recommendations for improvement. The report shall be drawn up after the first reporting of the data by Member States and every four years thereafter.

7.   By 31 March 2019, the Commission shall adopt implementing acts laying down the format for reporting the data referred to in paragraphs 1, 3, 4 and 5 of this Article. For the purposes of reporting on the implementation of points (a) and (b) of Article 11(2), Member States shall use the format established in Commission Implementing Decision of 18 April 2012 establishing a questionnaire for Member States reports on

the implementation of Directive 2008/98/EC of the European Parliament and of the Council on waste. For the purpose of reporting on food waste, the methodology developed under Article 9(8) shall be taken into account when developing the format for reporting. Those implementing acts shall be adopted in accordance with the examination procedure referred to in Article 39(2) of this Directive.

Article 38

Exchange of information and sharing of best practices, interpretation and adaptation to technical progress

1.   The Commission shall organise a regular exchange of information and sharing of best practices among Member States, including, where appropriate, with regional and local authorities, on the practical implementation and enforcement of the requirements of this Directive, including on:

(a)   the application of the calculation rules set out in Article 11a and the development of measures and systems to trace municipal waste streams from sorting to recycling;

(b)   adequate governance, enforcement, cross-border cooperation;

(c)   innovation in the field of waste management;

(d)   national by-product and end-of-waste criteria, as referred to in Article 5(3) and in Article 6(3) and (4), facilitated by a Union-wide electronic register to be established by the Commission;

(e)   the economic instruments and other measures used in accordance with Article 4(3) in order to boost the achievement of the objectives laid down in that Article;

(f)   measures laid down in Article 8(1) and (2);

(g)   prevention and the setting up of systems which promote re-use activities and the extension of life span;

(h)   the implementation of the obligations with regard to separate collection;

(i)   the instruments and incentives towards achieving the targets laid down in points (c), (d) and (e) of Article 11(2).

The Commission shall make the results of the exchange of information and sharing of best practices publicly available.

2.   The Commission may develop guidelines for the interpretation of the requirements set out in this Directive, including on the definition of waste, prevention, re-use, preparing for re-use, recovery, recycling, disposal, and on the application of the calculation rules set out in Article 11a.

The Commission shall develop guidelines on the definitions of municipal waste and backfilling.

The Commission is empowered to adopt delegated acts in accordance with Article 38a to amend this Directive by specifying the application of the formula for incineration facilities referred to in point R1 of Annex II. Local climatic conditions may be taken into account, such as the severity of the cold and the need for heating insofar as they influence the amounts of energy that can technically be used or produced in the form of electricity, heating, cooling or processing steam. Local conditions of the outermost regions as recognised in the third paragraph of Article 349 of the Treaty on the Functioning of the European Union and of the territories mentioned in Article 25 of the 1985 Act of Accession may also be taken into account.

3.   The Commission is empowered to adopt delegated acts in accordance with Article 38a to amend Annexes IV and V in the light of scientific and technical progress.

Article 38a

Exercise of the delegation

1.   The power to adopt delegated acts is conferred on the Commission subject to the conditions laid down in this Article.

2.   The power to adopt delegated acts referred to in Articles 7(1), 9(8), 11a(10), 27(1), 27(4), 38(2) and 38(3) shall be conferred on the Commission for a period of five years from 4 July 2018. The Commission shall draw up a report in respect of the delegation of power not later than nine months before the end of the five-year period. The delegation of power shall be tacitly extended for periods of an identical duration, unless the European Parliament or the Council opposes such extension not later than three months before the end of each period.

3.   The delegation of power referred to in Articles 7(1), 9(8), 11a(10), 27(1), 27(4), 38(2) and 38(3) may be revoked at any time by the European Parliament or by the Council. A decision to revoke shall put an end to the delegation of the power specified in that decision. It shall take effect the day following the publication of the decision in the Official Journal of the European Union or at a later date specified therein. It shall not affect the validity of any delegated acts already in force.

4.   Before adopting a delegated act, the Commission shall consult experts designated by each Member State in accordance with the principles laid down in the Interinstitutional Agreement of 13 April 2016 on Better Law Making[12].

5.   As soon as it adopts a delegated act, the Commission shall notify it simultaneously to the European Parliament and to the Council.

6.   A delegated act adopted pursuant to Articles 7(1), 9(8), 11a(10), 27(1), 27(4), 38(2) and 38(3) shall enter into force only if no objection has been expressed either by the European Parliament or by the Council within a period of two months of notification of that act to the European Parliament and to the Council or if, before the expiry of that period, the European Parliament and the Council have both informed the Commission that they will not object. That period shall be extended by two months at the initiative of the European Parliament or of the Council.

---

[12]   OJ L 123, 12.5.2016, p. 1.

Article 39

Committee procedure

1.   The Commission shall be assisted by a committee. That committee shall be a committee within the meaning of Regulation (EU) No 182/2011 of the European Parliament and of the Council[13].

2.   Where reference is made to this paragraph, Article 5 of Regulation (EU) No 182/2011 shall apply.

Where the committee delivers no opinion, the Commission shall not adopt the draft implementing act and the third subparagraph of Article 5(4) of Regulation (EU) No 182/2011 shall apply.

Article 40

Transposition

1.   Member States shall bring into force the laws, regulations and administrative provisions necessary to comply with this Directive by 12 December 2010.

When Member States adopt these measures, they shall contain a reference to this Directive or shall be accompanied by such reference on the occasion of their official publication. The methods of making such reference shall be laid down by Member States.

2.   Member States shall communicate to the Commission the text of the main provisions of national law which they adopt in the field covered by this Directive.

Article 41

Repeal and transitional provisions

Directives 75/439/EEC, 91/689/EEC and 2006/12/EC are hereby repealed with effect from 12 December 2010.

Article 42

Entry into force

This Directive shall enter into force on the twentieth day following that of its publication in the Official Journal of the European Union.

Article 43

Addressees

This Directive is addressed to the Member States.

---

[13]   Regulation (EU) No 182/2011 of the European Parliament and of the Council of 16 February 2011 laying down the rules and general principles concerning mechanisms for control by Member States of the Commission's exercise of implementing powers (OJ L 55, 28.2.2011, p. 13).

ANNEX I

DISPOSAL OPERATIONS

D 1   Deposit into or on to land (e.g. landfill, etc.)

D 2   Land treatment (e.g. biodegradation of liquid or sludgy discards in soils, etc.)

D 3   Deep injection (e.g. injection of pumpable discards into wells, salt domes or naturally occurring repositories, etc.)

D 4   Surface impoundment (e.g. placement of liquid or sludgy discards into pits, ponds or lagoons, etc.)

D 5   Specially engineered landfill (e.g. placement into lined discrete cells which are capped and isolated from one another and the environment, etc.)

D 6   Release into a water body except seas/oceans

D 7   Release to seas/oceans including sea-bed insertion

D 8   Biological treatment not specified elsewhere in this Annex which results in final compounds or mixtures which are discarded by means of any of the operations numbered D 1 to D 12

D 9   Physico-chemical treatment not specified elsewhere in this Annex which results in final compounds or mixtures which are discarded by means of any of the operations numbered D 1 to D 12 (e.g. evaporation, drying, calcination, etc.)

D 10   Incineration on land D 11 Incineration at sea (*)

D 12   Permanent storage (e.g. emplacement of containers in a mine, etc.)

D 13   Blending or mixing prior to submission to any of the operations numbered D 1 to D 12 (**)

D 14   Repackaging prior to submission to any of the operations numbered D 1 to D 13

D 15   Storage pending any of the operations numbered D 1 to D 14 (excluding temporary storage, pending collection, on the site where the waste is produced) (***)

(*)   This operation is prohibited by EU legislation and international conventions.

(**)   If there is no other D code appropriate, this can include preliminary operations prior to disposal including pre-processing such as, inter alia, sorting, crushing, compacting, pelletising, drying, shredding, conditioning or separating prior to submission to any of the operations numbered D1 to D12.

(***)   Temporary storage means preliminary storage according to point (10) of Article 3.

ANNEX II

RECOVERY OPERATIONS

R 1   Use principally as a fuel or other means to generate energy [*]

R 2   Solvent reclamation/regeneration

R 3   Recycling/reclamation of organic substances which are not used as solvents (including composting and other biological transformation processes) [**]

R 4   Recycling/reclamation of metals and metal compounds [***]

R 5   Recycling/reclamation of other inorganic materials [****]

---

[*]   This includes incineration facilities dedicated to the processing of municipal solid waste only where their energy efficiency is equal to or above:

—   0,60 for installations in operation and permitted in accordance with applicable Community legislation before 1 January 2009,

—   0,65 for installations permitted after 31 December 2008, using the following formula:

Energy efficiency = (Ep - (Ef + Ei))/(0,97 × (Ew + Ef)) In which:

Ep means annual energy produced as heat or electricity. It is calculated with energy in the form of electricity being multiplied by 2,6 and heat produced for commercial use multiplied by 1,1 (GJ/year)

Ef means annual energy input to the system from fuels contributing to the production of steam (GJ/year)

Ew means annual energy contained in the treated waste calculated using the net calorific value of the waste (GJ/year)

Ei means annual energy imported excluding Ew and Ef (GJ/year)

0,97 is a factor accounting for energy losses due to bottom ash and radiation.

This formula shall be applied in accordance with the reference document on Best Available Techniques for waste incineration.

The energy efficiency formula value will be multiplied by a climate correction factor (CCF) as shown below:

1.   CCF for installations in operation and permitted in accordance with applicable Union legislation before 1 September 2015.

CCF = 1 if HDD >= 3 350

CCF = 1,25 if HDD <= 2 150

CCF = – (0,25/1 200) × HDD + 1,698 when 2 150 < HDD < 3 350

2.   CCF for installations permitted after 31 August 2015 and for installations under 1 after 31 December 2029:

CCF = 1 if HDD >= 3 350

CCF = 1,12 if HDD <= 2 150

CCF = – (0,12/1 200) × HDD + 1,335 when 2 150 < HDD < 3 350
(The resulting value of CCF will be rounded at three decimal places).

The value of HDD (Heating Degree Days) should be taken as the average of annual HDD values for the incineration facility location, calculated for a period of 20 consecutive years before the year for which CCF is calculated. For the calculation of the value of HDD the following method established by Eurostat should be applied: HDD is equal to (18 °C – Tm) × d if Tm is lower than or equal to 15 °C (heating threshold) and is nil if Tm is greater than 15 °C; where Tm is the mean (Tmin + Tmax)/2 outdoor temperature over a period of d days. Calculations are to be executed on a daily basis (d = 1), added up to a year.

[**]   This includes preparing for re-use, gasification and pyrolysis using the components as chemicals and recovery of organic materials in the form of backfilling.

[***]   This includes preparing for re-use.

[****]   This includes preparing for re-use, recycling of inorganic construction materials, recovery of inorganic materials in the form of backfilling, and soil cleaning resulting in recovery of the soil.

R 6   Regeneration of acids or bases

R 7   Recovery of components used for pollution abatement

R 8   Recovery of components from catalysts

R 9   Oil re-refining or other reuses of oil

R 10   Land treatment resulting in benefit to agriculture or ecological improvement

R 11   Use of waste obtained from any of the operations numbered R 1 to R 10

R 12   Exchange of waste for submission to any of the operations numbered R 1 to R 11[*]

R 13   Storage of waste pending any of the operations numbered R 1 to R 12 (excluding temporary storage, pending collection, on the site where the waste is produced) [**]

---

[*]   If there is no other R code appropriate, this can include preliminary operations prior to recovery including pre-processing such as, inter alia, dismantling, sorting, crushing, compacting, pelletising, drying, shredding, conditioning, repackaging, separating, blending or mixing prior to submission to any of the operations numbered R1 to R11.

[**] Temporary storage means preliminary storage according to point (10) of Article 3.

ANNEX III

PROPERTIES OF WASTE WHICH RENDER IT HAZARDOUS

HP 1 'Explosive:' waste which is capable by chemical reaction of producing gas at such a temperature and pressure and at such a speed as to cause damage to the surroundings. Pyrotechnic waste, explosive organic peroxide waste and explosive self-reactive waste is included.

When a waste contains one or more substances classified by one of the hazard class and category codes and hazard statement codes shown in Table 1, the waste shall be assessed for HP 1, where appropriate and proportionate, according to test methods. If the presence of a substance, a mixture or an article indicates that the waste is explosive, it shall be classified as hazardous by HP 1.

Table 1: Hazard Class and Category Code(s) and Hazard statement Code(s) for waste constituents for the classification of wastes as hazardous by HP 1:

Hazard Class and Category Code(s)	Hazard statement Code(s)
Unst. Expl.	H 200
Expl. 1.1	H 201
Expl. 1.2	H 202
Expl. 1.3	H 203
Expl. 1.4	H 204
Self-react. A	H 240
Org. Perox. A	
Self-react. B	H 241
Org. Perox. B	

HP 2 'Oxidising:' waste which may, generally by providing oxygen, cause or contribute to the combustion of other materials.

When a waste contains one or more substances classified by one of the hazard class and category codes and hazard statement codes shown in Table 2, the waste shall be assessed for HP 2, where appropriate and proportionate, according to test methods. If the presence of a substance indicates that the waste is oxidising, it shall be classified as hazardous by HP 2.

Table 2: Hazard Class and Category Code(s) and Hazard statement Code(s) for the classification of wastes as hazardous by HP 2:

Hazard Class and Category Code(s)	Hazard statement Code(s)
Ox. Gas 1	H 270
Ox. Liq. 1	H 271
Ox. Sol. 1	
Ox. Liq. 2, Ox. Liq. 3	H 272
Ox. Sol. 2, Ox. Sol. 3	

HP 3   'Flammable:'

— flammable liquid waste liquid waste having a flash point below 60°C or waste gas oil, diesel and light heating oils having a flash point > 55°C and ≤ 75°C;

— flammable pyrophoric liquid and solid waste solid or liquid waste which, even in small quantities, is liable to ignite within five minutes after coming into contact with air;

— flammable solid waste solid waste which is readily combustible or may cause or contribute to fire through friction;

— flammable gaseous waste gaseous waste which is flammable in air at 20°C and a standard pressure of 101.3 kPa;

— water reactive waste which, in contact with water, emits flammable gases in dangerous quantities;

— other flammable waste flammable aerosols, flammable self-heating waste, flammable organic peroxides and flammable self- reactive waste.

When a waste contains one or more substances classified by one of the following hazard class and category codes and hazard statement codes shown in Table 3, the waste shall be assessed, where appropriate and proportionate, according to test methods. If the presence of a substance indicates that the waste is flammable, it shall be classified as hazardous by HP 3.

Table 3: Hazard Class and Category Code(s) and Hazard statement Code(s) for waste constituents for the classification of wastes as hazardous by HP 3:

Hazard Class and Category Code(s)	Hazard statement Code(s)
Flam. Gas 1	H220
Flam. Gas 2	H221
Aerosol 1	H222
Aerosol 2	H223
Flam. Liq. 1	H224
Flam. Liq.2	H225
Flam. Liq. 3	H226
Flam. Sol. 1	H228
Flam. Sol. 2	
Self-react. CD	H242
Self-react. EF	
Org. Perox. CD	
Org. Perox. EF	
Pyr. Liq. 1	H250
Pyr. Sol. 1	
Self-heat.1	H251
Self-heat. 2	H252
Water-react. 1	H260
Water-react. 2	H261
Water-react. 3	

HP 4   **'Irritant — skin irritation and eye damage:'** waste which on application can cause skin irritation or damage to the eye.

When a waste contains one or more substances in concentrations above the cut- off value, that are classified by one of the following hazard class and category codes and hazard statement codes and one or more of the following concentration limits is exceeded or equalled, the waste shall be classified as hazardous by HP 4.

The cut-off value for consideration in an assessment for Skin corr. 1A (H314), Skin irrit. 2 (H315), Eye dam. 1 (H318) and Eye irrit. 2 (H319) is 1 %.

If the sum of the concentrations of all substances classified as Skin corr. 1A (H314) exceeds or equals 1 %, the waste shall be classified as hazardous according to HP 4.

If the sum of the concentrations of all substances classified as H318 exceeds or equals 10 %, the waste shall be classified as hazardous according to HP 4.

If the sum of the concentrations of all substances classified H315 and H319 exceeds or equals 20 %, the waste shall be classified as hazardous according to HP 4.

Note that wastes containing substances classified as H314 (Skin corr.1A, 1B or 1C) in amounts greater than or equal to 5 % will be classified as hazardous by HP 8. HP 4 will not apply if the waste is classified as HP 8.

HP 5   **Specific Target Organ Toxicity (STOT)/Aspiration Toxicity:'** waste which can cause specific target organ toxicity either from a single or repeated expo-sure, or which cause acute toxic effects following aspiration.

When a waste contains one or more substances classified by one or more of the following hazard class and category codes and hazard statement codes shown in Table 4, and one or more of the concentration limits in Table 4 is exceeded or equalled, the waste shall be classified as hazardous according to HP 5. When sub-stances classified as STOT are present in a waste, an individual substance has to be present at or above the concentration limit for the waste to be classified as hazardous by HP 5.

When a waste contains one or more substances classified as Asp. Tox. 1 and the sum of those substances exceeds or equals the concentration limit, the waste shall be classified as hazardous by HP 5 only where the overall kinematic viscosity (at 40°C) does not exceed 20.5 mm^2/s.[14]

Table 4: Hazard Class and Category Code(s) and Hazard statement Code(s) for waste constituents and the corresponding concentration limits for the classifi-cation of wastes as hazardous by HP 5

Hazard Class and Category Code(s)	Hazard statement Code(s)	Concentration limit
STOT SE 1	H370	1 %
STOT SE 2	H371	10 %

---

[14]   The kinematic viscosity shall only be determined for fluids.

Hazard Class and Category Code(s)	Hazard statement Code(s)	Concentration limit
STOT SE 3	H335	20 %
STOT RE 1	H372	1 %
STOT RE 2	H373	10 %
Asp. Tox. 1	H304	10 %

HP 6 **'Acute Toxicity:'** waste which can cause acute toxic effects following oral or dermal administration, or inhalation exposure.

If the sum of the concentrations of all substances contained in a waste, classified with an acute toxic hazard class and category code and hazard statement code given in Table 5, exceeds or equals the threshold given in that table, the waste shall be classified as hazardous by HP 6. When more than one substance classified as acute toxic is present in a waste, the sum of the concentrations is required only for substances within the same hazard category.

The following cut-off values shall apply for consideration in an assessment:

— For Acute Tox. 1, 2 or 3 (H300, H310, H330, H301, H311, H331): 0.1 %;

— For Acute Tox. 4 (H302, H312, H332): 1 %.

Table 5: Hazard Class and Category Code(s) and Hazard statement Code(s) for waste constituents and the corresponding concentration limits for the classification of wastes as hazardous by HP 6

Hazard Class and Category Code(s)	Hazard statement Code(s)	Concentration limit
Acute Tox.1 (Oral)	H300	0,1 %
Acute Tox. 2 (Oral)	H300	0,25 %
Acute Tox. 3 (Oral)	H301	5 %
Acute Tox 4 (Oral)	H302	25 %
Acute Tox.1 (Dermal)	H310	0,25 %
Acute Tox.2 (Dermal)	H310	2,5 %
Acute Tox. 3 (Dermal)	H311	15 %
Acute Tox 4 (Dermal)	H312	55 %
Acute Tox 1 (Inhal.)	H330	0,1 %
Acute Tox.2 (Inhal.)	H330	0,5 %
Acute Tox. 3 (Inhal.)	H331	3,5 %
Acute Tox. 4 (Inhal.)	H332	22,5 %

HP 7 **'Carcinogenic:'** waste which induces cancer or increases its incidence.

When a waste contains a substance classified by one of the following hazard class and category codes and hazard statement codes and exceeds or equals one of the following concentration limits shown in Table 6, the waste shall be classified as hazardous by HP 7. When more than one substance classified as carcinogenic is present in a waste, an individual substance has to be present at or above the concentration limit for the waste to be classified as hazardous by HP 7.

Table 6: Hazard Class and Category Code(s) and Hazard statement Code(s) for waste constituents and the corresponding concentration limits for the classification of wastes as hazardous by HP 7

Hazard Class and Category Code(s)	Hazard statement Code(s)	Concentration limit
Carc. 1A	H350	0,1 %
Carc. 1B		
Carc. 2	H351	1,0 %

HP 8   **'Corrosive:'** waste which on application can cause skin corrosion.

When a waste contains one or more substances classified as Skin corr.1A, 1B or 1C (H314) and the sum of their concentrations exceeds or equals 5 %, the waste shall be classified as hazardous by HP 8.

The cut-off value for consideration in an assessment for Skin corr. 1A, 1B, 1C (H314) is 1.0 %.

HP 9   **'Infectious:'** waste containing viable micro-organisms or their toxins which are known or reliably believed to cause disease in man or other living organisms.

The attribution of HP 9 shall be assessed by the rules laid down in reference documents or legislation in the Member States.

HP 10   **Toxic for reproduction:'** waste which has adverse effects on sexual function and fertility in adult males and females, as well as developmental toxicity in the offspring.

When a waste contains a substance classified by one of the following hazard class and category codes and hazard statement codes and exceeds or equals one of the following concentration limits shown in Table 7, the waste shall be classified hazardous according to HP 10. When more than one substance classified as toxic for reproduction is present in a waste, an individual substance has to be present at or above the concentration limit for the waste to be classified as hazardous by HP 10.

Table 7: Hazard Class and Category Code(s) and Hazard statement Code(s) for waste constituents and the corresponding concentration limits for the classification of wastes as hazardous by HP 10

Hazard Class and Category Code(s)	Hazard statement Code(s)	Concentration limit
Repr. 1A	H360	0,3 %
Repr. 1B		
Repr. 2	H361	3,0 %

HP 11   **'Mutagenic:'** waste which may cause a mutation, that is a permanent change in the amount or structure of the genetic material in a cell.

When a waste contains a substance classified by one of the following hazard class and category codes and hazard statement codes and exceeds or equals one of the following concentration limits shown in Table 8, the waste shall be classified as hazardous according to HP 11. When more than one substance classified as mutagenic is present in a waste, an individual substance has to be present at or above the concentration limit for the waste to be classified as hazardous by HP 11.

Table 8: Hazard Class and Category Code(s) and Hazard statement Code(s) for waste constituents and the corresponding concentration limits for the classification of wastes as hazardous by HP 11

Hazard Class and Category Code(s)	Hazard statement Code(s)	Concentration limit
Muta. 1A,	H340	0,1 %
Muta. 1B		
Muta. 2	H341	1,0 %

HP 12   'Release of an acute toxic gas:' waste which releases acute toxic gases (Acute Tox. 1, 2 or 3) in contact with water or an acid.

When a waste contains a substance assigned to one of the following supplemental hazards EUH029, EUH031 and EUH032, it shall be classified as hazardous by HP 12 according to test methods or guidelines.

HP 13   'Sensitising:' waste which contains one or more substances known to cause sensitising effects to the skin or the respiratory organs.

When a waste contains a substance classified as sensitising and is assigned to one of the hazard statement codes H317 or H334 and one individual substance equals or exceeds the concentration limit of 10 %, the waste shall be classified as hazardous by HP 13.

HP 14   'Ecotoxic:' waste which presents or may present immediate or delayed risks for one or more sectors of the environment.

Waste which fulfils any of the following conditions shall be classified as hazardous by HP 14:

— Waste which contains a substance classified as ozone depleting assigned the hazard statement code H420 in accordance with Regulation (EC) No 1272/2008 of the European Parliament and of the Council[15] and the concentration of such a substance equals or exceeds the concentration limit of 0,1 %. [c(H420) ≥ 0,1 %]

— Waste which contains one or more substances classified as aquatic acute assigned the hazard statement code H400 in accordance with Regulation (EC) No 1272/2008 and the sum of the concentrations of those substances equals

---

[15] Regulation (EC) No 1272/2008 of the European Parliament and of the Council of 16 December 2008 on classification, labelling and packaging of substances and mixtures, amending and repealing Directives 67/548/EEC and 1999/45/EC, and amending Regulation (EC) No 1907/2006 (OJ L 353, 31.12.2008, p. 1).

or exceeds the concentration limit of 25 %. A cut-off value of 0,1 % shall apply to such substances.

[Σ c (H400) ≥ 25 %]

— Waste which contains one or more substances classified as aquatic chronic 1, 2 or 3 assigned to the hazard statement code(s) H410, H411 or H412 in accordance with Regulation (EC) No 1272/2008, and the sum of the concentrations of all substances classified as aquatic chronic 1 (H410) multiplied by 100 added to the sum of the concentrations of all substances classified as aquatic chronic 2 (H411) multiplied by 10 added to the sum of the concentrations of all substances classified as aquatic chronic 3 (H412) equals or exceeds the concentration limit of 25 %. A cut-off value of 0,1 % applies to substances classified as H410 and a cut-off value of 1 % applies to substances classified as H411 or H412.

[100 × Σc (H410) + 10 × Σc (H411) + Σc (H412) ≥ 25 %]

— Waste which contains one or more substances classified as aquatic chronic 1, 2, 3 or 4 assigned the hazard statement code(s) H410, H411, H412 or H413 in accordance with Regulation (EC) No 1272/2008, and the sum of the concentrations of all substances classified as aquatic chronic equals or exceeds the concentration limit of 25 %. A cut-off value of 0,1 % applies to substances classified as H410 and a cut-off value of 1 % applies to substances classified as H411, H412 or H413.

[Σ c H410 + Σ c H411 + Σ c H412 + Σ c H413 ≥ 25 %]

Where: Σ = sum and c = concentrations of the substances.

HP 15 'Waste capable of exhibiting a hazardous property listed above not directly displayed by the original waste'.

When a waste contains one or more substances assigned to one of the hazard statements or supplemental hazards shown in Table 9, the waste shall be classified as hazardous by HP 15, unless the waste is in such a form that it will not under any circumstance exhibit explosive or potentially explosive properties.

Table 9: Hazard statements and supplemental hazards for waste constituents for the classification of wastes as hazardous by HP 15

Hazard Statement(s)/Supplemental Hazard(s)	
May mass explode in fire	H205
Explosive when dry	EUH001
May form explosive peroxides	EUH019
Risk of explosion if heated under confinement	EUH044

In addition, Member States may characterise a waste as hazardous by HP 15 based on other applicable criteria, such as an assessment of the leachate.

Test methods:
The methods to be used are described in Commission Regulation (EC) No 440/2008[16] and in other relevant CEN notes or other internationally recognised test methods and guidelines.

---

[16] Commission Regulation (EC) No 440/2008 of 30 May 2008 laying down test methods pursuant to Regulation (EC) No 1907/2006 of the European Parliament and of the Council on the Registration, Evaluation, Authorisation and Restriction of Chemicals (REACH) (OJ L 142, 31.5.2008, p. 1).

ANNEX IV

EXAMPLES OF WASTE PREVENTION MEASURES REFERRED TO IN ARTICLE 29

Measures that can affect the framework conditions related to the generation of waste

1. The use of planning measures, or other economic instruments promoting the efficient use of resources.

2. The promotion of research and development into the area of achieving cleaner and less wasteful products and technologies and the dissemination and use of the results of such research and development.

3. The development of effective and meaningful indicators of the environmental pressures associated with the generation of waste aimed at contributing to the prevention of waste generation at all levels, from product comparisons at Community level through action by local authorities to national measures.

Measures that can affect the design and production and distribution phase

4. The promotion of eco-design (the systematic integration of environmental aspects into product design with the aim to improve the environmental performance of the product throughout its whole life cycle).

5. The provision of information on waste prevention techniques with a view to facilitating the implementation of best available techniques by industry.

6. Organise training of competent authorities as regards the insertion of waste prevention requirements in permits under this Directive and Directive 96/61/EC.

7. The inclusion of measures to prevent waste production at installations not falling under Directive 96/61/EC. Where appropriate, such measures could include waste prevention assessments or plans.

8. The use of awareness campaigns or the provision of financial, decision making or other support to businesses. Such measures are likely to be particularly effective where they are aimed at, and adapted to, small and medium sized enterprises and work through established business networks.

9. The use of voluntary agreements, consumer/producer panels or sectoral negotiations in order that the relevant businesses or industrial sectors set their own waste prevention plans or objectives or correct wasteful products or packaging.

10. The promotion of creditable environmental management systems, including EMAS and ISO 14001.

Measures that can affect the consumption and use phase

11. Economic instruments such as incentives for clean purchases or the institution of an obligatory payment by consumers for a given article or element of packaging that would otherwise be provided free of charge.

12. The use of awareness campaigns and information provision directed at the general public or a specific set of consumers.

13. The promotion of creditable eco-labels.

14. Agreements with industry, such as the use of product panels such as those being carried out within the framework of Integrated Product Policies or with retailers on the availability of waste prevention information and products with a lower environmental impact.

15. In the context of public and corporate procurement, the integration of environmental and waste prevention criteria into calls for tenders and contracts, in line with the Handbook on environmental public procurement published by the Commission on 29 October 2004.

16. The promotion of the reuse and/or repair of appropriate discarded products or of their components, notably through the use of educational, economic, logistic or other measures such as support to or establishment of accredited repair and reuse-centres and networks especially in densely populated regions.

ANNEX IVa

EXAMPLES OF ECONOMIC INSTRUMENTS AND OTHER MEASURES TO PROVIDE INCENTIVES FOR THE APPLICATION OF THE WASTE HIERARCHY REFERRED TO IN ARTICLE 4(3)[17]

1. Charges and restrictions for the landfilling and incineration of waste which incentivise waste prevention and recycling, while keeping landfilling the least preferred waste management option;

2. 'Pay-as-you-throw' schemes that charge waste producers on the basis of the actual amount of waste generated and provide incentives for separation at source of recyclable waste and for reduction of mixed waste;

3. Fiscal incentives for donation of products, in particular food;

4. Extended producer responsibility schemes for various types of waste and measures to increase their effectiveness, cost efficiency and governance;

5. Deposit-refund schemes and other measures to encourage efficient collection of used products and materials;

6. Sound planning of investments in waste management infrastructure, including through Union funds;

7. Sustainable public procurement to encourage better waste management and the use of recycled products and materials;

8. Phasing out of subsidies which are not consistent with the waste hierarchy;

9. Use of fiscal measures or other means to promote the uptake of products and materials that are prepared for re-use or recycled;

10. Support to research and innovation in advanced recycling technologies and remanufacturing;

11. Use of best available techniques for waste treatment;

12. Economic incentives for regional and local authorities, in particular to promote waste prevention and intensify separate collection schemes, while avoiding support to landfilling and incineration;

13. Public awareness campaigns, in particular on separate collection, waste prevention and litter reduction, and mainstreaming these issues in education and training;

14. Systems for coordination, including by digital means, between all competent public authorities involved in waste management;

15. Promoting continuous dialogue and cooperation between all stakeholders in waste management and encouraging voluntary agreements and company reporting on waste.

---

[17] While these instruments and measures may provide incentives for waste prevention, which is the highest step in the waste hierarchy, a comprehensive list of more specific examples of waste prevention measures is set out in Annex IV.

ANNEX IVb

IMPLEMENTATION PLAN TO BE SUBMITTED PURSUANT TO ARTICLE 11(3)

The implementation plan to be submitted pursuant to Article 11(3) shall contain the following:

1. assessment of the past, current and projected rates of recycling, landfilling and other treatment of municipal waste and the streams of which it is composed;

2. assessment of the implementation of waste management plans and waste prevention programmes in place pursuant to Articles 28 and 29;

3. reasons for which the Member State considers that it might not be able to attain the relevant target laid down in Article 11(2) within the deadline set therein and an assessment of the time extension necessary to meet that target;

4. measures necessary to attain the targets set out in Article 11(2) and (5) that are applicable to the Member State during the time extension, including appropriate economic instruments and other measures to provide incentives for the application of the waste hierarchy as set out in Article 4(1) and Annex IVa;

5. a timetable for the implementation of the measures identified in point 4, determination of the body competent for their implementation and an assessment of their individual contribution to attaining the targets applicable in the event of a time extension;

6. information on funding for waste management in line with the polluter-pays principle;

7. measures to improve data quality, as appropriate, with a view to better planning and monitoring performance in waste management.

ANNEX V

CORRELATION TABLE

[Not published herein]

II.5   European Waste List (EWL)

COMMISSION DECISION

of 3 May 2000[1]

replacing Decision 94/3/EC establishing a list of wastes pursuant to Article 1(a) of Council Directive 75/442/EEC on waste and Council Decision 94/904/EC establishing a list of hazardous waste pursuant to Article 1(4) of Council Directive 91/689/EEC on hazardous waste

(notified under document number C(2000) 1147)

(Text with EEA relevance)

(2000/532/EC)

THE COMMISSION OF THE EUROPEAN COMMUNITIES,

Having regard to the Treaty establishing the European Community

Having regard to Council Directive 75/442/EEC of 15 July 1975 on waste[2], as amended by Directive 91/156/EEC[3], and in particular Article 1(a) thereof,

Having regard to Council Directive 91/689/EEC of 12 December 1991 on hazardous waste[4], and in particular Article 1(4), second indent thereof,

Whereas:

(1)   Several Member States have notified a number of waste categories which they consider to display one or more of the properties listed in Annex III to Directive 91/689/EEC.

(2)   Article 1(4) of Directive 91/689/EEC requires the Commission to examine notifications from Member States with a view to amending the list of hazardous wastes laid down in Council Decision 94/904/EC[5].

(3)   Any waste inserted in the list of hazardous wastes must also be included in the European Waste Catalogue laid down in Commission Decision 94/3/EC[6]. It is appropriate, in order to increase the transparency of the listing system and to simplify existing provisions, to establish one Community list which integrates the list of

---

[1]   Consolidated version of Commission Decision (2000/532/EC), OJ, L 226, 6.9.2000, p. 3 as amended by Commission Decision 2014/955/EU of 18 December 2014, OJ, L 370, 30.12.2014, p. 44

[2]   OJ L 194, 25.7.1975, p. 47.

[3]   OJ L 78, 26.3.1991, p. 32.

[4]   OJ L 377, 31.12.1991, p. 20.

[5]   OJ L 356, 31.12.1994, p. 14.

[6]   OJ L 5, 7.1.1994, p. 15.

wastes laid down in Decision 94/3/EC and that of hazardous wastes laid down in Decision 94/904/EC.

(4)   The Commission is assisted in this task by the Committee established by Article 18 of Directive 75/442/EEC.

(5)   The measures laid down in this Decision are in accordance with the opinion expressed by the aforementioned Committee,

HAS ADOPTED THIS DECISION:

Article 1

The list in the Annex to this Decision is adopted.

Article 2 and 3 have been deleted

Article 4

Member States shall take the measures necessary to comply with this Decision not later than 1 January 2002.

Article 5

Decision 94/3/EC and Decision 94/904/EC are repealed with effect from 1 January 2002.

Article 6

This Decision is addressed to the Member States.

ANNEX

LIST OF WASTE REFERRED TO IN ARTICLE 7 OF DIRECTIVE 2008/98/EC

DEFINITIONS

For the purposes of this Annex, the following definitions shall apply:

1. 'hazardous substance' means a substance classified as hazardous as a consequence of fulfilling the criteria laid down in parts 2 to 5 of Annex I to Regulation (EC) No 1272/2008;

2. 'heavy metal' means any compound of antimony, arsenic, cadmium, chromium (VI), copper, lead, mercury, nickel, selenium, tellurium, thallium and tin, as well as these materials in metallic form, as far as these are classified as hazardous substances;

3. 'polychlorinated biphenyls and polychlorinated terphenyls' ('PCBs') means PCBs as defined in Article 2(a) of Council Directive 96/59/EC[7];

4. 'transition metals' means any of the following metals: any compound of scandium, vanadium, manganese, cobalt, copper, yttrium, niobium, hafnium, tungsten, titanium, chromium, iron, nickel, zinc, zirconium, molybdenum and tantalum, as well as these materials in metallic form, as far as these are classified as hazardous substances;

5. 'stabilisation' means processes which change the hazardousness of the constituents in the waste and transform hazardous waste into non-hazardous waste;

6. 'solidification' means processes which only change the physical state of the waste by using additives without changing the chemical properties of the waste;

7. partly stabilised wastes' means wastes containing, after the stabilisation process, hazardous constituents which have not been changed completely into non-hazardous constituents and could be released into the environment in the short, middle or long term.

ASSESSMENT AND CLASSIFICATION

1. Assessment of hazardous properties of waste

When assessing the hazardous properties of wastes, the criteria laid down in Annex III to Directive 2008/98/EC shall apply. For the hazardous properties HP 4, HP 6 and HP 8, cut-off values for individual substances as indicated in Annex III to Directive 2008/98/EC shall apply to the assessment. Where a substance is present in the waste below its cut-off value, it shall not be included in any calculation of a threshold. Where a hazardous property of a waste has been assessed by a test and by using the concentrations of hazardous substances as indicated in Annex III to Directive 2008/98/EC, the results of the test shall prevail.

---

[7] Council Directive 96/59/EC of 16 September 1996 on the disposal of polychlorinated biphenyls and polychlorinated terphenyls (PCB/PCT), OJ, L 243, 24.9.1996, p. 31).

2. Classification of waste as hazardous

Any waste marked with an asterisk (*) in the list of wastes shall be considered as hazardous waste pursuant to Directive 2008/98/EC, unless Article 20 of that Directive applies.

For those wastes for which hazardous and non-hazardous waste codes could be assigned, the following shall apply:

— An entry in the harmonised list of wastes marked as hazardous, having a specific or general reference to 'hazardous substances', is only appropriate to a waste when that waste contains relevant hazardous substances that cause the waste to display one or more of the hazardous properties HP 1 to HP 8 and/or HP 10 to HP 15 as listed in Annex III to Directive 2008/98/EC. The assessment of the hazardous property HP 9 'infectious' shall be made according to relevant legislation or reference documents in the Member States.

— A hazardous property can be assessed by using the concentration of substances in the waste as specified in Annex III to Directive 2008/98/EC or, unless otherwise specified in Regulation (EC) No 1272/2008, by performing a test in accordance with Regulation (EC) No 440/2008 or other internationally recognised test methods and guidelines, taking into account Article 7 of Regulation (EC) No 1272/2008 as regards animal and human testing.

— Wastes containing polychlorinated dibenzo-p-dioxins and dibenzofurans (PCDD/PCDF), DDT (1,1,1-trichloro-2,2-bis (4-chlorophenyl)ethane), chlordane, hexachlorocyclohexanes (including lindane), dieldrin, endrin, heptachlor, hexaclorobenzene, chlordecone, aldrine, pentachlorobenzene, mirex, toxaphene hexabromobiphenyl and/or PCB exceeding the concentration limits indicated in Annex IV to Regulation (EC) No 850/2004 of the European Parliament and of the Council[8] shall be classified as hazardous.

— The concentration limits defined in Annex III to Directive 2008/98/EC do not apply to pure metal alloys in their massive form (not contaminated with hazardous substances). Those waste alloys that are considered as hazardous waste are specifically enumerated in this list and marked with an asterisk (*).

— Where applicable, the following notes included in Annex VI to Regulation (EC) No 1272/2008 may be taken into account when establishing the hazardous properties of wastes:

— 1.1.3.1. Notes relating to the identification, classification and labelling of substances: Notes B, D, F, J, L, M, P, Q, R, and U.

— 1.1.3.2. Notes relating to the classification and labelling of mixtures: Notes 1, 2, 3 and 5.

— After assessing the hazardous properties for a waste according to this method, an appropriate hazardous or non-hazardous entry from the list of wastes shall be assigned.

All other entries in the harmonised list of wastes are considered non-hazardous.

---

[8] Regulation (EC) No 850/2004 of the European parliament and of the Council of 29 April 2004 on persistent organic pollutants and amending Directive 79/117/EEC (OJ L 158, 30.4.2004, p. 7).

LIST OF WASTE

The different types of waste in the list are fully defined by the six-digit code for the waste and the respective two-digit and four-digit chapter headings. This implies that the following steps should be taken to identify a waste in the list:

— Identify the source generating the waste in Chapters 01 to 12 or 17 to 20 and identify the appropriate six-digit code of the waste (excluding codes ending with 99 of these chapters). Note that a specific production unit may need to classify its activities in several chapters. For instance, a car manufacturer may find its wastes listed in Chapters 12 (wastes from shaping and surface treatment of metals), 11 (inorganic wastes containing metals from metal treatment and the coating of metals) and 08 (wastes from the use of coatings), depending on the different process steps.

— If no appropriate waste code can be found in Chapters 01 to 12 or 17 to 20, the Chapters 13, 14 and 15 must be examined to identify the waste.

— If none of these waste codes apply, the waste must be identified according to Chapter 16.

— If the waste is not in Chapter 16 either, the 99 code (wastes not otherwise specified) must be used in the section of the list corresponding to the activity identified in step one.

INDEX

[Not published herein]

See appendix 2 for a commented list of waste

II.6   EU Regulation 1418/2007

COMMISSION REGULATION (EC) No 1418/2007

of 29 November 2007[1]

concerning the export for recovery of certain waste listed in Annex III or IIIA to Regulation (EC) No 1013/2006 of the European Parliament and of the Council to certain countries to which the OECD Decision on the control of transboundary movements of wastes does not apply

(Text with EEA relevance)

THE COMMISSION OF THE EUROPEAN COMMUNITIES,

Having regard to the Treaty establishing the European Community,

Having regard to Regulation (EC) No 1013/2006 of the European Parliament and of the Council of 14 June 2006 on shipments of waste[2], and in particular the third sub-paragraph of Article 37(2) thereof,

After consultation of the countries concerned,

Whereas:

(1)   In accordance with Article 37(1) of Regulation (EC) No 1013/2006 the Commission has sent a written request to each country to which Decision C(2001)107/Final of the OECD Council concerning the revision of Decision C(92)39/Final on control of transboundary movements of wastes destined for recovery operations does not apply, seeking confirmation in writing that waste which is listed in Annex III or IIIA to that Regulation and the export of which is not prohibited under its Article 36 may be exported from the Community for recovery in that country and requesting an indication as to which control procedure, if any, would be followed in the country of destination.

(2)   In those requests, each country was asked to indicate if it had opted for a prohibition or a procedure of prior written notification and consent, or if it would exercise no control, in respect of such waste.

(3)   Pursuant to the first subparagraph of Article 37(2) of Regulation (EC) No 1013/2006, and before the date of application of that Regulation, the Commission was required to adopt a Regulation taking into account all the replies received. The

---

[1]   Consolidated version of Regulation (EC) No 1418/2007 concerning the export for recovery of certain waste to certain non-OECD countries, OJ, L 316 4.12.2007, p. 6, as last amended by Commission Regulation (EU) No 2021/1840 of 20 October 2021, OJ, L 373, 21.10.2021, p. 1

[2]   OJ L 190, 12.7.2006, p. 1

Commission duly adopted Regulation (EC) No 801/2007 of 6 July 2007[3]. However, further replies and clarifications received since that date provide a better understanding of how the replies of the countries of destination should be taken into account.

(4)   The Commission has now received replies to its written requests from Algeria, Andorra, Argentina, Bangladesh, Belarus, Benin, Botswana, Brazil, Chile, China, Chinese Taipei, Costa Rica, Croatia, Cuba, Egypt, Georgia, Guyana, Hong Kong (China), India, Indonesia, Israel, Ivory Coast, Kenya, Kyrgyzstan, Lebanon, Liechtenstein, Macau (China), Malawi, Mali, Malaysia, Moldova, Morocco, Oman, Pakistan, Paraguay, Peru, Philippines, Russian Federation, Seychelles, South Africa, Sri Lanka, Thailand, Tunisia, Vietnam.

(5)   Certain countries have not issued a confirmation in writing that the waste may be exported to them from the Community for recovery. Therefore, in accordance with the second subparagraph of Article 37(2) of Regulation (EC) No 1013/2006, those countries are to be regarded as having chosen a procedure of prior written notification and consent.

(6)   Certain countries have in their replies made known their intention to follow control procedures applicable under national law that are distinct from those provided for in Article 37(1) of Regulation (EC) No 1013/2006. In addition, and in accordance with Article 37(3) of Regulation (EC) No 1013/2006, Article 18 of that Regulation should apply mutatis mutandis to such shipments, unless a waste is also subject to the prior notification and consent procedure.

(7)   Regulation (EC) No 801/2007 should be amended accordingly. For the sake of clarity, given the number of changes required, it is appropriate to repeal that Regulation and replace it by this Regulation. However, waste classified in Regulation (EC) No 801/2007 as subject to no control in the country of destination but which in this Regulation is shown as requiring prior notification and consent should continue to be classified as subject to no control in the country of destination during a transitional period of 60 days after entry into force,

HAS ADOPTED THIS REGULATION:

Article 1

Export for recovery of waste listed in Annex III or IIIA to Regulation (EC) No 1013/2006, which is not prohibited under Article 36 of that Regulation, to certain countries to which Decision C(2001)107/Final of the OECD Council concerning the revision of Decision C(92)39/Final on control of transboundary movements of wastes destined for recovery operations does not apply shall be governed by the procedures set out in the Annex.

---

[3]   OJ L 179, 7.7.2007, p. 6

442

Article 1a

The replies received following a written request by the Commission in accordance with the first subparagraph of Article 37 (1) of Regulation (EC) No 1013/2006 are listed in the Annex.

Where it is indicated in the Annex that a country, with regard to certain shipments of waste, does not prohibit them or apply the procedure of prior written notification and consent as described in Article 35 of that Regulation, Article 18 of that Regulation shall apply mutatis mutandis to such shipments.

Article 2

Regulation (EC) No 801/2007 is repealed.

Article 3

This Regulation shall enter into force on the fourteenth day following that of its publication in the Official Journal of the European Union.

It shall apply from the date of entry into force.

However, Regulation (EC) No 801/2007 shall continue to apply for 60 days after that date to waste listed in column (c) of the Annex to that Regulation which is listed in column (b), or in columns (b) and (d), of the Annex to the present Regulation.

This Regulation shall be binding in its entirety and directly applicable in all Member States.

Done at Brussels, 29 November 2007.

EU Regulation 1418/2007

ANNEX

The headings of the columns in this Annex refer to the following:

(a)   prohibition;

(b)   prior written notification and consent as described in Article 35 of Regulation (EC) No 1013/2006;

(c)   no control in the country of destination;

(d)   other control procedures will be followed in the country of destination under applicable national law. As for the waste included in column (c), the general information requirements laid down in Article 18 of Regulation (EC) No 1013/2006 apply mutatis mutandis unless a waste is also included in column (b).

Where two codes are separated by a hyphen, this is to be understood as covering the two codes and all codes in between them.

Where two codes are separated by a semicolon, this is to be understood as covering the two codes in question.

Where column (b) and column (d) are both designated for the same entry, that means that control procedures in the country of destination are applicable in addition to those laid down in Article 35 of Regulation (EC) No 1013/2006.

Where a particular waste or mixture of wastes is not indicated for a given country, this means that this country has not issued a sufficiently clear confirmation that this waste or mixture of wastes may be exported for recovery in that country and as to which control procedure, if any, would be followed in that country. Pursuant to Article 37(2) of Regulation (EC) No 1013/2006, the procedure of prior written notification and consent as described in Article 35 of that Regulation is applicable in such cases.

[List not published herein]

Please cf. webside of European Commission[4]

---

[4]   https://eur-lex.europa.eu/legal-content/EN/TXT/?uri=CELEX%3A02007R1418-20140718

II.7   Bamako Convention

BAMAKO CONVENTION ON THE BAN OF THE IMPORT INTO AFRICA AND THE CONTROL OF TRANSBOUNDARY MOVEMENT AND MANAGEMENT OF HAZARDOUS WASTES WITHIN AFRICA

of 30 January 1991[1]

PREAMBLE

The Parties to this Convention,

1.   Mindful of the growing threat to human health and the environment posed by the increased generation and the complexity of hazardous wastes,

2.   Further mindful that the most effective way of protection human health and the environment from the dangers posed by such wastes is the reduction of their generation to a minimum in term of quantity and/or hazard potential,

3.   Aware of the risk of damage to human health and the environment caused by transboundary movements of hazardous wastes,

4.   Reiterating that States should ensure that the generator should carry out his responsibilities with regard to the transport and disposal of hazardous wastes in a manner that is consistent with the protection of human health and environment, whatever the place of disposal,

5.   Recalling relevant chapters of the Charter of the Organization of African Unity (OAU) on environmental protection the African Charter for Human and Peoples' Rights, Chapter IX of the Lagos Plan of Action and other Recommendations adopted by the Organization of African Unity on the environment,

6.   Further recognizing the sovereignty of States to ban the importation into, and the transit through, their territory, of hazardous wastes and substances for human health and environmental reasons,

7.   Recognizing also the increasing mobilization in Africa for the prohibition of transboundary movements of hazardous wastes and their disposal in African countries,

8.   Convinced that hazardous wastes should, as far as is compatible with environmentally sound and efficient management be disposed in the State where they were generated,

9.   Convinced that the effective control and minimization of transboundary movements of hazardous wastes will act as an incentive, in Africa and elsewhere, for the reduction of the volume of the generation of such wastes,

---

[1]   Adopted in Bamako, Mali on 30 January 1991, entered into force on 22. April 1998

10.   Noting that a number of international and regional agreement deal with the problem of the protection and preservation of the environment with regard to the transit of dangerous goods,

11.   Taking into account the Declaration of the United Nations Conference on the Human Environment (Stockholm, 1972), the Cairo Guidelines and Principles for the Environmentally Sound Management of Hazardous Wastes adopted by the Governing Council of the United Nations Environment Programme (UNEP) by Decision 14/30 of 17 June, 1987, the Recommendations of the United Nations Committee of Experts on the Transport of Dangerous Goods (formulated in 1957 and updated biennially), the Charter of Human Rights, relevant recommendations, declarations, instruments and regulations adopted within the United Nations System, the relevant articles of the 1989 Basel Convention on the Control of Transboundary Movements of Hazardous Wastes and their Disposal which allow for the establishment of regional agreements which may be equal to or stronger than its own provisions, Article 39 of the Lomé IV Convention relating to the international movement of hazardous wastes and radioactive wastes, African intergovernmental organizations and the work and studies done within other international and regional organizations,

12.   Mindful of the spirit, principles, aims and functions of the African Convention on the Conservation of Nature and Natural Resources adopted by the African Heads of State and Government in Algiers (1968) and the World Charter for Nature adopted by the General Assembly of the United Nations at its Thirty-seventh Session (1982) as the rule of ethics in respect of the protection of the human environment and the conservation of natural resources,

13.   Concerned by the problem of transboundary traffic in hazardous wastes,

14.   Recognizing the need to promote the development of clean production methods, including clean technologies, for the sound management of hazardous wastes produced in Africa, in particular, to avoid, minimize and eliminate the generation of such wastes,

15.   Recognizing also that where necessary hazardous wastes should be transportted in accordance with relevant international conventions and recommendations,

16.   Determined to protect, by strict control, the human health of the African population and the environment against the adverse effects which may result from the generation of hazardous wastes,

17.   Affirming a commitment also to responsibly address the problem of hazardous wastes originating within the Continent of Africa,

HAVE AGREED AS FOLLOWS:

Article 1

Definitions

For the purpose of this Convention:

1.   "Wastes" are substances or materials which are disposed of, or are intended to be disposed of, or are required to be disposed of by the provisions of national law;

2.   "Hazardous wastes" means wastes as specified in Article 2 of this Convention;

3.   "Management" means the prevention and reduction of hazardous wastes and the collection, transport, storage, and treatment either for the reuse or disposal, of hazardous wastes including after-care of disposal sites;

4.   "Transboundary movement" means any movement of hazardous wastes from an area under the national jurisdiction of any State to or through an area under the national jurisdiction of another State, or to or through an area not under the national jurisdiction of another State, provided at least two States are involved in the movement;

5.   "Clean production methods" means production or industrial systems which avoid, or eliminate the generation of hazardous wastes and hazardous products in conformity-with Article 4, section 3 (f) and (g) of this Convention;

6.   "Disposal" means any operation specified in Annex III to this Convention;

7.   "Approved site or facility" means a site or facility for the disposal of hazardous wastes which is authorized or permitted to operate for this purpose by a relevant authority of the State where the site or facility is located;

8.   "Competent authority" means one governmental authority designated by a Party to be responsible, within such geographical areas as the Party may think fit, for receiving the notification of a transboundary movement of hazardous wastes and any information related to it, and for responding to such a notification, as provided in Article 6 of this Convention;

9.   "Focal point" means the entity of a Party referred to in Article 5 of this Convention responsible for receiving and submitting information as provided for in Articles 13 and 16;

10.   "Environmentally sound management of hazardous wastes" means taking all practicable steps to ensure that hazardous wastes are managed in a manner which will protect human health and the environment against the adverse effects which may result from such wastes;

11.   "Area under the national jurisdiction of a State" means any land, marine area or airspace within which a State exercises administrative and regulatory responsibility in accordance with international law in regard to the protection of human health or the environment;

12.   "State of export" means a State from which a transboundary movement of hazardous wastes is planned to be initiated or is initiated;

13.   "State of import" means a State to which a transboundary movement is planned

or takes place for the purpose of disposal therein or for the purpose of loading prior to disposal in an area not under the national jurisdiction of any State;

14. "State of transit" means any State, other than the State of export or import, through which a movement of hazardous wastes is planned or takes place;

15. "States concerned" means States of export or import, or transit states, whether or not Parties;

16. "Person" means any natural or legal person;

17. "Exporter" means any person under the jurisdiction of the State of export who arranges for hazardous wastes to be exported;

18. "Importer" means any person under the jurisdiction of the State of import who arranges for hazardous wastes to be imported;

19. "Carrier" means any person who carries out the transport of hazardous wastes;

20. "Generator" means any person whose activity produces hazardous wastes, or, if that person is not known, the person who is in possession and/or control of those wastes;

21. "Disposer" means any person to whom hazardous wastes are shipped and who carries out the disposal of such wastes;

22. "Illegal traffic" means any transboundary movement of hazardous wastes as specified in Article 9 of this Convention;

23. "Dumping at sea" means the deliberate disposal of hazardous wastes at sea from vessels, aircraft, platforms or other man-made structures at sea, and includes ocean incineration and disposal into the seabed and sub-seabed.

Article 2

Scope of the Convention

1. The following substances shall be "hazardous wastes" for the purposes of this convention:

    (a) Wastes that belong to any category contained in Annex I of this Convention;

    (b) Wastes that are not covered under paragraph (a) above but are defined as, or are considered to be, hazardous wastes by the domestic legislation of the State of export, import or transit;

    (c) Wastes which possess any of the characteristics contained in Annex II of this Convention;

    (d) Hazardous substances which have been banned, cancelled or refused registration by government regulatory action, or voluntarily withdrawn from registration in the country of manufacture, for human health or environmental reasons.

2.   Wastes which, as a result of being radioactive, are subject to any international control systems, including international instruments, applying specifically to radioactive materials, are included in the scope of this Convention.

3.   Wastes which derive from the normal operations of a ship, the discharge of which is covered by another international instrument, shall not fall within the scope of this convention.

Article 3

National Definitions of Hazardous Wastes

1.   Each State shall, within six months of becoming a Party to this Convention, inform the Secretariat of the Convention of the wastes, other than those listed in Annex I of this Convention, considered or defined as hazardous under its national legislation and of any requirements concerning transboundary movement procedures applicable to such wastes.

2.   Each Party shall subsequently inform the Secretariat of any significant changes to the information it has provided pursuant to Paragraph 1 of this Article.

3.   The Secretariat shall forthwith inform all Parties of the information it has received pursuant to paragraphs 1 and 2 of this Article.

4.   Parties shall be responsible for making the information transmitted to them by the Secretariat under Paragraph 3 of this Article available to their exporters and other appropriate bodies.

Article 4

General Obligations

1.   **Hazardous Waste Import Ban**

All Parties shall take appropriate legal, administrative and other measures within the area under their jurisdiction to prohibit the import of all hazardous wastes, for any reason, into Africa from non-Contracting Parties. Such import shall be deemed illegal and a criminal act. All Parties shall:

(a)   Forward as soon as possible, all information relating to such illegal hazardous waste import activity to the Secretariat who shall distribute the information to all Contracting Parties;

(b)   Cooperate to ensure that no imports of hazardous wastes from a non-Party enter a Party to this Convention. To this end, the Parties shall, at the Conference of the Contracting Parties, consider other enforcement mechanisms.

2.   **Ban on Dumping of Hazardous Wastes at Sea and Internal Waters**

(a)   Parties in conformity with related international conventions and instruments shall, in the exercise of their jurisdiction within their internal waters, territorial seas, exclusive economic zones and continental shelf, adopt legal, administrative and other appropriate measures to control all carriers from non-Parties, and prohibit the dumping at sea of hazardous wastes, including their incineration at sea and their disposal in the seabed and sub-seabed. Any dumping of hazardous wastes at sea, including incineration at sea as well as seabed and sub-seabed disposal, by Contracting Parties, whether in internal waters, territorial seas, exclusive economic zones or high seas shall be deemed to be illegal;

(b)   Parties shall forward, as soon as possible, all information relating to dumping of hazardous wastes to the Secretariat which shall distribute the information to all Contracting Parties.

3.   **Waste Generation in Africa**

Each Party Shall:

(a)   Ensure that hazardous waste generators submit to the Secretariat reports regarding the wastes that they generate in order to enable the Secretariat of the Convention to produce a complete hazardous waste audit;

(b)   Impose strict, unlimited liability as well as joint and several liability on hazardous waste generators;

(c)   Ensure that the generation of hazardous wastes within the area under its jurisdiction is reduced to a minimum taking into account social, technological and economic aspects;

(d)   Ensure the availability of adequate treatment and/or disposal facilities, for the environmentally sound management of hazardous wastes which shall be located, to the extent possible, within its jurisdiction;

(e)   Ensure that persons involved in the management of hazardous wastes within its jurisdiction take such steps as are necessary to prevent pollution arising from such wastes and, if such pollution occurs, to minimize the consequence thereof for human health and the environment;

The Adoption of Precautionary Measures:

(f)   Each Party shall strive to adopt and implement the preventive, precautionary approach to pollution problems which entails, inter-alia, preventing the release into the environment of substances which may cause harm to humans or the environment without waiting for scientific proof regarding such harm. The Parties shall cooperate with each other in taking the appropriate measures to implement the precautionary principle to pollution prevention through the application of clean production methods, rather than the pursuit of a permissible emissions approach based on assimilative capacity assumptions;

(g)   In this respect Parties shall promote clean production methods applicable to entire product life cycles including:

    —   raw material selection, extraction and processing;

    —   product conceptualization, design, manufacture and assemblage;

- materials transport during all phases;

- industrial and household usage;

- reintroduction of the product into industrial systems or nature when it no longer serves a useful function;

Clean production shall not include "end-of-pipe" pollution controls such as filters and scrubbers, or chemical, physical or biological treatment measures which reduce the volume of waste by incineration or concentration, mask the hazard by dilution, or transfer pollutants from one environmental medium to another, are also excluded;

(h) The issue of preventing the transfer to Africa of polluting technologies shall be kept under systematic review by the Secretariat of the Conference and periodic reports shall be made to the Conference of the Parties;

Obligations in the Transport and Transboundary Movement of Hazardous Wastes from Contracting Parties:

(i) Each Party shall prevent the export of hazardous wastes to States which have prohibited by their legislation or international agreement all such imports, or if it has reason to believe that the wastes in question will not be managed in an environmentally sound manner, according to criteria to be decided on by the Parties at their first meeting;

(j) A Party shall not permit hazardous wastes to be exported to a State which does not have the facilities for disposing of them in an environmentally sound manner;

(k) Each Party shall ensure that hazardous wastes to be exported are managed in an environmentally sound manner in the State of import and transit. Technical guidelines for the environmentally sound management of wastes subject to this Convention shall be decided by the Parties at their first meeting;

(l) The Parties agree not to allow the export of hazardous wastes for disposal within the area South of 60 degrees South Latitude, whether or not such wastes are subject to transboundary movement;

(m) Furthermore, each Party shall:

(i) Prohibit all persons under its national jurisdiction from transporting, storing or disposing of hazardous wastes unless such persons are authorized or allowed to perform such operations;

(ii) Ensure that hazardous wastes that are to be the subject of a transboundary movement are packaged, labelled, and transported in conformity with generally accepted and recognized international rules and standards in the field of packaging, labelling, and transport, and that due account is taken of relevant internationally recognized practices;

(iii) Ensure that hazardous wastes be accompanied by a movement document, containing information specified in Annex IV B, from the point at which a transboundary movement commences to the point of disposal;

(n) Parties shall take the appropriate measures to ensure that the transboundary movements of hazardous wastes only are allowed if:

   (i) The State of export does not have the technical capacity and the necessary facilities, capacity or suitable disposal sites in order to dispose of the wastes in question in an environmentally sound and efficient manner; or

   (ii) The transboundary movement in question is in accordance with other criteria to be decided by the Parties, provided those criteria do not differ from the objectives of this Convention;

(o) Under this Convention, the obligation of States in which hazardous wastes are generated, requiring that those wastes are managed in an environmentally sound manner, may not under any circumstances be transferred to the States of import or transit;

(p) Parties shall undertake to review periodically the possibilities for the reduction of the amount and/or the pollution potential of hazardous wastes which are exported to other States;

(q) Parties exercising their right to prohibit the import of hazardous wastes for disposal shall inform the other Parties of their decision pursuant to Article 13 of this Convention;

(r) Parties shall prohibit or shall not permit the export of hazardous wastes to States which have prohibited the import of such wastes, when notified by the secretariat or any competent authority pursuant to sub-paragraph (q) above;

(s) Parties shall prohibit or shall not permit the export of hazardous wastes if the State of import does not consent in writing to the specific import, in the case where that State of import has not prohibited the import of such wastes;

(t) Parties shall ensure that the transboundary movement of hazardous wastes is reduced to the minimum consistent with the environmentally sound and efficient management of such wastes, and is conducted in a manner which will protect human health and the environment against the adverse effects which may result from such movement;

(u) Parties shall require that information about a proposed transboundary movement of hazardous wastes be provided to the States concerned, according to Annex IV A of this Convention, and clearly state the potential effects of the proposed movement on human health and the environment.

4. **Furthermore**

(a) Parties shall undertake to enforce the obligations of this Convention against offenders and infringements according to relevant national laws and/or international law;

(b) Nothing in this Convention shall prevent a Party from imposing additional requirements that are consistent with the provisions of this Convention, and are in accordance with the rules of international law, in order to better protect human health and the environment;

(c) This Convention recognizes the sovereignty of States over their territorial sea, waterways, and air space established in accordance with international law, and jurisdiction which States have in their exclusive economic zone and their continental shelves in accordance with international law, and the exercise by ships and aircraft of all States of navigation rights and freedoms as provided for in international law and as reflected in relevant international instruments.

Article 5

Designation of Competent Authorities, Focal Point and Dumpwatch

To facilitate the implementation of this Convention, the Parties shall:

1. Designate or establish one or more competent authorities and one focal point. One competent authority shall be designated to receive the notification in case of a State of transit.

2. Inform the Secretariat, within three months of the date of the entry into force of this Convention for them, which agencies they have designated as their focal point and their competent authorities.

3. Inform the Secretariat, within one month of the date of decision, of any changes regarding the designations made by them under paragraph 2 above.

4. Appoint a national body to act as a Dumpwatch. In such capacity as a Dumpwatch, the designated national body only will be required to coordinate with the concerned governmental and non-governmental bodies.

Article 6

Transboundary Movement and Notification Procedures

1. The State of export shall notify, or shall require the generator or exporter to notify, in writing, through the channel of the competent authority of the State of export, the competent authority of the States concerned of any proposed transboundary movement of hazardous wastes. Such notification shall contain the declarations and information specified in Annex IV A of this Convention, written in a language acceptable to the State of import. Only one notification needs to be sent to each State concerned.

2. The Party of import shall respond to the notifier in writing consenting to the movement with or without conditions, denying permission for the movement, or requesting additional information. A copy of the final response of the State of import shall be sent to the competent authorities of the States concerned that are Parties to this Convention.

3.   The State of export shall not allow the transboundary movement until it has received:

(a)   written consent of the State of import; and

(b)   from the State of import, written confirmation of the existence of a contract between the exporter and the disposer specifying environmentally sound management of the wastes in question.

4.   Each State of transit which is a Party to this Convention shall promptly acknowledge to the notifier receipt of the notification. It may subsequently respond to the notifier in writing, within 60 days, consenting to the movement with or without conditions, denying permission for the movement, or requesting additional information. The State of export shall not allow the transboundary movement to commence until it has received the written consent of the State of transit.

5.   In the case of a transboundary movement of hazardous wastes where the wastes are legally defined as or considered to be hazardous wastes only:

(a)   By the State of export, the requirements of paragraph 8 of this Article that apply to the importer or disposer and the State of import shall apply mutatis mutandis to the exporter and State of export, respectively;

(b)   By the Party of import, or by the States of import and transit which are Parties to this Convention, the requirements of paragraphs 1, 3, 4 and 6 of this Article that apply to the exporter and State of export shall apply mutatis mutandis to the importer or disposer and Party of import, respectively; or

(c)   By any State of transit which is a Party to this Convention, the provisions of paragraph 4 of this Article shall apply to such State.

6.   The State of export shall use a shipment specific notification even where hazardous wastes having the same physical and chemical characteristics are shipped regularly to the same disposer via the same customs office of entry of the State of import, and in the case of transit, via the same customs office of entry and exit of the State or States of transit; specific notification of each and every shipment shall be required and contain the information in Annex IV A of this Convention.

7.   Each Party to this Convention shall limit their points or ports of entry and notify the Secretariat to this effect for distribution to all Contracting Parties. Such points and ports shall be the only ones permitted for the transboundary movement of hazardous wastes.

8.   The Parties to this Convention shall require that each person who takes charge of a transboundary movement of hazardous wastes sign the movement document either upon delivery or receipt of the wastes in question. They shall also require that the disposer inform both the exporter and the competent authority of the State of export of receipt by the disposer of the wastes in question and, in due course, of the completion of disposal as specified in the notification. If no such information is received within the State of export, the competent authority of the State of export or the exporter shall so notify the State of import.

10.   Any transboundary movement of hazardous wastes shall be covered by insurance, bond or other guarantee as may be required by the State of import, or any State of transit which is a Party to this Convention.

Article 7

Transboundary Movement from a Party through States which are not Parties

Paragraph 2 and 4 of Article 6 of this Convention shall apply mutatis mutandis to, transboundary movements of hazardous wastes from a Party through a State or States which are not Parties.

Article 8

Duty to Re-import

When a transboundary movement of hazardous wastes to which the consent of the States concerned has been given, subject to the provisions of this Convention, cannot be completed in accordance with the terms of the contract, the State of export shall ensure that the wastes in question are taken back into the State of export, by the exporter, if alternative arrangements cannot be made for their disposal in an environmentally sound manner within a maximum of 90 days from the time that the importing State informed the State of export and the Secretariat. To this end, the State of export and any State of transit shall not oppose, hinder or prevent the return of those waste to the State of export.

Article 9

Illegal traffic

1.   For the purpose of this Convention, any transboundary movement of hazardous wastes under the following situations shall be deemed to be illegal traffic:

   (a)   if carried out without notification, pursuant to the provisions of this Convention, to all States concerned; or

   (b)   if carried out without the consent, pursuant to the provisions of this Convention, of a State concerned; or

   (c)   if consent is obtained from States concerned through falsification, misrepresentation or fraud; or

   (d)   if it does not conform in a material way with the documents; or

   (e)   if it results in deliberate disposal of hazardous wastes in contravention of this Convention and of general principles of international law.

2.   Each Party shall introduce appropriate national legislation for imposing criminal penalties on all persons who have planned, carried out, or assisted in such illegal imports. Such penalties shall be sufficiently high to both punish and deter such conduct.

3.   In case of a transboundary movement of hazardous wastes deemed to be illegal traffic as the result of conduct on the part of the exporter or generator, the State of export shall ensure that the wastes in question are taken back by the exporter or generator or if necessary by itself into the State of export, within 30 days from the time the State of export has been informed about the illegal traffic. To this end the States concerned shall not oppose, hinder or prevent the return of those wastes to the State of export and appropriate legal action shall be taken against the contravenor(s).

4.   In the case of a transboundary movement of hazardous wastes deemed to be illegal traffic as the result of conduct on the part of the importer or disposer, the State of import shall ensure that the wastes in question are returned to the exporter by the importer and that legal proceedings according to the provisions of this Convention are taken against the contravenor(s).

Article 10

Intra-African Co-operation

1.   The Parties to this Convention shall cooperate with one another and with relevant African organizations, to improve and achieve the environmentally sound management of hazardous wastes.

2.   To this end, the Parties shall:

(a)   Make available information, whether on a bilateral or multilateral basis, with a view to promoting clean production methods and the environmentally sound management of hazardous wastes, including harmonization of technical standards and practices for the adequate management of hazardous wastes;

(b)   Cooperate in monitoring the effects of the management of hazardous wastes on human health and the environment;

(c)   Cooperate, subject to their national laws, regulations and policies, in the development and implementation of new environmentally sound clean production technologies and the improvement of existing technologies with a view to eliminating, as far as practicable, the generation of hazardous wastes and achieving more effective and efficient methods of ensuring their management in an environmentally sound manner, including the study of the economic, social and environmental effects of the adoption of such new and improved technologies;

(d)   Cooperate actively, subject to their national laws, regulations and policies, in the transfer of technology and management systems related to the environmentally sound management of hazardous wastes. They shall also cooperate in developing the technical capacity among Parties, especially those which may need and request technical assistance in this field;

(e)   Cooperate in developing appropriate technical guidelines and/or codes of practice;

(f)   Cooperate in the exchange and dissemination of information on the movement of hazardous wastes in conformity with Article 13 of this Convention.

Article 11

International Cooperation Bilateral, Multilateral and Regional Agreements

1.   Parties to this Convention may enter into bilateral, multilateral, or regional agreements or arrangements regarding the transboundary movement and management of hazardous wastes generated in Africa with Parties or non-Parties provided that such agreements or arrangements do not derogate from the environmentally sound management of hazardous wastes as required by this Convention. These agreements or arrangements shall stipulate provisions which are no less environmentally sound than those provided for by this Convention.

2.   Parties shall notify the Secretariat of any bilateral, multilateral or regional agreements or arrangements referred to in paragraph 1 of this Article and those which they have entered into prior to the entry into force of this Convention for them, for the purpose of controlling transboundary movements of hazardous wastes which take place entirely among the Parties to such agreements. The provisions of this Convention shall not affect transboundary movements of hazardous wastes generated in Africa which take place pursuant to such agreements provided that such agreements are compatible with the environmentally sound management of hazardous wastes as required by this Convention.

3.   Each Contracting Party shall prohibit vessels flying its flag or aircraft registered in its territory from carrying out activities in contravention of this Convention.

4.   Parties shall use appropriate measures to promote South-South cooperation in the implementation of this Convention.

5.   Taking into account the needs of developing countries, co-operation between international organizations is encouraged in order to promote, among other things, public awareness, the development of rational management of hazardous waste, and the adoption of new and non/less polluting technologies.

Article 12

Liabilities and Compensation

The Conference of Parties shall set up an Ad Hoc expert organ to prepare a draft Protocol setting out appropriate rules and procedures in the field of liabilities and compensation for damage resulting from the transboundary movement of hazardous wastes.

Article 13

Transmission of Information

1.   The Parties shall ensure that in the case of an accident occurring during the transboundary movement of hazardous wastes or their disposal which is likely to present risks to human health and the environment in other States, those States are immediately informed.

2.   The States shall inform each other, through the Secretariat, of:

   (a)   Changes regarding the designation of competent authorities and/or focal points, pursuant to Article 5 of this Convention;

   (b)   Changes in their national definition of hazardous wastes, pursuant to Article 3 of this Convention;

   (c)   Decisions made by them to limit or ban the import of hazardous wastes;

   (d)   Any other information required pursuant to paragraph 4 of this Article.

3.   The Parties, consistent with national laws and regulations, shall set up information collection and dissemination mechanisms on hazardous wastes. They shall transmit such information through the Secretariat, to the Conference of the Parties established under Article 15 of this Convention, before the end of each calendar year, in a report on the previous calendar year, containing the following information:

   (a)   Competent authorities, Dumpwatch, and focal points that have been designated by them pursuant to Article 5 of this Convention;

   (b)   Information regarding transboundary movements of hazardous wastes in which they have been involved, including:

   (i)   The quantity of hazardous wastes exported, their category, characteristics, destination, any transit country and disposal method as stated in the notification;

   (ii)   The amount of hazardous wastes imported, their category, characteristics, origin, and disposal methods;

   (iii) Disposal which did not proceed as intended;

   (iv) Efforts to achieve a reduction of the amount of hazardous wastes subject to transboundary movement;

   (c)   Information on the measures adopted by them in the implementation of this Convention;

   (d)   Information on available qualified statistics - which have been compiled by them on the effects on human health and the environment of the generation, transportation, and disposal of hazardous wastes - as part of the information required in conformity with Article 4 Section 3 (a) of this Convention;

   (e)   Information concerning bilateral, multilateral and regional agreements and arrangements entered into pursuant to Article 11 of this Convention;

(f)   Information on accidents occurring during the transboundary movements, treatment and disposal of hazardous wastes and on the measures undertaken to deal with them;

(g)   Information on treatment and disposal options operated within the area under their national jurisdiction;

(h)   Information on measures undertaken for the development of clean production methods, including clean production technologies, for the reduction and/or elimination of the production of hazardous wastes; and

(i)   Such other matters as the Conference of the Parties shall deem relevant.

4.   The Parties, consistent with national laws and regulations, shall ensure that copies of each notification concerning any given transboundary movement of hazardous wastes, and the response to it, are sent to the Secretariat.

Article 14

Financial Aspects

1.   The regular budget of the Conference of Parties, as required in Article 15 and 16 of this Convention, shall be prepared by the Secretariat and approved by the Conference.

2.   Parties shall, at the first meeting of the Conference of the Parties, agree on a scale of contributions to the recurrent budget of the Secretariat.

3.   The Parties shall also consider the establishment of a revolving fund to assist, on an interim basis, in case of emergency situations to minimize damage from disasters or accidents arising from transboundary movements of hazardous wastes or during the disposal of such wastes.

4.   The Parties agree that, according to the specific needs of different regions and sub-regions, regional or sub-regional centres for training and technology transfers regarding the management of hazardous wastes and the minimization of their generation should be established, as well as appropriate funding mechanisms of a voluntary nature.

Article 15

Conference of the Parties

1.   A Conference of the Parties, made up of ministers having the environment as their mandate, is hereby established. The first meeting of the Conference of the Parties shall be convened by the secretary-General of the OAU not later than one year after the entry into force of this Convention. Thereafter, ordinary meetings of the Conference of the Parties shall be held at regular intervals to be determined by the Conference at its first meeting.

2.   The Conference of the Parties to this Convention shall adopt Rules of Procedure for itself and for any subsidiary body it may establish, as well as financial rules

to determine in particular the financial participation of the Parties to this Convention.

3.   The Parties to this Convention at their first meeting shall consider any additional measures needed to assist them in fulfilling their responsibilities with respect to the protection and the preservation of the marine and inland waters environments in the context of this Convention.

4.   The Conference of the Parties shall keep under continued review and evaluation the effective implementation of this Convention, and in addition, shall:

(a)   promote the harmonization of appropriate policies, strategies and measures for minimizing harm to human health and the environment by hazardous wastes;

(b)   consider and adopt amendments to this Convention and its annexes, taking into consideration, inter alia, available scientific, technical, economic and environmental information;

(c)   consider and undertake any additional action that may be required for the achievement of the purpose of this Convention in the light of experience gained in its operation and in the operation of the agreements and arrangements envisaged in Article 11 of this Convention;

(d)   consider and adopt protocols as required;

(e)   establish such subsidiary bodies as are deemed necessary for the implementation of this Convention; and

(f)   make decisions for the peaceful settlement of disputes arising from the transboundary movement of hazardous wastes, if need be, according to international law.

5.   Organizations may be represented as observers at meetings of the Conference of the Parties. Any body or agency, whether national or international, governmental or non-governmental, qualified in fields relating to hazardous wastes which has informed the Secretariat, may be represented as an observer at a meeting of the Conference of the Parties. The admission and participation of observers shall be subject to the rules of procedure adopted by the Conference of the Parties.

Article 16

Secretariat

1.   The functions of the Secretariat shall be:

(a)   To arrange for, and service, meetings provided for in Article 15 and 17 of this Convention;

(b)   To prepare and transmit reports based upon information received in accordance with Articles 3, 4, 6, 11, and 13 of this Convention as well as upon information derived from meetings of subsidiary bodies established under Article 15 of this Convention as well as upon, as appropriate, information provided by relevant inter-governmental and non-governmental entities;

(c)   To prepare reports on its activities carried out in the implementation of its functions under this Convention and present them to the Conference of the Parties;

(d)   To ensure the necessary coordination with relevant international bodies, and in particular to enter into such administrative and contractual arrangements as may be required for the effective discharge of its functions;

(e)   To communicate with focal points, competent authorities and Dumpwatch established by the Parties in accordance with Article 5 of this Convention as well as appropriate inter-governmental and non-governmental organizations which may provide assistance in the implementation of this Convention;

(f)   To compile information concerning approved national sites and facilities of Parties to this Convention available for the disposal and treatment of their hazardous wastes and to circulate this information;

(g)   To receive and convey information from and to Parties on:

  –   sources of technical assistance and training;

  –   available technical and scientific know-how;

  –   sources of advice and expertise; and

  –   availability of resources;

This information will assist them in,

  –   the management of the notification system of this Convention;

  –   environmentally sound clean production methods relating to hazardous wastes, such as clean production technologies;

  –   the assessment of disposal capabilities and sites;

  –   the monitoring of hazardous wastes; and

  –   emergency responses;

(h)   To provide Parties to this Convention with information on consultants or consulting firms having the necessary technical competence in the field, which can assist them with examining a notification for a transboundary movement, the concurrence of a shipment of hazardous wastes with the relevant notification, and/or whether the proposed disposal facilities for hazardous wastes are environmentally sound, when they have reason to believe that the wastes in question will not be managed in an environmentally sound manner. Any such examinations would not be at the expense of the Secretariat;

(i)   To assist Parties to this Convention in their identification of cases of illegal traffic and to circulate immediately to the Parties concerned any information it has received regarding illegal traffic;

(j)   To cooperate with Parties to this Convention and with relevant and competent international organizations and agencies in the provision of experts and equipment for the purpose of rapid assistance to States in the event of an emergency situation; and

(k)   To perform such other functions relevant to the purposes of this Convention as may be determined by the Conference of the Parties to this Convention.

2.   The Secretariat's functions shall be carried out on an interim basis by the Organization of African Unity (OAU) jointly with the United Nations Economic Commission for Africa (ECA) until the completion of the first meeting of the Conference of the Parties held pursuant to Article 15 of this Convention. At this meeting, the Conference of the Parties shall also evaluate the implementation by the interim Secretariat of the functions assigned to it, in particular under paragraph 1 above, and decide upon the structures appropriate for those functions.

Article 17

Amendment of the Convention and of Protocols

1.   Any Party may propose amendments to this Convention and any Party to a Protocol may propose amendments to that Protocol. Such amendments shall take due account, inter alia, of relevant scientific, technical, environmental and social considerations.

2.   Amendments to this Convention shall be adopted at a meeting of the Conference of the Parties. Amendments to any Protocol shall be adopted at a meeting of the Parties to the Protocol in question. The text of any proposed amendment to this Convention or to any Protocol, except as may otherwise be provided in such Protocol, shall be communicated to the Parties by the Secretariat at least six months before the meeting at which it is proposed for adoption. The Secretariat shall also communicate proposed amendments to the Signatories to this Convention for their information.

3.   The Parties shall make every effort to reach agreement on any proposed amendment to this Convention by consensus. If all efforts at consensus have been exhausted, and no agreement reached, the amendment shall, as a last resort, be adopted by a two-thirds majority vote of the Parties present and voting at the meeting. It shall then be submitted by the Depository to all Parties for ratification, approval, formal confirmation or acceptance.

Amendment of Protocols to this Convention

4.   The procedure specified in paragraph 3 above shall apply to amendments to any protocol, except that a two-thirds majority of the Parties to that Protocol present and voting at the meeting shall suffice for their adoption.

GENERAL PROVISIONS

5.   Instruments of ratification, approval, formal confirmation or acceptance of

amendments shall be deposited with the Depository. Amendments adopted in accordance with paragraph 3 or 4 above shall enter into force between Parties having accepted them, on the ninetieth day after the receipt by the Depository of the instrument of ratification, approval, formal confirmation or acceptance by at least two-thirds of the Parties who accepted the amendments to the Protocol concerned, except as may otherwise be provided in such Protocol. The amendments shall enter into force for any other Party on the ninetieth day after that Party deposits its instrument of ratification, approval, formal confirmation or acceptance of the amendments.

6.   For the purpose of this Article, "Parties present and voting" means Parties present and casting an affirmative or negative vote.

Article 18

Adoption and Amendment of Annexes

1.   The annexes to this Convention or to any Protocol shall form an integral part of this Convention or of such Protocol, as the case may be and, unless expressly provided otherwise, a reference to this Convention or its protocols constitutes at the same time a reference to any annexes thereto. Such annexes shall be restricted to scientific, technical and administrative matters.

2.   Except as may be otherwise provided in any Protocol with respect to its annexes, the following procedures shall apply to the proposal, adoption and entry into force of additional annexes to this Convention or of annexes to a protocol:

(a)   Annexes to this Convention and its Protocols shall be proposed and adopted according to the procedure laid down in Article 17, paragraphs 1, 2, 3, and 4 of this Convention;

(b)   Any Party that is unable to accept an additional annex to this Convention or an annex to any Protocol to which it is Party shall so notify the Depository, in writing, within six months from the date of the communication of the adoption by the Depository. The Depository shall without delay notify all Parties of any such notification received. A Party may at any time substitute an acceptance for a previous declaration of objection and the annexes shall thereupon enter into force for that Party;

(c)   Upon the expiration of six months from the date of the circulation of the communication by the Depository, the annex shall become effective for all Parties to this Convention or to any Protocol concerned, which have not submitted a notification in accordance with the provision of subparagraph (b) above.

3.   The proposal, adoption and entry into force of amendments to annexes to this Convention or to any Protocol shall be subject to the same procedure as for the proposal, adoption and entry into force of annexes to the Convention or annexes to a Protocol. Annexes and amendments thereto shall take due account, inter alia, of relevant scientific and technical considerations.

4.   If an additional annex or an amendment to an annex involves an amendment

to this Convention or to any Protocol, the additional annex or amended annex shall not enter into force until such time as the amendment to this Convention or to the Protocol enters into force.

Article 19

Verification

Any Party which has reason to believe that another Party is acting or has acted in breach of its obligations under this Convention must inform the Secretariat thereof, and in such an event, shall simultaneously and immediately inform, directly or through the Secretariat, the Party against whom the allegations are made. The Secretariat shall carry out a verification of the substance of the allegation and submit a report thereof to all the Parties to this Convention.

Article 20

Settlement of Disputes

1. In case of dispute between Parties as to the interpretation or application of, or compliance with, this Convention or any Protocol thereto, the Parties shall seek a settlement of the dispute through negotiations or any other peaceful means of their own choice.

2. If the Parties concerned cannot settle their dispute as provided in paragraph 1 of this Article, the dispute shall be submitted either to an Ad Hoc organ set up by the Conference for this purpose, or to the International Court of Justice.

3. The conduct of arbitration of disputes between Parties by the Ad Hoc organ provided for in paragraph 2 of this Article shall be as provided in Annex V of this Convention.

Article 21

Signature

This Convention shall be open for signature by Member States of the OAU in Bamako and Addis Ababa for a period of six months from 30 January 1991 to 31 July 1991.

Article 22

Ratification, Acceptance, Formal Confirmation or Approval

1. This Convention shall be subject to ratification, acceptance, formal confirmation, or approval by Member States of the OAU. Instruments of ratification, acceptance, formal confirmation, or approval shall be deposited with the Depository.

2. Parties shall be bound by all obligations of this Convention.

Article 23

Accession

This Convention shall be open for accession by Member States of the OAU from the day after the date on which the Convention is closed for signature. The instruments of accession shall be deposited with the Depository.

Article 24

Right to Vote

Each Contracting Party to this Convention shall have one vote.

Article 25

Entry into Force

1.   This Convention shall enter into force on the ninetieth day after the date of deposit of the tenth instrument of ratification from Parties signatory to this Convention.

2.   For each State which ratifies this Convention or accedes thereto after the date of the deposit of the tenth instrument of ratification, it shall enter into force on the ninetieth day after the date of deposit by such State of its instrument of accession or ratification.

Article 26

Reservations and Declarations

1.   No reservations or exception may be made to this Convention.

2.   Paragraph 1 of this Article does not preclude a State when signing, ratifying, or acceding to this Convention, from making declarations or statements, however phrased or named, with a view, inter alia, to the harmonization of its laws and regulations with the provisions of this Convention, provided that such declarations or statements do not purport to exclude or to modify the legal effects of the provisions of the Convention in their application to that State.

Article 27

Withdrawal

1.   At any time after three years from the date on which this Convention has entered into force for a Party, that Party may withdraw from the Convention by giving written notification to the Depository.

2.   Withdrawal shall be effective one year after receipt of notification by the Depository, or on such later date as may be specified in the notification.

3.   Withdrawal shall not exempt the withdrawing Party from fulfilling any obligations it might have incurred under this Convention.

Article 28

Depositary

The Secretary-General of the Organization of African Unity shall be the Depository for this Convention and of any Protocol thereto.

Article 29

Registration

This Convention, as soon as it enters into force, shall be registered with the Secretary-General of the United Nations organization (UNO) in conformity with Article 102 of the Charter of the UNO.

Article 30

Authentic Texts

The Arabic, English, French and Portuguese texts of this Convention are equally authentic.

ANNEX I

CATEGORIES OF WASTES WHICH ARE HAZARDOUS WASTES

Waste Streams:

Y0 All wastes containing or contaminated by radionuclides the concentration or properties of which result from human activity

Y1 Clinical wastes from medical care in hospitals, medical centers and clinics

Y2 Wastes from the production and preparation of pharmaceutical products

Y3 Waste pharmaceuticals, drugs and medicines

Y4 Wastes from the production, formulation and use of biocide and phytopharmaceuticals

Y5 Wastes from the manufacture, formulation and use of wood preserving chemicals

Y6 Wastes from the production, formulation and use of organic solvents

Y7 Wastes from heat treatment and tempering operation containing cyanides

Y8 Waste mineral oils unfit for their originally intended use

Y9 Waste oils/water, hydrocarbons/water mixtures, emulsions

Y10 Waste substances and articles containing or contaminated with polychlorinated biphenyls (PCBS) and/or polychlorinated terphenyls (PCTS) and/or polybrominated biphenyls (PBBs)

Y11 Waste tarry residues arising from refining, distillation and any pyrolytic treatment

Y12 Wastes from production, formulation and use of inks, dyes pigments, paints, lacquers, varnish

Y13 Wastes from production, formulation and use of resins latex, plasticizers, glues/adhesives

Y14 Waste chemical substances arising from research and development or teaching activities which are not identified and/or are new and whose effects on man and/or the environment are not known

Y15 Wastes of an explosive nature not subject to other legislation

Y16 Wastes from production, formulation and use of photographic chemicals and processing materials

Y17 Wastes resulting from surface treatment of metals and plastics

Y18 Residues arising from industrial waste disposal operations

Y46 Wastes collected from households, including sewage and sewage sludges

Y47 Residues arising from the incineration of household wastes

Bamako Convention

Wastes having as constituents:

Y19 Metal carbonyls

Y20 Beryllium; beryllium compounds

Y21 Hexavalent chromium compounds

Y22 Copper compounds

Y23 Zinc compounds

Y24 Arsenic; arsenic compounds

Y25 Selenium; selenium compounds

Y26 Cadmium; cadmium compounds

Y27 Antimony; antimony compounds

Y28 Tellurium; tellurium compounds

Y29 Mercury; mercury compounds

Y30 Thallium; thallium compounds

Y31 Lead; lead compounds

Y32 Inorganic fluorine compounds excluding calcium fluoride

Y33 Inorganic cyanides

Y34 Acidic solutions or acids in solid form

Y35 Basic solutions or bases in solid form

Y36 Asbestos (dust and fibres)

Y37 Organic phosphorous compounds

Y38 Organic cyanides

Y39 Phenols; phenolcompounds including chlorophenols

Y40 Ethers

Y41 Halogenated organic solvents

Y42 Organic solvents excluding halogenated solvents

Y43 Any congener of polychlorinated dibenzo-furan

Y44 Any congener of polychlorinated dibenzo-p-dioxin

Y45 Organohalogen compounds other than substances referred to in this Annex
(e.g., Y39, Y41, Y42, Y43, Y44)

ANNEX II

LIST OF HAZARDOUS CHARACTERISTICS

UN Class*   **Code Characteristics**

1   **H1 Explosive**

An explosive substance or waste is a solid or liquid substance or waste (or mixture of substances or wastes) which is in itself capable by chemical reaction or producing gas at such a temperature and pressure and at such a speed as to cause damage to the surroundings.

3   **H3 Flammable liquids**

The word "flammable" has the same meaning as "inflammable". Flammable liquids are liquids, or mixtures of liquids, or liquids containing solids in solution or suspension (for example paints, varnishes, lacquers, etc., but not including substances or wastes otherwise classified on account of their dangerous characteristics) which give off a flammable vapour at temperatures of not more than 60.5 degrees C, closed-cup test, or not more than 65.6 degrees C, open-cup test. (Since the results of open-cup tests and of closed-cup tests are not strictly comparable and even individual results by the same test are often variable, regulations varying from the above figures to make allowance for such difference would be within the spirit of this definition).

4.1   **H4.1 Flammable solids**

Solids, or waste solids, other than those classed as explosives, which under conditions encountered in transport are readily combustible, or may cause or contribute to fire through friction.

4.2   **H4.2 Substances or wastes liable to spontaneous combustion**

Substances or wastes which are liable to spontaneous heating under normal conditions encountered in transport, or to heating up on contact with air, and being then liable to catch fire.

4.3   **H4.3 Substances or wastes which, in contact with water emit flammable gases**

Substances or wastes which, by interaction with water, are liable to become spontaneously flammable or to give off flammable gases in dangerous quantities

5.1   **H5.1 Oxidizing**

Substances or wastes which, while in themselves not necessarily combustible, may, generally by yielding oxygen, cause or contribute to the combustion of other materials.

---

*   Corresponds to the hazardous classification system included in the United Nations Recommendations on the transport of Dangerous Goods (ST/SG/AC.10/I/Rev.5, United Nations, New York, 1988).

5.2	**H5.2 Organic peroxides**

Organic substances or wastes which contain the bivalent-0-0-structure are thermally unstable substances which may undergo exothermic self-accelerating decomposition.

6.1	**H6.1 Poisonous (Acute)**

Substances or wastes liable either to cause death or serious injury or to harm human health if swallowed or inhaled or by skin contact.

6.2	**H6.2 Infectious substances**

Substances or wastes containing viable microorganisms or their toxins which are known or suspected to cause disease in animals or humans.

8	**H8 Corrosives**

Substances or wastes which, by chemical action, will cause severe damage when in contact with living tissue, or in the case of leakage, will materially damage, or even destroy, other goods or the means of transport; they may also cause other hazards.

9	**H10 Liberation of toxic gases in contact with air or water**

Substances or wastes which, by interaction with air or water, are liable to give off toxic gases in dangerous quantities.

9	**H11 Toxic (Delayed or chronic)**

Substances or wastes which, if they are inhaled or ingested or if they penetrate the skin, may involve delayed or chronic effects, including carcinogenicity.

9	**H12 Ecotoxic**

Substances or wastes which if released present or may present immediate or delayed adverse impacts to the environment by means of bioaccumulation and/or toxic effects upon biotic systems.

9	**H13**

Capable, by any means, after disposal, of yielding another material, e.g., leachate, which possesses any of the characteristics listed above.

ANNEX III

DISPOSAL OPERATIONS

D1   Deposit into or onto land, (e.g., landfill, etc.)

D2   Land treatment, (e.g., biodegradation of liquid or sludgy discards in soils, etc.)

D3   Deep injection, (e.g., injection of pumpable discards into wells, salt domes or naturally occurring repositories, etc.)

D4   Surface impoundments, (e.g., placement of liquid or sludge discards into pits, ponds, or lagoons, etc.)

D5   Specially engineered landfill, (e.g., placement into lined discrete cells which are capped and isolated from one another and the environment, etc.)

D6   Release into a water body except seas/oceans

D7   Release into seas/oceans including sea-bed insertion

D8   Biological treatment not specified elsewhere in this Annex which results in final compounds or mixtures which are discarded by means of any of the operations in Annex III

D9   Physico-chemical treatment not specified elsewhere in this Annex which results in final compounds or mixtures which are discarded by means of any of the operations in Annex III, (e.g., evaporation, drying, calcination, neutralisation precipitation, etc.)

D10   Incineration on land

D11   Incineration at sea

D12   Permanent storage, (e.g., emplacement of containers in a mine, etc.)

D13   Blending or mixing prior to submission to any of the operations in Annex III

D14   Repackaging prior to submission to any of the operations in Annex III

D15   Storage pending any of the operations in Annex III

D16   Use as a fuel (other than in direct incineration) or other means to generate energy

D17   Solvent reclamation/regeneration

D18   Recycling/reclamation of organic substances which are not used as solvents

D19   Recycling/reclamation of metals and metal compounds

D20   Recycling/reclamation of other inorganic materials

D21   Regeneration of acids and bases

D22   Recovery of components used for pollution abatement

D23   Recovery of components from catalysts

D24   Used oil re-refining or other reuses of previously used oil

D25   Land treatment resulting in benefit to agriculture or ecological improvement

D26   Uses of residual materials obtained from any of the operations numbered D1-D25

D27   Exchange of wastes for submission to any of the operations numbered D1-D26

D28   Accumulation of material intended for any operation in Annex III

ANNEX IV A

INFORMATION TO BE PROVIDED ON NOTIFICATION

1. Reason for waste export
2. Exporter of the waste 1/
3. Generator(s) of the waste and site of generation 1/
4. Importer and Disposer of the waste and actual site of disposal 1/
5. Intended carrier(s) of the waste or their agents, if known 1/
6. Country of export of the waste Competent authority 2/
7. Countries of transit Competent authority 2/
8. Country of import of the waste Competent authority 2/
9. Projected date of shipment and period of time over which waste is to be exported and proposed itinerary (including point of entry and exit)
10. Means of transport envisaged (road, rail, sea, air, inland waters)
11. Information relating to insurance 3/
12. Designation and physical description of the waste including Y number and UN number and its composition 4/ and information on any special handling requirements including emergency provisions in case of accidents.
13. Type of packaging envisaged (e.g., bulk, drummer, tanker)
14. Estimated quantity in weight/volume
15. Process by which the waste is generated 5/
16. Waste classifications from Annex II of this Convention:

    Hazardous characteristics, H number, and UN class
17. Method of disposal as per Annex III of this Convention
18. Declaration by the generator and exporter that the information is correct
19. Information transmitted (including technical description of the plant) to the exporter or generator from the disposer of the waste upon which the latter has based his assessment that there was no reason to believe that the wastes will not be managed in an environmentally sound manner in accordance with the laws and regulations of the country of import
20. Information concerning the contract between the exporter and disposer

Notes
1/   Full name and address, telephone, telex or telefax number and the name, address, telephone, telex, or telefax number of the person to be contacted.
2/   Full name and address, telephone, telex or telefax number.
3/   Information to be provided on relevant insurance requirements and how they are met by exporter, carrier, and disposer.
4/   The nature and the concentration of the most hazardous components, in terms of toxicity and other dangers presented by the waste both in handling and in relation to the proposed disposal method.
5/   Insofar as this is necessary to assess the hazard and determine the appropriateness of the proposed disposal operation.

ANNEX IV B

INFORMATION TO BE PROVIDED ON THE MOVEMENT DOCUMENT

1. Exporter of the waste 1/

2. Generator(s) of the waste and site of generation 1/

3. Disposer of the waste and actual site of disposal 1/

4. Carrier(s) of the waste 1/ or his agent(s)

5. The date the transboundary movement started and date(s) and signature on receipt by each person who takes charge of the waste

6. Means of transport (road, rail, inland waterway, sea, air) including countries of export, transit and import, also point of entry and exit where these have been designated

7. General description of the waste (physical state, proper UN shipping name and class, UN number, Y number and H number as applicable)

8. Information on special handling requirements including emergency provisions in case of accidents

9. Type and number of packages

10. Quantity in weight/volume

11. Declaration by the generator or exporter that the information is correct

12. Declaration by the generator or exporter indicating no objection from the competent authorities of all States concerned

13. Certification by disposer of receipt at designated disposal facility and indication of method of disposal and of the appropriate date of disposal

Notes

The information required on the movement document shall where possible be integrated into one document with that required under transport rules. Where this is not possible, the information should complement rather than duplicate that required under the transport rules. The movement document shall carry instructions as to who is to provide information and fill-out any form.

1/ Full name and address, telephone, telex or telefax number and the name, address, telephone, telex or telefax number of the person to be contacted in case of emergency.

ANNEX V

ARBITRATION

Article 1

Unless the agreement referred to in Article 20 of the Convention provides otherwise, the arbitration procedure shall be conducted in accordance with Articles 2 to 10 below.

Article 2

The claimant Party shall notify the Secretariat that the Parties have agreed to submit the dispute to arbitration pursuant to paragraph 1 or paragraph 2 of Article 20 of this Convention and include, in particular, the Articles of the Convention, and the interpretation or application of which are at issue. The Secretariat shall forward the information thus received to all Parties to the Convention.

Article 3

The arbitral tribunal shall consist of three members. Each of the Parties to the dispute shall appoint an arbitrator, and the two arbitrators so appointed shall designate by common agreement the third arbitrator, who shall be the chairman of the tribunal. The latter shall not be a national of one of the parties to the dispute, nor have his usual place of residence in one of the Parties, nor be employed by any of them, nor have dealt with the case in any other capacity.

Article 4

1.   If the chairman of the arbitral tribunal has not been designated within two months of the appointment of the second arbitrator, the Secretary-General of the OAU shall, at the request of either Party, designate him within a further two months period.

2.   If one of the Parties to the dispute does not appoint an arbitrator within two months of the receipt of the request, the other Party may inform the Secretary-General of the OAU who shall designate the chairman of the arbitral tribunal within a further two months period. Upon designation, the chairman of the arbitral tribunal shall request the Party which has not appointed an arbitrator to do so within two months.

After such period, he shall inform the Secretary-General of the OAU who shall make this appointment within a further two month's period.

Article 5

1.   The arbitral tribunal shall render its decision in accordance with international law and in accordance with the provisions of this Convention.

2.   Any arbitral tribunal constituted under the provisions of this Annex shall draw up its own rules of procedure.

Article 6

1.   The decisions of the arbitral tribunal both on procedure and on substance, shall be taken by majority vote of its members.

2.   The tribunal may take all appropriate measures in order to establish the facts. It may, at the request of one of the Parties, recommend essential interim measures of protection.

3.   The Parties to the dispute shall provide all facilities necessary for the effective conduct of the proceedings.

4.   The absence or default of a Party in the dispute shall not constitute an impediment to the proceedings.

Article 7

The tribunal may hear and determine counter-claims arising directly out of the subject-matter of the dispute.

Article 8

Unless the arbitral tribunal determines otherwise because of the particular circumstances of the case, the expenses of the tribunal, including the remuneration of its members, shall be borne by the Parties to the dispute in equal shares. The tribunal shall keep a record of all its expenses, and shall furnish a final statement thereof to the Parties.

Article 9

Any Party that has an interest of a legal nature in the subject-matter of the dispute which may be affected by the decision in the case, may intervene in the proceedings with the consent of the tribunal.

Article 10

1.   The tribunal shall render its award within five months of the date on which it is established unless it finds it necessary to extend the time-limit for a period which should not exceed five months.

2.   The award of the arbitral tribunal shall be accompanied by a statement of reasons. It shall be final and binding upon the Parties to the dispute.

3.   Any dispute which may arise between the Parties concerning the interpretation or execution of the award may be submitted by either Party to the arbitral tribunal which made the award or, if the latter cannot be seized thereof, to another tribunal constituted for this purpose in the same manner as the first.

II.8   Waigani Convention

WAIGANI CONVENTION

**CONVENTION TO BAN THE IMPORTATION INTO FORUM ISLAND
COUNTRIES OF HAZARDOUS AND RADIOACTIVE WASTES AND TO
CONTROL THE TRANSBOUNDARY MOVEMENT AND MANAGEMENT OF
HAZARDOUS WASTES WITHIN THE SOUTH PACIFIC REGION**

of 16 September 1995[1]

PREAMBLE:

The Parties to this Convention:

Conscious of their responsibility to protect, preserve and improve the environment of the South Pacific for the good health, benefit and enjoyment of present and future generations of the people of the South Pacific;

Concerned about the growing threat to human health and the environment posed by the increasing generation of hazardous wastes and the disposal of such wastes by environmentally unsound methods;

Concerned also about the dangers posed by radioactive wastes to the people and environment of the South Pacific;

Aware that their responsibilities to protect, preserve and improve the environment of the South Pacific can be met only by cooperative effort among all peoples of the South Pacific based on an understanding of the needs and capacities of all Parties;

Taking full account of the Programme of Action for the Sustainable Development of Small Island Developing States adopted in Barbados on 6 May 1994;

Noting with concern that a number of approaches have been made to certain Island Countries of the South Pacific by unscrupulous foreign waste dealers for the importation into and the disposal within the South Pacific of hazardous wastes generated in other countries;

Concerned by the slowness of progress towards a satisfactory resolution of the issues surrounding international trade in goods which have been banned, cancelled or refused registration in the country of **manufacture for human health or environmental reasons;**

Recalling their commitments under existing regional treaties and arrangements for the protection and preservation of the environment of the South Pacific, including the Convention for the Protection of the Natural Resources and Environment of the South Pacific Region, signed in Noumea on 24 November 1986, the Protocol

---

[1]   Entry into force: 21 October 2001; Last updated: 28 November 2018

concerning Cooperation in Combating Pollution Emergencies in the South Pacific Region, adopted by Parties on 25 November 1986, and the South Pacific Nuclear Free Zone Treaty, signed in Rarotonga on 6 August 1985;

Further Recalling the Basel Convention on the Control of Transboundary Movements of Hazardous Wastes and their Disposal adopted by the Conference of the Plenipotentiaries on 22 March 1989, and noting decisions of its Conference of the Parties including Decision II 12 of 25 March 1994;

Desiring to conclude an agreement under Article 11 of the Basel Convention;

Mindful of the International Atomic Energy Agency (IAEA) Code of Practice on the International Transboundary Movement of Radioactive Waste and recognising the need for its strict observance in the South Pacific Region;

Noting as well the preliminary negotiations on a Convention on the Safe Management of Nuclear Waste;

Further Recalling the Declaration of the United Nations Conference on the Human Environment (Stockholm, 1972), the Cairo Guidelines and Principles for the Environmentally Sound Management of Hazardous Wastes adopted by the Governing Council of the United Nations Environment Programme (UNEP) by Decision 14/30 of 17 June 1987 and the Recommendations of the United Nations Committee of Experts on the Transport of Dangerous Goods (formulated in 1957 and updated biennially);

Recalling also Agenda 21 adopted by the United Nations Conference on Environment and Development in Rio de Janeiro on 14 June 1992, which reaffirms that effective control of the generation, storage, treatment, recycling and reuse, transport, recovery, and disposal of hazardous wastes is of paramount importance for proper health, environmental protection and natural resources management and sustainable development;

Resolving to prohibit the importation of hazardous wastes into Pacific Island Developing Parties, and to regulate and facilitate the environmentally sound management of such wastes generated within the Convention Area; and

Resolving also to prohibit the importation of all radioactive wastes into Pacific Island Developing Parties while at the same time recognising that the standards, procedures and the authorities responsible for the environmentally sound management of radioactive wastes will differ from those in respect of hazardous wastes.

Have agreed as follows:

ARTICLE 1

Definitions

For the purposes of this Convention:

"Approved site or facility" means a site or facility for the disposal of hazardous wastes which is authorised or permitted to operate for this purpose by a relevant authority of the Party where the site or facility is located;

"Area under the jurisdiction of a Party" means any land, marine area or airspace within which a Party exercises administrative and regulatory responsibility in accordance with international law in regard to the protection of human health or the environment;

"Authorised transboundary movement" means a transboundary movement of hazardous wastes to which the consent of the Parties concerned has been given in accordance with the provisions of this Convention;

"Basel Convention" means the Convention on the Control of Transboundary Movements of Hazardous Wastes and their Disposal, 1989;

"Carrier" means any person who carries out the transport of hazardous wastes;

"Cleaner production" means the conceptual and procedural approach to production that demands that all phases of the life-cycle of a product or process should be addressed, with the objective of prevention or minimisation of short and long-term risks to humans and to the environment;

"Competent authority" means any one governmental authority designated by a Party to be responsible within such geographical areas as the Party may think fit for receiving the notification of a transboundary movement of hazardous wastes and any information related to it, and for responding to such a notification, as provided in Article 6 of this Convention;

"Convention Area" shall comprise:

(i) the land territory, internal waters, territorial sea, continental shelf, archipelagic waters and exclusive economic zones established in accordance with international law of:

— **American Samoa**	— **The Commonwealth of**
— **Australia**	**Northern Mariana Islands**
— **Cook Islands**	— **Republic of Palau**
— **Federated States of Micronesia**	— **Papua New Guinea**
— **Fiji**	— **Pitcairn**
— **French Polynesia**	— **Solomon Islands**
— **Guam**	— **Tokelau**
— **Kiribati**	— **Tonga**
— **Republic of Marshall Islands**	— **Tuvalu**
— **Nauru**	— **Vanuatu**
— **New Caledonia and Dependencies**	— **Wallis and Futuna**
— **New Zealand**	— **Western Samoa**
— **Niue**	

(ii)   those areas of high seas which are enclosed from all sides by the exclusive economic zones referred to in sub-paragraph (i);

(iii)   areas of the Pacific Ocean which have been included in the Convention Area pursuant to Article 2.6;

"Countries concerned" means countries of export, import or transit whether or not Parties to this Convention;

"Days" means calendar days unless otherwise specified;

"Disposal" means any operation specified in Annex V to this Convention;

"Disposer" means any person for whom hazardous wastes are destined and who carries out the actual disposal of such wastes;

"Domestically prohibited goods" means substances or products which have been banned, cancelled or refused registration by government regulatory action, or voluntarily withdrawn from registration in the country of manufacture, for human health or environmental reasons;

"Environmentally sound management of hazardous wastes" means taking all practicable steps to ensure that hazardous wastes are managed in a manner which will protect human health and the environment against the adverse effects which may result from such wastes;

"Exporter" means any person under the jurisdiction of the exporting Party who arranges for hazardous wastes to be exported;

"Exporting Party" means a Party from which a transboundary movement of hazardous wastes is planned to be initiated or is initiated;

"Focal point" means the entity of a Party referred to in Article 5 of this Convention responsible for receiving and submitting information as provided for in Articles 7 and 14;

"Forum Island Countries" means all Members of the South Pacific Forum with the exception of Australia and New Zealand;

"Generator" means any person whose activity produces hazardous wastes or, if that person is not known, the person who is in possession and/or control of those wastes;

"Hazardous wastes" means wastes as specified in Article 2 of this Convention;

"IAEA" means the International Atomic Energy Agency;

"Illegal traffic" means any transboundary movement of hazardous wastes as specified in Article 9 of this Convention;

"Importer" means any person under the jurisdiction of the importing Party who arranges for hazardous wastes to be imported;

"Importing Party" means a Party to which transboundary movement of hazardous wastes is planned or takes place for the purpose of disposal therein or for the purpose of loading prior to disposal in an area not under the national jurisdiction of any State;

"London Convention" means the Convention on the Prevention of Marine Pollution by Dumping of Wastes and Other Matter, 1972;

"Management" means the prevention and reduction of hazardous wastes and the collection, transport, storage, and treatment or disposal, of hazardous wastes including after-care of disposal sites;

"Other Party" means a Party listed in Annex IV or any Party which is accepted by the Conference of the Parties to be an Other Party in accordance with the procedures established pursuant to Article 13.4(g);

"Pacific Island Developing Party" means a Party listed in Annex III or any Party which is accepted by the Conference of the Parties to be a Pacific Island Developing Party in accordance with the procedures established pursuant to Article 13.4(g);

"Party" means a Party to this Convention;

"Person" means any natural or legal person;

"Precautionary principle" means the principle that in order to protect the environment, the precautionary approach shall be widely applied by Parties according to their capabilities. Where there are threats of serious or irreversible damage, lack of full scientific certainty shall not be used as a reason for postponing cost-effective measures to prevent environmental degradation;

"Radioactive wastes" means wastes which, as a result of being radioactive, are subject to other international control systems, including international instruments, applying specifically to radioactive materials;

"Secretariat" means the Secretariat established pursuant to Article 14 of this Convention;

"SPREP" means the South Pacific Regional Environment Programme;

"Transboundary movement" means any movement of hazardous wastes from an area under the jurisdiction of any Party, to or through an area under the jurisdiction of another Party, or to or through an area not under the jurisdiction of another Party, provided at least two Parties are involved in the movement;

"Transit Party" means any Party, other than the exporting Party or importing Party, through which a movement of hazardous wastes is planned or takes place;

"Vessels" and **"Aircraft"** mean waterborne or airborne craft of any type whatsoever. This expression includes air cushioned craft and floating craft, whether self-propelled or not;

"Wastes" means substances or materials which are disposed of, or are intended to be disposed of, or are required to be disposed of, by provisions of national legislation.

ARTICLE 2

Scope of the Convention and Area of Coverage

Scope of the Convention

1.   The following substances shall be "hazardous wastes" for the purposes of this Convention:

> (a)   Wastes that belong to any category contained in Annex I of this Convention, unless they do not possess any of the characteristics contained in Annex II of this Convention; and

> (b)   Wastes that are not covered under sub-paragraph (a) above, but which are defined as, or are considered to be, hazardous wastes by the national legislation of the exporting, importing or transit Party to, from or through which such wastes are to be sent.

2.   Radioactive wastes are excluded from the scope of this Convention except as specifically provided for in Articles 4.1, 4.2, 4.3, and 4.5 of this Convention.

3.   Wastes which derive from the normal operations of a vessel, the discharge of which is covered by another international instrument, shall not fall within the scope of this Convention.

4.   Nothing in this Convention shall affect in any way the sovereignty of States over their territorial sea, the sovereign rights and jurisdiction that States have in their exclusive economic zones and continental shelves, and the exercise by vessels and aircraft of all States of navigational rights and freedoms, as provided for in international law and as reflected in the 1982 United Nations Convention on the Law of the Sea and other relevant international instruments.

5.   Nothing in this Convention shall affect in any way the rights and obligations of any Party under international law including under other international agreements in force. Such agreements include the London Convention as amended; the 1982 United Nations Convention on the Law of the Sea, including in particular Articles 31, 210 and 236 thereof; the South Pacific Nuclear Free Zone Treaty, 1985, including in particular Article 7 thereof; and the International Convention for the Prevention of Pollution from Ships, 1973.

Area of Coverage

6.   A Party may add areas under its jurisdiction within the Pacific Ocean between the Tropic of Cancer and 60 degrees South latitude and between 130 degrees East longitude and 120 degrees West longitude to the Convention Area. Such addition shall be notified to the Depositary who shall promptly notify the other Parties and the Secretariat. Such areas shall be incorporated within the Convention Area ninety days after notification to the Parties by the Depositary, provided there has been no objection to the proposal to add new areas by any Party. If there is any such objection the Parties concerned will consult with a view to resolving the matter.

ARTICLE 3

National Definitions of Hazardous Wastes

1.   Each Party shall, within six months of becoming a Party to this Convention, inform the Secretariat of the wastes, other than those listed in Annex I of this Convention, considered or defined as hazardous under its national legislation and of any requirements concerning transboundary movement procedures applicable to such wastes.

2.   Each Party shall subsequently inform the Secretariat of any significant changes to the information it has provided pursuant to paragraph 1 of this Article.

3.   The Secretariat shall forthwith inform all Parties of the information it has received pursuant to paragraphs 1 and 2 of this Article.

4.   Parties shall be responsible for making the information transmitted to them by the Secretariat under paragraph 3 of this Article available to their exporters, importers and other appropriate bodies.

ARTICLE 4

General Obligations

1.   Hazardous Wastes and Radioactive Wastes Import and Export Ban

   (a)   Each Pacific Island Developing Party shall take appropriate legal, administrative and other measures within the area under its jurisdiction to ban the import of all hazardous wastes and radioactive wastes from outside the Convention Area. Such import shall be deemed an illegal and criminal act; and

   (b)   Each Other Party shall take appropriate legal, administrative and other measures within the area under its jurisdiction to ban the export of all hazardous wastes and radioactive wastes to all Forum Island Countries, or to territories located in the Convention Area with the exception of those that have the status of Other Parties in accordance with Annex IV. Such export shall be deemed an illegal and criminal act.

2.   To facilitate compliance with paragraph 1 of this Article, all Parties:

   (a)   Shall forward in a timely manner all information relating to illegal hazardous wastes and radioactive wastes import activity within the area under its jurisdiction to the Secretariat who shall distribute the information as soon as possible to all Parties; and

   (b)   Shall cooperate to ensure that no illegal import of hazardous wastes and radioactive wastes from a non-Party enters areas under the jurisdiction of a Party to this Convention.

3.   Ban on Dumping of Hazardous Wastes and Radioactive Wastes at Sea

   (a)   Each Party which is a Party to the London Convention, the South Pacific Nuclear Free Zone Treaty, 1985, the 1982 United Nations Convention on the Law of the Sea or the Protocol for the Prevention of Pollution of the South Pacific Region by Dumping, 1986, reaffirms the commitments under those instruments which require it to prohibit dumping of hazardous wastes and radioactive wastes at sea; and

(b) Each Party which is not a Party either to the London Convention or the Protocol for the Prevention of Pollution of the South Pacific Region by Dumping, 1986, should consider becoming a Party to both of those instruments.

4. Wastes Located in the Convention Area

Each Party shall:

(a) Ensure that within the area under its jurisdiction, the generation of hazardous wastes is reduced at its source to a minimum taking into account social, technological and economic needs;

(b) Take appropriate legal, administrative and other measures to ensure that within the area under its jurisdiction, all transboundary movements of hazardous wastes generated within the Convention Area are carried out in accordance with the provisions of this Convention;

(c) Ensure the availability of adequate treatment and disposal facilities for the environmentally sound management of hazardous wastes, which shall be located, to the extent practicable, within areas under its jurisdiction, taking into account social, technological and economic considerations. However, where Parties are for geographic, social or economic reasons unable to dispose safely of hazardous wastes within those areas, cooperation should take place as provided for under Article 10 of this Convention;

(d) In cooperation with SPREP, participate in the development of programmes to manage and simplify the transboundary movement of hazardous wastes which cannot be disposed of in an environmentally sound manner in the countries in which they are located. Provided that such programmes do not derogate from the environmentally sound management of hazardous wastes as required by this Convention, they may be registered as arrangements under Article 11 of this Convention;

(e) Develop a national hazardous wastes management strategy which is compatible with the SPREP South Pacific Regional Pollution Prevention, Waste Minimization and Management Programme;

(f) Submit to the Secretariat such reports as the Conference of the Parties may require regarding the hazardous wastes generated in the area under its jurisdiction in order to enable the Secretariat to produce a regular hazardous wastes report;

(h) Take appropriate legal, administrative and other measures to prohibit vessels flying its flag or aircraft registered in its territory from carrying out activities in contravention of this Convention.

5. Radioactive Wastes

(a) Parties shall give active consideration to the implementation of the IAEA Code of Practice on the International Transboundary Movement of Radioactive Wastes and such other international and national standards which are at least as stringent; and

(b) Subject to available resources, Parties shall actively participate in the development of the Convention on the Safe Management of Nuclear Waste.

6. Domestically Prohibited Goods:

(a) Subject to available resources, Parties shall endeavour to participate in relevant international fora to find an appropriate global solution to the problems associated with the international trade of domestically prohibitted goods; and

(b) Nothing in this Convention shall be interpreted as limiting the sovereign right of Parties to act individually or collectively, consistent with their international obligations, to ban the importation of domestically prohibited goods into areas under their jurisdiction.

ARTICLE 5

Competent Authorities and Focal Points

1. To facilitate the implementation of this Convention, each Party shall designate or establish one competent authority and one focal point. A Party need not designate or establish new or separate authorities to perform the functions of the competent authority and the focal point.

2. The competent authority shall be responsible for the implementation of notification procedures for transboundary movement of hazardous wastes in accordance with the provisions of Article 6 of this Convention.

3. The focal point shall be responsible for transmitting and receiving information in accordance with the provisions of Article 7 of this Convention.

4. The Parties shall inform the Secretariat, within three months of the date of the entry into force of this Convention for them, which authorities they have designated or established as the competent authority and the focal point.

ARTICLE 6

Notification Procedures for Transboundary Movements of Hazardous Wastes between Parties

1. The exporting Party shall notify, or shall require the generator or exporter to notify, in writing, through its competent authority, the competent authority of the countries concerned of any proposed transboundary movement of hazardous wastes. Such notification shall contain the declarations and information specified in Annex VI A of this Convention, written in a language acceptable to the importing Party. Only one notification needs to be sent to each country concerned.

2.   The importing Party shall acknowledge within reasonable time, which in the case of Other Parties shall not exceed fourteen working days, the receipt of the notification referred to in paragraph 1 of this Article. The importing Party shall have sixty days after issuing the acknowledgement to inform the notifier that it is consenting to the movement, with or without conditions, denying permission for the movement or requesting additional information. In the event that additional information has been sought, a new period of twenty one days recommences from the time of receipt of the additional information.

3.   The exporting Party shall not allow the transboundary movement until it has received:

   (a)   Written consent of the importing Party;

   (b)   Written consent from every transit Party;

   (c)   Written consent of every non-Party country of transit;

   (d)   Written confirmation from the importing Party of the existence of a contract between the exporter and the disposer specifying the environmentally sound management of the wastes in question; and

   (e)   Written confirmation from the exporter of the existence of adequate insurance, bond or other guarantee satisfactory to the exporting Party.

4.   Each transit Party shall acknowledge within reasonable time, which in the case of Other Parties shall not exceed fourteen working days, the receipt of the notification referred to in paragraph 1 of this Article. Each transit Party shall have sixty days after issuing the acknowledgement to inform the notifier that it is consenting to the movement, with or without conditions, denying permission for the movement or requesting additional information. In the event that additional information has been sought, a new period of

5   In the case of a transboundary movement of hazardous wastes, where the wastes are legally defined as or are considered to be hazardous wastes only:

   (a)   By the exporting Party, the requirement at paragraph 10 of this Article, that any transboundary movement shall be covered by insurance, bond or other guarantee shall be as required by the exporting Party; or

   (b)   By the importing Party, or the transit Party, the requirements of paragraphs 1, 3, 4, and 6 of this Article that apply to the exporter and exporting Party, shall apply mutatis mutandis to the importer or disposer and importing Party, respectively; or

   (c)   By any transit Party, the provisions of paragraph 4 of this Article shall apply to such Party.

6.   The exporting Party may, subject to the written consent of the countries concerned, allow the generator or the exporter to use a general notification where hazardous wastes having the same physical and chemical characteristics are shipped regularly to the same disposer via the same customs office of exit of the exporting Party, via the same customs office of entry of the importing Party, and, in the case of transit, via the same customs office of entry and exit of the Party or Parties of

transit.

7.   The countries concerned may make their written consent to the use of the general notification referred to in paragraph 6 of this Article subject to the supply of certain information, such as the exact quantities or periodical lists of hazardous wastes to be shipped.

8.   The general notification and written consent referred to in paragraphs 6 and 7 of this Article may cover multiple shipments of hazardous wastes during a maximum period of twelve months.

9.   Each transboundary movement of hazardous wastes shall be accompanied by a movement document which includes the information listed in Annex VI B. The Parties to this Convention shall require that each person who takes charge of a transboundary movement of hazardous wastes sign the movement document either upon delivery or receipt of the wastes in question. They shall also require the disposer to inform both the exporter and the competent authority of the exporting Party of receipt by the disposer of the wastes in question and, in due course, of the completion of disposal as specified in the notification. If no such information is received by the exporting Party, the competent authority of the exporting Party or the exporter shall so notify the importing Party.

10.   Any transboundary movement of hazardous wastes shall be covered by insurance, bond or other guarantee as may be required or agreed to by the importing Party or any transit Party.

ARTICLE 7

Transmission of Information

1.   The Parties shall ensure that in the case of an accident occurring during the transboundary movement of hazardous wastes or their disposal which is likely to present risks to human health and the environment in other States and Parties, those States and Parties and the Secretariat are immediately informed.

2.   The Parties shall inform one another, through the Secretariat, of:

   (a)   Changes regarding the designation of competent authorities and/or focal points, pursuant to Article 5 of this Convention; and

   (b)   Changes in their national definition of hazardous wastes, pursuant to Article 3 of this Convention.

3.   The Parties, consistent with national laws and regulations, shall set up information collection and dissemination mechanisms on hazardous wastes to enable the Secretariat to fulfil the functions listed in Article 14.

ARTICLE 8

Duty to Re-import

1.   The exporting Party shall adopt appropriate administrative and legal measures

to ensure that when an authorised transboundary movement of hazardous wastes cannot be completed in accordance with the terms of the contract or of this Convention, the wastes in question are returned to it by the exporter. To this end, the importing Party and the transit Party or Parties shall not oppose, hinder or prevent the return of those wastes to the exporting Party.

2.   Notwithstanding the provisions of paragraph 1 of this Article, where an authorised transboundary movement of hazardous wastes cannot be completed within the terms of the contract or the terms of this Convention, the exporting Party need not re-import those wastes provided that alternative arrangements are made for the disposal of the wastes in a manner which is compatible with the environmentally sound management of hazardous wastes as required by this Convention and other international legal obligations. Such disposal shall take place within ninety days from the time that the importing Party informed the exporting Party and the Secretariat, or such other period of time as the Parties concerned agree.

ARTICLE 9

Illegal Traffic

1.   For the purpose of this Convention, any transboundary movement of hazardous wastes shall be deemed to be illegal traffic if:

(a)   Carried out without notification, pursuant to the provisions of this Convention, to all countries concerned;

(b)   Carried out without the consent, pursuant to the provisions of this Convention, of a country concerned;

(c)   Consent is obtained from countries concerned through falsification, misrepresentation or fraud;

(d)   The contents do not conform in a material way with the supporting documentation;

(e)   It results in deliberate disposal of hazardous wastes in contravention of this Convention, other relevant international instruments and of general principles of international law; or

(f)   It is in contravention of the import or export bans established by Article 4.1.

2.   Each Party shall introduce or adopt appropriate national legislation to prevent and punish illegal traffic. The Parties shall cooperate with a view to achieving the objects of this Article.

3.   (a)   In the case of a transboundary movement of hazardous wastes deemed to be illegal traffic as the result of conduct on the part of the exporter or generator, the exporting Party shall ensure that, within thirty days from the time the exporting Party has been informed about the illegal traffic or such other period of time the countries concerned may agree, the wastes in question are either:

      (i)   taken back by the exporter or generator or if necessary by itself into the exporting Party; or, if impracticable,

      (ii)   otherwise disposed of in accordance with the provisions of this Convention;

   (b)   In the case of paragraph 3(a)(i) of this Article, the Parties concerned shall not oppose, hinder or prevent the return of those wastes to the exporting Party.

4.   In the case of a transboundary movement of hazardous wastes deemed to be illegal traffic as a result of conduct on the part of the importer or disposer, the importing Party shall ensure that the wastes in question are disposed of in an environmentally sound manner by the importer or disposer or, if necessary, by itself within thirty days from the time the illegal traffic has come to the attention of the importing Party or such time as the countries concerned may agree. To this end, the importing Party and the exporting Party shall cooperate, as necessary, in the disposal of the wastes in an environmentally sound manner.

5.   In cases where the responsibility for the illegal traffic cannot be assigned either to the exporter or generator or to the importer or disposer, the Parties concerned or any other Parties, as appropriate, shall ensure through cooperation that the wastes in question are disposed of as soon as possible in an environmentally sound manner either in the exporting Party or the importing Party or elsewhere as appropriate.

6.   The Secretariat shall undertake the necessary coordination with the Secretariat of the Basel Convention in relation to the effective prevention and monitoring of illegal traffic in hazardous wastes. Such coordination shall include:

   (a)   Exchanging information on incidents or alleged incidents of illegal traffic in the Convention Area and on the appropriate steps to remedy such incidents; and

   (b)   Providing assistance in the field of capacity building including development of national legislation and of appropriate infrastructure in the Pacific Island Developing Parties with a view to the prevention and penalisation of illegal traffic of hazardous wastes.

ARTICLE 10

Cooperation among Parties and International Cooperation

1.   The Parties to this Convention shall cooperate with one another, non-Parties and relevant regional and international organisations, to facilitate the availability of adequate treatment and disposal facilities and to improve and achieve the environmentally sound management of hazardous wastes. Such facilities shall be located within the Convention Area to the extent practicable taking into account social, technological and economic considerations.

2.   To this end, the Parties shall:

   (a)   Upon request, make information available, whether on a bilateral or regional basis, with a view to promoting the environmentally sound management of hazardous wastes, including harmonisation of relevant technical standards and practices;

(b) Cooperate in monitoring the effects of hazardous wastes and their management on human health and the environment;

(c) Cooperate, subject to their national laws and policies, in the development and implementation of new environmentally sound and cleaner production technologies and the improvement of existing technologies. Such cooperation shall be with a view to eliminating, as far as practicable, the generation of hazardous wastes and achieving more effective and efficient methods of ensuring their management in an environmentally sound manner, including the study of the economic, social and environmental impacts of the adoption of such new and improved technologies;

(d) Cooperate, subject to their national laws and policies, actively in the transfer of technology and management systems related to the environmentally sound management of hazardous wastes. They shall also cooperate in developing the technical capacity and infrastructure of Parties, especially those which may need and request technical assistance in this field; and

(e) Cooperate in developing appropriate technical guidelines and/or codes of practice.

3.   The Secretariat shall encourage Other Parties and other concerned developed countries to take all practicable steps to promote, facilitate and finance, as appropriate, the transfer of, or access to, environmentally sound technologies and know-how to Pacific Island Developing Parties, to enable them to implement the provisions of this Convention. Other Parties undertake to cooperate with the Secretariat in this regard.

4.   Taking into account the needs of developing countries, Parties shall encourage cooperation with international organisations in order to promote, among other things, public awareness, the development of rational management of hazardous wastes, and the adoption of new technologies which are environmentally sound, including cleaner production technologies.

ARTICLE 11

Bilateral, Regional or Multilateral Agreements or Arrangements

1.   Notwithstanding the provisions of Article 4.4(g), Parties to this Convention may enter into bilateral, regional or multilateral agreements or arrangements with non-Parties regarding the transboundary movement and management of hazardous wastes provided that such agreements or arrangements do not derogate from the provisions of Article 4.1 or from the environmentally sound management of such wastes as required by this Convention.

2.   The Parties shall notify the Secretariat of any bilateral, regional or multilateral

agreements or arrangements referred to in paragraph 1 of this Article and those which they have entered into prior to the entry into force of this Convention for them, for the purpose of controlling transboundary movements of hazardous wastes which take place entirely among the parties to such agreements or arrangements.

3.   The provisions of this Convention shall not affect transboundary movements of hazardous wastes which take place pursuant to such agreements or arrangements provided that such agreements or arrangements are compatible with the environmentally sound management of hazardous wastes as required by this Convention.

ARTICLE 12

Liabilities and Compensation

The Conference of the Parties shall consider the preparation and adoption of appropriate arrangements in the field of liability and compensation arising from transboundary movements of hazardous wastes in the Convention Area without prejudice to the application and further development of relevant rules of international law.

ARTICLE 13

Conference of the Parties

1.   A Conference of the Parties to this Convention is hereby established. The first meeting of the Conference of the Parties shall be convened not later than one year after the entry into force of this Convention. Thereafter, ordinary meetings of the Conference of the Parties shall be held at regular intervals to be determined by the Conference at its first meeting. The quorum for meetings of the Conference of the Parties shall be two thirds of the Parties.

2.   The Conference of the Parties shall adopt by consensus at its first ordinary meeting, or as soon as practicable thereafter, Rules of Procedure. It shall also adopt by consensus financial rules, including the scale of contributions of the Parties to this Convention to the regular budget.

3.   The first meeting of the Conference of the Parties shall consider the adoption of any additional measures in accordance with the Precautionary principle relating to the implementation of this Convention.

4.   The Conference of the Parties shall keep under continuous review and evaluation the effective implementation of this Convention, and in particular, shall:

    (a)   Promote the harmonisation, at high levels of protection, of appropriate legislation, policies, strategies and measures for minimising harm to human health and the environment;

    (b)   Consider and adopt, where necessary, amendments to this Convention, and its annexes, taking into consideration, inter alia, available scientific, technical, economic and environmental information;

    (c)   Examine and approve the regular budget prepared by the Secretariat in accordance with Article 14;

(d) Consider and undertake any additional action that may be necessary for the achievement of the purposes of this Convention in the light of experience gained in the operation of the Convention and developments elsewhere;

(e) Consider and adopt protocols as necessary;

(f) Establish and/or designate such subsidiary bodies or agencies as are deemed necessary for the implementation of this Convention; and

(g) Determine and adopt appropriate rules and procedures for the acceptance of new Parties to this Convention in accordance with Article 23 and Annexes III and IV.

5. Any State which is eligible to become a Party to this Convention may be represented as an observer at meetings of the Conference of the Parties. Any other State or any body or agency, whether national, regional or international, governmental or nongovernmental, with an interest in the subject matter of this Convention which has informed the Secretariat of its wish to be represented as an observer at a meeting of the Conference of the Parties, may be admitted unless at least one-third of the Parties present object. The admission and participation of observers shall be subject to the rules of procedure adopted by the Conference of the Parties.

ARTICLE 14

Secretariat

1. A Secretariat for this Convention is hereby established. The functions of the Secretariat shall be to:

(a) Arrange and service meetings of the Parties to this Convention;

(b) Prepare the regular budget of the Conference of the Parties, as required by this Convention;

(c) Prepare and transmit reports based upon information received in accordance with Articles 3, 4, 7, and 11 of this Convention;

(d) Prepare and transmit information derived from meetings of subsidiary bodies and agencies established under Article 13 of this Convention or provided by relevant intergovernmental and Non-Governmental entities;

(e) Ensure coordination with the Secretariat of the Basel Convention and other relevant international and regional bodies, and in particular to enter into such administrative arrangements as may be required for the effective discharge of its functions;

(f) Communicate with the competent authorities and focal points established by the Parties in accordance with Article 5 of this Convention as well as appropriate intergovernmental and Non-Governmental Organisations which may provide financial and/or technical assistance in the implementation of this Convention;

(g) Compile information concerning approved sites and facilities available for the disposal of hazardous wastes and means of transport to these sites and facilities and to circulate this information;

(h) Receive and convey on request to Parties information on available sources of technical and scientific expertise;

(i) Receive and convey on request to Parties information on consultants or consulting firms having the necessary technical competence in the field which can assist them with examining a notification for a transboundary movement of hazardous wastes, the concurrence of a shipment of hazardous wastes with the relevant notification, and/or whether the proposed disposal facilities for hazardous wastes are environmentally sound, when they have reason to believe that the wastes in question will not be managed in an environmentally sound manner;

(j) Assist Parties to this Convention in their identification of cases of illegal traffic and to circulate immediately to the Parties concerned any information it has received regarding illegal traffic, and to undertake the necessary coordination with the Secretariat of the Basel Convention as provided for in Article 9.6;

(k) To cooperate with countries concerned and with relevant and competent international organisations and agencies in the provision of experts and equipment for the purpose of rapid assistance in the event of an emergency situation in the Convention Area;

(l) To report the information prescribed in paragraph 2 of this Article, to the Parties to this Convention, before the end of each calendar year; and

(m) To perform such other functions relevant to the purposes of this Convention as may be determined by the Conference of the Parties.

2.   The Secretariat shall transmit to the Parties, before the end of each calendar year, a report taking into account material provided by Parties under Articles 4.4(f) and 7.3 on the previous calendar year, containing the following:

(a) Information regarding transboundary movement of hazardous wastes in which Parties have been involved, including:

(i)   the quantity of hazardous wastes exported, their category, characteristics, destination, any transit country and disposal method as stated in the notification;

(ii)   the amount of hazardous wastes imported, their category, characteristics, origin, and disposal methods;

(iii) disposals which did not proceed as intended; and

(iv)   efforts to achieve a reduction of the amount of hazardous wastes subject to transboundary movement.

(b)   Information on measures adopted by Parties in the implementation of this Convention;

(c)   Information where it is available on the effects on human health and the environment from the generation, transportation and disposal of hazardous wastes in the Convention Area. The information may take the form of statistical data;

(d)   Information on accidents occurring during transboundary movements, treatment and disposal of hazardous wastes and on measures undertaken to deal with them;

(e)   Information on environmentally sound treatment and disposal options operated by Parties; and

(f)   Information on measures undertaken by Parties for the development of cleaner production technologies for the reduction and/or elimination of the production of hazardous wastes.

3.   The Secretariat's functions shall be carried out by SPREP.

ARTICLE 15

Revolving Fund

The Conference of the Parties shall consider the establishment of a revolving fund to assist on an interim basis in case of emergency situations to minimise damage from disasters or accidents arising from transboundary movement or disposal of hazardous wastes within the Convention Area.

ARTICLE 16

Amendments to this Convention

1.   Any Party may propose amendments to this Convention.

2.   Amendments to this Convention may be adopted only at a meeting of the Conference of the Parties at which at least two-thirds of the Parties are represented. The text of any proposed amendment to this Convention shall be communicated to the Parties by the Secretariat at least six months before the meeting at which it is proposed for adoption. The Secretariat shall also communicate proposed amendments to the Signatories to this Convention and to the Depositary for their information.

3.   The Parties shall make every effort to reach agreement on any proposed amendment to this Convention by consensus. If all efforts at consensus have been exhausted, and no agreement reached, the amendment shall, as a last resort, be adopted by a two-thirds majority vote of Parties present and voting, each Party having one vote, and shall be submitted by the Depositary to all Parties for ratification, approval or acceptance.

4.   Instruments of ratification, acceptance or approval of amendments shall be de-

posited with the Depositary. Amendments shall enter into force between Parties having accepted such amendments on the ninetieth day following the date of receipt by the Depositary of the instruments of at least three-fourths of the Parties to this Convention. Thereafter the amendments shall enter into force for any other Party on the ninetieth day after the date on which that Party deposits its instrument

5.   For the purpose of this Article, "Parties present and voting" means Parties present and casting an affirmative or negative vote.

ARTICLE 17

Protocols to this Convention

1.   The Conference of the Parties may, at any ordinary meeting, adopt protocols to this Convention.

2.   The text of any proposed protocol shall be communicated to the Parties by the Secretariat at least six months before the meeting at which it is proposed for adoption.

3.   The procedure specified in Article 16.3 shall apply to the adoption of, and any amendments to, any protocol.

4.   The requirements for the entry into force of any protocol or subsequent amendments to such protocol shall be established by that protocol.

5.   Decisions under any protocol shall be taken only by the Parties to that protocol.

ARTICLE 18

Adoption and Amendment of Annexes

1.   The annexes to this Convention shall form an integral part of this Convention and, unless expressly provided otherwise, a reference to this Convention constitutes, at the same time a reference to any annexes thereto. Such annexes shall be restricted to scientific, technical and administrative matters.

2.   The following procedures shall apply to the proposal, adoption and entry into force of additional annexes, or amendments to annexes, to this Convention:

   (a)   Such additional annexes or amendments to annexes shall be proposed and adopted according to the procedure laid down in Articles 16.1, 16.2 and 16.3 of this Convention;

   (b)   Any Party that is unable to accept such additional annexes or amendments to annexes, shall so notify the Depositary, in writing, within six months from the date of the communication of the adoption by the Depositary. The Depositary shall without delay notify all Parties of any such notification received. A Party may at any time substitute an acceptance for a previous declaration of objection and the annexes or amendments to annexes shall thereupon enter into force for that Party; and

(c)   Upon the expiration of six months from the date of the circulation of the communication by the Depositary, the annexes or amendments to annexes shall enter into force for all Parties to this Convention, which have not submitted a notification in accordance with the provisions of sub-paragraph (b) above.

3.   If an additional annex or an amendment to an annex involves an amendment to this Convention or to any protocol, the additional annex or amended annex shall not enter into force until such time as the amendment to this Convention or to the protocol enters into force.

ARTICLE 19

Verification

1.   Any Party which has reason to believe that another Party is acting or has acted in breach of its obligations under this Convention may inform the Secretariat thereof, and in such an event, shall simultaneously and immediately inform, directly or through the Secretariat, the Party against whom the allegations are made. All relevant information should be submitted by the Secretariat to the Parties.

2.   The Conference of the Parties shall consider the adoption of a protocol dealing with detailed procedures and arrangements for the verification of alleged breaches of obligations under this Convention.

ARTICLE 20

Settlement of Disputes

1.   In case of a dispute between Parties as to the interpretation or application of, or compliance with, this Convention or any protocol thereto, the Parties concerned shall seek a settlement of the dispute through negotiation, mediation or any other peaceful means of their own choice.

2.   If the Parties concerned cannot settle their dispute through the means mentioned in paragraph 1 of this Article, the dispute, if the Parties to the dispute agree, shall be submitted to arbitration under the conditions set out in Annex VII of this Convention or to the International Court of Justice. However, failure to reach common agreement on submission of the dispute to arbitration or to the International Court of Justice shall not absolve the Parties from the responsibility of continuing to seek to resolve it by the means referred to in paragraph 1.

3.   When ratifying, accepting, approving or acceding to this Convention, or at any time thereafter, a Party may declare that it recognises as compulsory ipso facto and without special agreement, in relation to any Party accepting the same obligation:

(a)   Arbitration in accordance with the procedures set out in Annex VII; and/or

(b)   Submission of the dispute to the International Court of Justice.

Such declaration shall be notified in writing to the Secretariat which shall communicate it to the Parties.

ARTICLE 21

Signature

1.   This Convention shall be open for signature by the Members of the South Pacific Forum at Waigani, Papua New Guinea, on 16 September 1995.

2.   This Convention shall remain open for signature by the Members of the South Pacific Forum from 22 September 1995 until 21 March 1996 at the South Pacific Forum Secretariat, Suva.

ARTICLE 22

Ratification, Acceptance or Approval

This Convention shall be subject to ratification, acceptance or approval by Members of the South Pacific Forum. Instruments of ratification, acceptance or approval shall be deposited with the Depositary.

ARTICLE 23

Accession

1.   This Convention shall be open for accession by Members of the South Pacific Forum from the day after the date on which the Convention is closed for signature. The instruments of accession shall be deposited with the Depositary.

2.   Other States not members of the South Pacific Forum which have territories in the Convention Area may accede to the Convention. In addition, other States which do not have territories in the Convention Area may also accede to the Convention pursuant to a decision of the Conference of the Parties under Article 13.4(g).

ARTICLE 24

Entry into Force

This Convention shall enter into force thirty days from the date of deposit of the tenth instrument of ratification, acceptance, approval or accession and thereafter for each State thirty days after the deposit of its instrument of ratification, acceptance, approval or accession.

ARTICLE 25

Reservations and Declarations

1.   No reservations or exceptions shall be made to this Convention.

2.   Paragraph 1 of this Article does not preclude a signatory or Party when signing, ratifying or acceding to this Convention, from making declarations or statements, however phrased or named, with a view, inter alia, to the harmonisation of its laws and regulations with the provisions of this Convention, provided that such declarations or statements do not purport to exclude or to modify the legal effect of the provisions of this Convention in their application to that Party.

ARTICLE 26

Withdrawal

1.   At any time after three years from the date on which this Convention has entered into force for a Party, that Party may withdraw by giving written notification to the Depositary.

2.   Withdrawal shall be effective one year after receipt of notification by the Depositary, or on such later date as may be specified in the notification.

3.   Withdrawal shall not exempt any withdrawing Party from fulfilling any obligations it might have incurred under this Convention, whilst a Party to this Convention.

ARTICLE 27

Depositary

The Secretary General of the South Pacific Forum Secretariat shall be the Depositary of this Convention and of any protocols thereto.

ARTICLE 28

Registration

This Convention, as soon as it enters into force, shall be registered by the Depositary with the Secretary-General of the United Nations in conformity with Article 102 of the Charter of the United Nations.

ANNEX I

CATEGORIES OF WASTES WHICH ARE HAZARDOUS WASTES

Wastes Streams:

Y1:   Clinical wastes from medical care in hospitals, medical centres and clinics.

Y2:   Wastes from the production and preparation of pharmaceutical products.

Y3:   Waste pharmaceuticals, drugs and medicines.

Y4:   Wastes from the production, formulation and use of biocides and phytophar-maceuticals.

Y5:   Wastes from the manufacture, formulation and use of wood preserving chemicals.

Y6:   Wastes from the production, formulation and use of organic solvents.

Y7:   Wastes from heat treatment and tempering operations containing cyanides.

Y8:   Waste mineral oils unfit for their originally intended use.

Y9:   Waste oils/water, hydrocarbons/water mixtures, emulsions.

Y10: Waste substances and articles containing or contaminated with polychlorinated biphenyls (PCBs) and/or polychlorinated terphenyls (PCTs) and/or polybrominated biphenyls (PBBs).

Y12: Wastes from production, formulation and use of inks, dyes, pigments, paints, lacquers, varnish.

Y13: Wastes from production, formulation and use of resins, latex, plasticisers, glues/adhesives.

Y14: Waste chemical substances arising from research and development or teaching activities which are not identified and/or are new and whose effects on human health and/or the environment are not known.

Y15: Wastes of an explosive nature not subject to other legislation.

Y16: Wastes from production, formulation and use of photographic chemicals and processing materials.

Y17: Wastes resulting from surface treatment of metals and plastics.

Y18: Residues arising from industrial waste disposal operations.

Y46: Wastes collected from households, including sewage and sewage sludges with the exception of clean sorted recyclable wastes which do not possess any of the hazardous characteristics defined in Annex II.

Y47: Residues arising from the incineration of household wastes.

Wastes having as constituents:

Y19: Metal carbonyls.

Y20: Beryllium; beryllium compounds.

Y21: Hexavalent chromium compounds.

Y22: Copper compounds.

Y23: Zinc compounds.

Y24: Arsenic; arsenic compounds

Y25: Selenium; selenium compounds.

Y26: Cadmium; cadmlum compounds.

Y27: Antimony; antimony compounds.

Y28: Tellurium; tellurium compounds

Y29: Mercury; mercury compounds.

Y30: Thallium; thallium compounds.

Y31: Lead; lead compounds.

Y32: Inorganic fluorine compounds excluding calcium fluoride.

Y33: Inorganic cyanides.

Y34: Acidic solutions or acids in solid form.

Y35: Basic solutions or bases in solid form.

Y36: Asbestos (dust and fibres).

Y37: Organic phosphorus compounds.

Y38: Organic cyanides.

Y39: Phenols; phenol compounds including chlorophenols.

Y40: Ethers.

Y41: Halogenated organic solvents.

Y42: Organic solvents excluding halogenated solvents.

Y43: Any congenor of polychlorinated dibenzo-furan.

Y44: Any congenor of polychlorinated dibenzo-p-dioxin.

Y45: Organohalogen compounds other than substances referred to in this Annex (e.g. Y39, Y41, Y42, Y43, Y44).

ANNEX II

LIST OF HAZARDOUS CHARACTERISTICS

UN CLASS*	CODE	CHARACTERISTICS

1 H1 **Explosive**

An explosive substance or waste is a solid or liquid substance or waste (or mixture of substances or wastes) which is in itself capable by chemical reaction of producing gas at such a temperature and pressure and at such speed as to cause damage to the surroundings.

3 H3 **Flammable liquids**

The word "flammable" has the same meaning as "inflammable". Flammable liquids are liquids, or mixtures of liquids, or liquids containing solids in solution or suspension (for example, paints, varnishes, lacquers, etc., but not including substances or wastes otherwise classified on account of their dangerous characteristics) which give off a flammable vapour at temperatures of not more than 60.5 degrees C, closed-cup test, or not more than 65.6 degrees C, open-cup test. (Since the results of open-cup tests and of closed-cup tests are not strictly comparable and even individual results by the same test are often variable, regulations varying from the above figures to make allowance for such differences would be within the spirit of this definition).

4.1 H4.1 **Flammable solids**

Solids, or waste solids, other than those classed as explosives, which under conditions encountered in transport are readily combustible, or may cause or contribute to fire through friction.

4.2 H4.2 **Substances or wastes liable to spontaneous combustion**

Substances or wastes which are liable to spontaneous heating under normal conditions encountered in transport, or to heating up on contact with air, and being then liable to catch fire.

4.3 H4.3 **Substances or wastes which, in contact with water, emit flammable gases**

Substances or wastes which, by interaction with water, are liable to become spontaneously flammable or to give off flammable gases in dangerous quantities.

5.1 H5.1 **Oxidizing**

Substances or wastes which, while in themselves not necessarily combustible, may, generally by yielding oxygen cause, or contribute to, the combustion of other materials.

5.2 H5.2 **Organic peroxides**

Organic substances or wastes which contain the bivalent-O - O-structure are thermally unstable substances which may undergo exothermic self-accelerating decomposition.

| 6.1 | H6.1 | **Poisonous (Acute)** |

Substances or wastes liable either to cause death or serious injury or to harm human health if swallowed or inhaled or by skin contact.

| 6.2 | H6.2 | **Infectious substances** |

Substances or wastes containing viable microorganisms or their toxins which are known or suspected to cause disease in animals or humans.

| 8 | H8 | **Corrosives** |

Substances or wastes which, by chemical action, will cause severe damage when in contact with living tissue, or in the case of leakage, will materially damage, or even destroy, other goods or the means of transport; they may also cause other hazards.

| 9 | H10 | **Liberation of toxic gases in contact with air or water** |

Substances or wastes which, if they are inhaled or ingested or if they penetrate the skin, may involve delayed or chronic effects, including carcinogenicity

| 9 | H11 | **Toxic (Delayed or chronic)** |

Substances or wastes which, if they are inhaled or ingested or if they penetrate the skin, may involve delayed or chronic effects, including carcinogenicity

| 9 | H12 | **Ecotoxic** |

Substances or wastes which, if released, present or may present immediate or delayed adverse impacts to the environment by means of bioaccumulation and/or toxic effects upon biotic systems.

| 9 | H13 | Capable, by any means, after disposal, of yielding another material, e.g. leachate, which possesses any of the characteristics listed above. |

Tests

The potential hazards posed by certain types of wastes are not yet fully documented; tests to define quantitatively these hazards do not exist. Further research is necessary in order to develop means to characterise potential hazards posed to human health and the environment by these wastes. Standardised tests have been derived with respect to pure substances and materials. Many countries have developed national tests which can be applied to materials listed in Annex I, in order to decide if these materials exhibit any of the characteristics listed in this Annex.

ANNEX III

PACIFIC ISLAND DEVELOPING PARTIES

1.   The following Members of the South Pacific Forum, on becoming party to this Convention, shall be considered to be Pacific Island Developing Parties for the purposes of the Convention:

- **Cook Islands**

- **Federated States of Micronesia**

- **Fiji**

- **Kiribati**

- **Republic of Marshall Islands**

- **Nauru**

- **Niue**

- **Republic of Palau**

- **Papua New Guinea**

- **Solomon Islands**

- **Tonga**

- **Tuvalu**

- **Vanuatu**

- **Western Samoa**

2.   The Conference of the Parties may, in accordance with Article 13.4(g), and upon agreement with such prospective party, accept the status of any new Party to this Convention as a Pacific Island Developing Party.

ANNEX IV

OTHER PARTIES

1. The following Members of the South Pacific Forum, on becoming party to this Convention, shall be considered to be Other Parties for the purposes of the Convention:

    – **Australia**

    – **New Zealand.**

2. (a) The Conference of the Parties may, in accordance with Article 13.4(g), and upon agreement with such prospective party, accept the status of any new Party to this Convention as an Other Party; and

    (b) An Other Party may designate a territory located within the Convention Area to which, upon agreement by the Conference of the Parties, the provisions of Article 4.1 of this Convention shall be applied mutatis mutandis in the same manner as they apply to a Pacific Island Developing Party.

Waigani Convention

ANNEX V

DISPOSAL OPERATIONS

[Not published herein]

Cf. Annex IV of Basel Convention

ANNEX VI A

INFORMATION TO BE PROVIDED ON NOTIFICATION

[Not published herein]

Cf. Annex VA of Basel Convention

ANNEX VI B

INFORMATION TO BE PROVIDED ON THE MOVEMENT DOCUMENT

[Not published herein]

Cf. Annex VB of Basel Convention

ANNEX VII

ARBITRATION

ARTICLE 1

Unless the agreement referred to in Article 20 of this Convention provides otherwise, the arbitration procedure shall be conducted in accordance with Articles 2 to 10 below.

ARTICLE 2

The claimant party shall notify the Secretariat that the Parties have agreed to submit the dispute to arbitration pursuant to Articles 20.2 or 20.3 of this Convention and include, in particular, the articles of this Convention, the interpretation or application of which, are at issue. The Secretariat shall forward the information thus received to all Parties to this Convention.

ARTICLE 3

The arbitral tribunal shall consist of three members. Each of the Parties to the dispute shall appoint an arbitrator, and the two arbitrators so appointed shall designate by common agreement the third arbitrator, who shall be the president of the arbitral tribunal. The latter shall not be a national of one of the Parties to the dispute, nor have their usual place of residence in the territory of one of the Parties, nor be employed by any of them, nor have dealt with the case in any other capacity.

ARTICLE 4

1. If the president of the arbitral tribunal has not been designated within two months of the appointment of the second arbitrator, the Secretary General of the Forum Secretariat in consultation with the Director of SPREP shall, at the request of either Party, designate that person within a further two months period.

2. If one of the Parties to the dispute does not appoint an arbitrator within two months of the receipt of the request, the other Party may inform the Secretary General of the Forum Secretariat who shall, in consultation with the Director of SPREP, designate the president of the arbitral tribunal within a further two months period. Upon designation, the president of the arbitral tribunal shall request the Party which has not appointed an arbitrator to do so within two months. After such period, the president shall inform the Secretary General of the Forum Secretariat who shall make this appointment within a further two months period, in consultation with the Director of SPREP.

ARTICLE 5

1. The arbitral tribunal shall render its decision in accordance with international law and in accordance with the provisions of this Convention.

2. Any arbitral tribunal established under the provisions of this Annex shall decide its own rules of procedure.

ARTICLE 6

1.   The decisions of the arbitral tribunal, both on procedure and on substance, shall be taken by majority vote of its members.

2.   The arbitral tribunal may take all appropriate measures in order to establish the facts. It may, at the request of one of the Parties, recommend essential interim measures of protection.

3.   The Parties to the dispute shall provide all facilities necessary for the effective conduct of the proceedings.

4.   The absence or default of a Party in the dispute shall not constitute an impediment to the proceedings.

ARTICLE 7

The arbitral tribunal may hear and determine counter-claims arising directly out of the subject matter of the dispute.

ARTICLE 8

Unless the arbitral tribunal determines otherwise because of the particular circumstances of the case, the expenses of the arbitral tribunal, including the remuneration of its members, shall be borne by the Parties to the dispute in equal shares. The arbitral tribunal shall keep a record of all its expenses, and shall furnish a final statement thereof to the Parties.

ARTICLE 9

Any Party that has an interest of a legal nature in the subject matter of the dispute which may be affected by the decision in the case, can intervene in the proceedings with the consent of the arbitral tribunal.

ARTICLE 10

1.   The arbitral tribunal shall render its award within five months of the date on which it is established unless it finds it necessary to extend the time-limit for a period which should not exceed five months.

2.   The award of the arbitral tribunal shall be accompanied by a statement of reasons. It shall be final and binding upon the Parties to the dispute.

3.   Any dispute which may arise between the Parties concerning the interpretation or execution of the award may be submitted by either Party to the arbitral tribunal which made the award or, if the latter cannot be seized thereof, to another tribunal constituted for this purpose in the same manner as the first.

II.9   Canada-US Agreement

Agreement between the Government of Canada and the Government of the United States of America Concerning the Transboundary Movement of Hazardous Waste

of 28 October 1986[1]

The Government of Canada (Canada), and the Government of the United States of America (the United States), hereinafter called "The Parties":

Recognizing that severe health and environmental damage may result from the improper treatment, storage, and disposal of hazardous waste and other waste;

Seeking to ensure that the treatment, storage, and disposal of hazardous waste and other waste are conducted so as to reduce the risks to public health, property and environmental quality;

Recognizing that the close trading relationship and the long common border between the United States and Canada engender opportunities for a generator of hazardous waste and other waste to benefit from using the nearest appropriate disposal facility, which may involve the transboundary shipment of hazardous waste and other waste;

Recognizing further that the most effective and efficient means of achieving environmentally sound management procedures for hazardous waste crossing the United States-Canada border is through cooperative, efforts and coordinated regulatory schemes;

Believing that a bilateral agreement is needed to facilitate the control of transboundary shipments of hazardous waste and other waste between the United States and Canada;

Reaffirming Principle 21 of the 1972 Declaration of the United Nations Conference on the Human Environment, adopted at Stockholm, which asserts that states have, in accordance with the Charter of the United Nations and the principles of international law, the sovereign right to exploit their own resources pursuant to their own environmental policies and the responsibility to ensure that activities within their jurisdiction or control do not cause damage to the environment of other states or of areas beyond the limits of national jurisdiction;

Taking into account Organization for Economic Co-operation and Development (OECD) Council Decisions and Recommendations on transfrontier movements of hazardous wastes, the United Nations Environment Program Cairo Guidelines and

---

[1]   This is a consolidated text of the 1986 Agreement and its 1992 Amendment for public use only, and does not constitutes an "official" text

Principles for the Environmentally Sound Management of Hazardous Waste, and resolutions of the London Dumping Convention,

Have agreed as follows:

Article 1 - Definitions

For the purposes of this Agreement:

(a)   **Designated** Authority means, in the case of the US Environmental Protection Agency (US EPA) and, in the case of Canada, the Department of the Environment.

(b)   **Hazardous Waste** means with respect to Canada, hazardous waste, and with respect to the United States, hazardous waste subject to a manifest requirement in the United States, as defined by their respective national legislations and implementing regulations.

(c)   **Country of Export** means the country from which the shipment of hazardous waste originated.

(d)   **Country of Import** means the country to which hazardous waste and other waste is sent for the purpose of treatment, storage (with the exception of short-term storage incidental to transportation) or disposal.

(e)   **Country of Transit** means the country which is neither the country of export nor the country of import, through whose land territory or internal waters hazardous waste and other waste is transported, or in whose ports such waste is unloaded for further transportation.

(f)   **Consignee** means the treatment, storage (with the exception of short-term storage incidental to transportation) or disposal facility in the country of import and the name of the person operating the facility.

(g)   **Exporter** means, in the case of the United States, the person defined as exporter, and in the case of Canada, the person defined as consignor, under their respective national laws and regulations governing hazardous waste and other waste.

(h)   **Other Waste** means Municipal Solid Waste (MSW) that is sent for final disposal or for incineration with energy recovery, and residues arising from the incineration of such waste, as defined by the Parties' respective national legislation and implementing regulations, but excluding waste covered under paragraph (b) of this Article.

Article 2 - General Obligation

The parties shall permit the export, import, and transit of hazardous waste and other waste across their common border for treatment, storage, or disposal pursuant to the terms of their domestic laws, regulations and administrative practices, and the provisions of this Agreement.

Article 3 - Notification to the Importing Country

(a)   The designated authority of the country of export shall notify the designated authority of the country of import of proposed transboundary shipments of hazardous waste and other waste.

(b) The notice referred to in paragraph (a) of this article may cover an individual shipment or a series of shipments extending over a twelve month or lesser period and shall contain the following information:

(i) The exporter's name, address and telephone number, and if required in the country of export, the identification number.

(ii) for each hazardous waste and other waste type and for each consignee:

1. A description of the hazardous waste and other waste to be exported, as identified by the waste identification number, the classification and the shipping name as required on the manifest in the country of export;

2. The estimated frequency or rate at which such waste is to be exported and the period of time over which such waste is to be exported;

3. The estimated total quantity of the hazardous waste and other waste in units as specified by the manifest required in the country of export;

4. The point of entry into the country of import;

5. The name and address of the transporter(s) and the means of transportation, such as the mode of transportation (air, highway, rail, water, etc.) and type(s) of container (drums, boxes, tanks, etc.);

6. A description of the manner in which the waste will be treated, stored or disposed of in the importing country;

7. The name and site address of the consignee;

8. An approximate date of the first shipment to each consignee, if available.

(c) The designated authority of the country of import shall have 30 days from the date of receipt of the notice provided pursuant to the date of receipt of the notice provided pursuant to paragraphs (a) and (b) of this article to respond to such notice, indicating its consent (conditional or not) or its objection to the export. Such response will be transmitted to the designated authority of the country of export. The date of receipt of the notice will be identified in an acknowledgement of receipt made immediately by the designated authority of the country of import to the country of export.

(d) If no response is received by the designated authority of the country of export within the 30 day period referred to in paragraph (c) of this article, the country of import shall be considered as having no objection to the export of hazardous waste and other waste described in the notice and the export may take place conditional upon the persons importing the hazardous waste and other waste complying with all the applicable laws of the country of import.

(e) The country of import shall have the right to amend the terms of the proposed shipment(s) as described in the notice.

(f)   The consent of the country of import, whether express, tacit, or conditional, provided pursuant to paragraphs (c) and (d) of this article, may be withdrawn or modified for good cause. The Parties will withdraw or modify such consent insofar as possible at the most appropriate time for the persons concerned.

(g)   For the purposes of this Article and Article 5, manifest-related requirements may, with respect to other waste, be substituted by alternative tracking requirements.

Article 4 - Notification to the Transit Country

(a)   The designated authority of the country of export shall notify the designated authority of the country of transit of the proposed shipment of hazardous waste and other waste at least 7 days prior to the date of the shipment. The notice shall include the information specified in paragraph (b) of Article 3, with the following exceptions:

(i)   The points of entry into and departure from the country of transit shall be provided in lieu of the entry point(s) into the country of import; and

(ii)   A description of the approximate length of time the hazardous waste and other waste will remain in the country of transit and the nature of its handling while there shall be submitted instead of a description of the treatment, storage, or disposal of the waste in the country of import.

Article 5 - Cooperative Efforts

1.   The Parties will cooperate to ensure, to the extent possible, that all transboundary shipments of hazardous waste and other waste comply with the manifest requirements of both countries.

2.   The Parties will cooperate in monitoring and spot-checking transboundary shipments of hazardous waste and other waste to ensure, to the extent possible, that such shipments conform to the requirements of the applicable legislation and of this Agreement.

3.   To the extent any implementing laws and regulations are necessary to comply with this Agreement, the Parties will act expeditiously to issue such regulations consistent with domestic law. Pending such issuance, the Parties will make best efforts to provide notification in accordance with this Agreement where current regulatory authority is insufficient. The Parties will provide each other with a diplomatic note upon the issuance and the coming into effect of any such laws and regulations.

Article 6 - Readmission of Exports

The country of export shall readmit any shipment of hazardous waste and other waste that may be returned by the country of import or transit.

Article 7 - Enforcement

The Parties shall ensure, to the extent possible, that within their respective jurisdictions, their domestic laws and regulations are enforced with respect to the transportation, storage, treatment and disposal of transboundary shipments of hazardous waste and other waste.

Article 8 - Protection of Confidential Information

If the provision of technical information pursuant to articles 3 and 4 would require the disclosure of information covered by agreement(s) of confidentiality between a Party and an exporter, the country of export shall make every effort to obtain the consent of the concerned person for the purpose of conveying any such information to the country of import or transit. The country of import or transit shall make every effort to protect the confidentiality of such information conveyed.

Article 9 - Insurance

The Parties may require, as a condition of entry, that any transboundary movement of hazardous waste and other waste be covered by insurance or other financial guarantee in respect to damage to third parties caused during the entire movement of hazardous waste and other waste, including loading and unloading.

Article 10 - Effects on International Agreements

Nothing in this Agreement shall be deemed to diminish the obligations of the Parties with respect to disposal of hazardous waste and other waste at sea contained in the 1972 London Dumping Convention.

Article 11 - Domestic Law

The provisions of this Agreement shall be subject to the applicable laws and regulations of the Parties.

Article 12 - Amendment

This Agreement may be amended by mutual written consent of the Parties or their authorized representatives.

Article 13 - Entry into Force

This Agreement shall enter into force on November 8, 1986 and continue in force for five years. It will automatically be renewed for additional five year periods unless either Party gives written notice of termination to the other at least three months prior to the expiration of any five year period. In any five year period, this Agreement may be terminated upon one year written notice given by one Party to the other.

II.10 Mexico-US Agreement

**AGREEMENT OF COOPERATION BETWEEN THE UNITED STATES OF
AMERICA AND THE UNITED MEXICAN STATES REGARDING THE
TRANSBOUNDARY SHIPMENTS OF HAZARDOUS WASTES AND
HAZARDOUS SUBSTANCES[1]**

of November, 1986

PREAMBLE

The Government of the United States of America ("the United States"), and the Government of the United Mexican States ("Mexico") ("the Parties"),

Recognizing that health and environmental damage may result from improper activities associated with hazardous waste;

Realizing the potential risks to public health, property and the environment associated with hazardous substances;

Seeking to ensure that activities associated with the transboundary shipment of hazardous waste are conducted so as to reduce or prevent the risks to public health, property and environmental quality, by effectively cooperating in regard to their export and import;

Seeking also to safeguard the quality of public health, property and environment from unreasonable risks by effectively regulating the export and import of hazardous substances;

Considering that transboundary shipments of hazardous waste and hazardous substances between the Parties, if carried out illegally and thus without the supervision and control of the competent authorities, or if improperly managed could endanger the public health, property and environment, particularly in the United States/Mexico border area;

Recognizing that the close trading relationship and the long common border between the Parties make it necessary to cooperate regarding transboundary shipments of hazardous waste and hazardous substances without unreasonably affecting the trade of goods and services;

Reaffirming Principle 21 of the 1972 Declaration of the United Nations Conference on the Human Environment, adopted at Stockholm, which provides that States have, in accordance with the Charter of the United Nations and the principles of international law, the sovereign right to exploit their own resources pursuant to their own environmental policies and the responsibility to ensure that activities within their jurisdiction or control do not cause damage to the environment of other States or of areas beyond the limits of national jurisdiction;

---

[1] Annex III to the Agreement Between the United States of America and the United Mexican States of Cooperation for the Protection and Improvement of the Environment in the Border Area

Recognizing that Article 3 of the Agreement between the Parties on Cooperation for the Protection and Improvement of the Environment in the Border Area of 1983 provides that the Parties may conclude specific arrangements for the solution of common problems in the border area as annexes to that Agreement;

Have agreed as follows:

ARTICLE I

Definitions

1. **"Designated Authority"** means, in the case of the United States, the Environmental Protection Agency and, in the case of Mexico, the Secretariat of Urban Development and Ecology through the Subsecretariat of Ecology.

2. **"Hazardous waste"** means any waste, as designated or defined by the applicable designated authority pursuant to national policies, laws or regulations, which if improperly dealt with in activities associated with them, may result in health or environmental damage.

3. **"Hazardous substance"** means any substance, as designated or defined by the applicable national policies, laws or regulations, including pesticides or chemicals, which when improperly dealt with in activities associated with them, may produce harmful effects to public health, property or the environment, and is banned or severely restricted by the applicable designated authority.

4. **"Activities"** associated with hazardous waste or hazardous substances means, as applicable, their handling, transportation, treatment, recycling, storage, application, distribution, reuse or other utilization.

5. **"Country of export"** means the Party from which the transboundary movement of hazardous waste or hazardous substances is to be initiated.

6. **"Country of Import"** means the Party to which the hazardous waste or hazardous substances are to be sent. This does not include "transit", as meaning transport of hazardous waste or hazardous substances through the territory of a Party without being imported through its Customs under applicable laws and regulations.

7. **"Consignee"** means the facility in the country of import which will ultimately receive the hazardous waste or hazardous substances.

8. **"Exporter"** means the physical or juridical person, whether public or private, acting on his behalf or as a contractor or subcontractor expressly or implicitly defined as exporter under the national laws and regulations of the country of export which specifically govern hazardous waste or hazardous substances.

9. **"Banned or severely restricted"** means final regulatory action, as designated or defined by the applicable designated authority, pursuant to national policies, laws or regulations.

   (a) Prohibiting, cancelling or suspending all or virtually all registered uses of a pesticide for human health or environmental reasons.

(b) Prohibiting or severely limiting the manufacture, processing, distribution or use of a chemical for human health or environmental reasons.

ARTICLE II

General Obligations

1.   Transboundary shipments of hazardous waste and hazardous substances across the common border of the Parties shall be governed by the terms of this Annex and their domestic laws and regulations.

2.   Each Party shall ensure, to the extent practicable, that its domestic laws and regulations are enforced with respect to transboundary shipments of hazardous waste and hazardous substances, and other substances as the Parties may mutually agree through appendices to this Annex, that pose dangers to public health, property and the environment.

3.   Each Party shall cooperate in monitoring and spot-checking transboundary shipments across the common border of hazardous waste and hazardous substances to ensure, to the extent practicable, that such shipments conform to the requirements of this Annex and its national laws and regulations. To this effect, a program of cooperation in this area should be concluded through an Appendix to this Annex, including the exchange of information resulting from the monitoring and spot-checking of transboundary shipments which may be useful to the other Party.

HAZARDOUS WASTE

ARTICLE III

Notification to the Importing Country

1.   The designated authority of the country of export shall notify the designated authority of the country of import of transboundary shipments of hazardous waste for which the consent of the country of import is required under the laws or regulations of the country of export, with a copy of the notification simultaneously sent through diplomatic channels.

2.   The notification referred to in paragraph 1 of this Article shall be given at least 45 days in advance of the planned date of export and may cover an individual shipment or a series of shipments extending over a twelve-month or lesser period and shall contain the following information for each shipment:

    (a) The exporter's name, address, telephone number, identification number and other relevant data required in the country of export.

    (b) By consignee, for each hazardous waste type:

        (i)   A description of the hazardous waste to be exported, as identified by the waste identification number(s) and the shipping description(s) required in the country of export.

        (ii)   The estimated frequency or rate at which such waste is to be exported and the period time over which such waste is to be exported.

(iii) The estimated total quantity of the hazardous waste in units as specified by the manifest or documents required in the country of export.

(iv) The point of entry into the country of import.

(v) The means of transportation, including the mode of transportation and the type of container involved.

(vi) A description of the treatment or storage to which the waste will be subjected in the country of import.

(vii) The name and site address of the consignee.

3.   In order to facilitate compliance with the requirements of the importing country for the exporter to provide information and documents additional to those described in paragraph 2 of this Article, the designated authority of the exporting country will cooperate by making such requirements for information and documents known to the exporter. To that end, the country of import may list such additional required information and documents in appendices to this Annex.

4.   The designated authority of the country of import shall have 45 days from the date of acknowledgement of receipt of the notification provided in paragraph 1 of this Article within which to respond to such notification, indicating its consent, with or without conditions, or its objection to the export.

5.   The country of import shall have the right to amend the terms of the proposed shipment contained in the notification in order to give its consent.

6.   The consent of the country of import provided pursuant to paragraphs 4 and 5 of this Article, may be withdrawn or modified at any time, pursuant to the national policies, laws or regulations of the country of import.

7.   Whenever the designated authority of a country of export requires notification of or is otherwise aware of a transboundary shipment that will be transported through the territory of the other Party, it shall, in accordance with its national laws and regulations, notify that Party.

ARTICLE IV

Readmission of Exports

The country of export shall readmit any shipment of hazardous waste that may be returned for any reason by the country of import.

HAZARDOUS SUBSTANCES

ARTICLE V

Notification of Regulatory Actions

1.   When a Party has banned or severely restricted a pesticide or chemical, its

designated authority shall notify the designated authority of the other Party that such action has been taken either directly or through an appropriate intergovernmental organization.

2.   The notice referred to in paragraph 1 of this Article shall contain the following information, if available:

(a) the name of the pesticide or chemical that is the object of the regulatory action;

(b) a concise summary of the regulatory action taken, including the timetable for any further actions that are planned. If the regulatory action bans or restricts certain uses but allows other uses, such information should be included;

(c) a concise summary of the reason for the regulatory action, including an indication of the potential risks to human health or the environment that are the grounds for the action;

(d) information concerning registered pesticides or substitute chemicals that could be used in lieu of the banned or severely restricted pesticide or chemical;

(e) the name and address of the contact point to which a request for further information should be addressed.

ARTICLE VI

Notification of Exports

1.   If the country of export becomes aware that an export of a hazardous substance to the country of import is occurring, the designated authority of the country of export shall notify the designated authority of the country of import.

2.   The purpose of such notice shall be to remind the country of import of the notification regarding regulatory action provided pursuant to Article 5 and to alert it to the fact that the export is occurring.

3.   The notice referred to in paragraph 1 of this Article shall contain the following information, if available:

(a) the name of the exported hazardous substance;

(b) for banned or severely restricted chemicals, approximate date(s) of the export;

(c) a copy of, or reference to, the information provided at the time of the notification of the regulatory action;

(d) name and address of the contact point for further information.

ARTICLE VII

Timing of the Notifications

1.   Notification of regulatory actions, required pursuant to Article 5, shall be transmitted as soon as practicable after the regulatory action has been taken, and in any event not later than 90 days following the taking of such action.

2.   When a Party has banned or severely restricted chemicals or pesticides prior to the entry into force of this Annex, its designated authority shall provide an inventory of such prior regulatory actions to the designated authority of the other Party.

3.   Notification of exports required pursuant to Article 6, shall be provided at the time the first export of a hazardous substance is occurring to the Country of import following the regulatory action and should recur at the time of the first export of the hazardous substance each subsequent year to that country.

4.   When the hazardous substance being exported has been banned or severely restricted prior to the entry into force of this Annex, the first export following the regulatory action shall be considered to be the first export following the provision of the inventory referred to in paragraph 2 of this Article.

ARTICLE VIII

Compliance with Requirements in the Importing Country

In order to facilitate compliance with the requirements in the importing country for the import of hazardous substances, the designated authority of the country of export will cooperate by making such requirements, including expected information and documents, known to the exporter. To that end, the country of import may list such requirements, information and documents in appendices to this Annex.

ARTICLE IX

Readmission of Exports

The country of export shall readmit any shipment of hazardous substances that was not lawfully imported into the country of import.

GENERAL PROVISIONS

ARTICLE X

Additional Arrangements

1.   The Parties shall consider and, as appropriate, establish additional arrangements to mitigate or avoid adverse effects on health, property and the environment from improper activities associated with hazardous waste and hazardous substances. Such arrangements may include the sharing of research data as well as the

definition of criteria regarding imminent and substantial endangerment and emergency responses, and may be included in appendices to this Annex.

2.   The Parties shall consult regarding experience with transboundary shipments of hazardous wastes and hazardous substances and, as problems are identified in the special circumstances of the United States-Mexico border relationship may include through appendices to this Annex, additional cooperation and mutual obligations aimed at achieving when necessary a more stringent control of transboundary shipments, such as provisions to bring uniformity in those relating to both hazardous wastes and hazardous substances regarding compulsory notification to and consent by the importing country for each transboundary shipment, as may become permitted by new national laws and regulations adopted by the Parties.

ARTICLE XI

Hazardous Waste Generated From Raw materials

Admitted In-Bond Hazardous waste generated in the processes of economic production, manufacturing, processing or repair, for which raw materials were utilized and temporarily admitted, shall continue to be readmitted by the country of origin of the raw materials in accordance with applicable national policies, laws and regulations.

ARTICLE XII

Information Exchange and Assistance

1.   The Parties shall, to the extent practicable, provide to each other, mutual assistance designed to increase the capability of each Party to enforce its laws applicable to transboundary shipments of hazardous waste or hazardous substances and to take appropriate action with respect to violators of its laws.

    (a) Such assistance may generally include:

        (i)   the exchange of information;

        (ii)   the provision of documents, records and reports;

        (iii) the facilitating of on-site visits to treatment, storage, or disposal facilities;

        (iv) assistance provided or required pursuant to any international agreements or treaties in force with respect to the Parties, or pursuant to any arrangement or practice that might otherwise be applicable;

        (v)   emergency notification of hazardous situations; and

        (vi) other forms of assistance mutually agreed upon by the Parties.

    (b) Save in exceptional circumstances, requests for assistance made pursuant to this Article shall be submitted in writing and translated into the language of the requested State.

(c) The requested State shall provide the requesting State with copies of publicly available records of government departments and agencies in the requested State.

(d) The requested State may provide any record or information in the possession of a government office or agency, but not publicly available, to the same extent and under the same conditions as it would be available to its own administrative, law enforcement, or judicial authorities.

2.   The Parties may establish in an appendix to this Annex a cooperative program relating to the exchange of scientific, technical, and other information for purposes of the development of their own respective regulatory mechanisms controlling hazardous waste and hazardous substances

ARTICLE XIII

Protection of Confidential Information

The Parties shall adopt procedures to protect the confidentiality of proprietary or sensitive information conveyed pursuant to this Annex, when such procedures do not already exist.

ARTICLE XIV

Damages

1.   The country of import may require, as a condition of entry, that any transboundary shipment of hazardous waste or hazardous substances be covered by insurance, bond or other appropriate and effective guarantee.

2.   Whenever a transboundary shipment of hazardous waste or hazardous substances is carried out in violation of this Annex, of the national laws and regulations of the Parties, or of the conditions to which the authorization for import was subject, or whenever the hazardous waste or hazardous substances produce damages to public health, property or the environment in the country of import, the competent authorities of the country of export shall take all practicable measures and initiate and carry out all pertinent legal actions that they are legally competent to undertake, so that when applicable in accordance with its national laws and regulations the physical or juridical persons involved:

(a) return the hazardous waste or hazardous substances to the country of export;

(b) return in as much as practicable the status quo ante of the affected ecosystem;

(c) repair, through compensation, the damages caused to persons, property or the environment.

The country of import shall also take, for the same purposes, all practicable measures and initiate and carry out all pertinent legal actions that its authorities are legally competent to undertake.

The country of export shall report to the country of import all measures and legal actions undertaken in the framework of this paragraph, and shall cooperate with the country of import, on the basis of this Annex or of other bilateral treaties and agreements in force between the Parties, and to the extent permitted by its national laws and regulations, to seek in its courts the satisfaction of those matters covered in subparagraphs a) to c) of this paragraph.

3.   The provisions of this Annex shall not be deemed to abridge or prejudice the Parties' national laws concerning transboundary shipments, or liability or compensation for damages resulting from activities associated with hazardous waste and hazardous substances.

ARTICLE XV

Effect on Other Instruments

1.   Nothing in this Annex shall be construed to prejudice other existing or future agreements concluded between the Parties, or affect the rights or obligations of the Parties under international agreements to which they are Party.

2.   The provisions of this Annex shall, in particular not be deemed to prejudice or otherwise affect the functions entrusted to the International Boundary and Water Commission, in accordance with the 1944 Treaty on the Utilization of Waters of the Colorado and Tijuana Rivers and of the Rio Grande.

ARTICLE XVI

Appendices

Any appendices to this Annex may be added through an exchange of diplomatic notes and shall form an integral part of this Annex.

ARTICLE XVII

Amendment

This Annex, and any appendices added hereto, may be amended by mutual agreement of the Parties through an exchange of diplomatic notes.

ARTICLE XVIII

Review

The Parties shall meet at least every two years from the date of entry into force of this Annex, at a time and place to be mutually agreed upon, in order to review the effectiveness of its implementation and to agree on whatever individual and joint measures are necessary to improve such effectiveness.

ARTICLE XIX

Entry into Force

This Annex shall enter into force upon an exchange of diplomatic notes between the Parties stating that each Party has completed its necessary internal procedures.

Article XX

Termination

This Annex shall remain in force indefinitely, unless one of the Parties notifies the other in writing through diplomatic channels of its desire to terminate it, in which case the Annex shall terminate six months after the date of such written notification. Unless otherwise agreed, such termination shall not affect the validity of any agreements made under this Annex.

II.11 Contract for Shipments of Waste

Example template for the contract concluded according to the Basel Convention (BC), the OECD-Decision or the EU Waste Shipment Regulation (EU-WSR).

Contract

concerning the shipment of *[waste description]* from *[country of export/dispatch]* to *[country of import/destination]* for *[recovery/disposal]* at the Facility in relation to notification number: *[notification number]*

between   **The Exporter/Notifier**

[Company name]

[Contact information]

and   **The Consignee/Importer/Facility**

[Company name]

[Contact information]

hereinafter, jointly referred to as "Contractors"

The Contractors agree on the following terms:

1. The Exporter/Notifier agrees to deliver the waste to the Importer/Consignee for the [recovery/disposal] thereof and the Importer/Consignee agrees to receive the waste and [recover/dispose] it according to the *[BC] [OECD-Decision] [EU-WSR]** and *[applicable national law]**. Shipments shall be carried out in accordance with the information provided in the notification subject to the conditions of consent and the approved financial guarantee and this contract.

2. The Exporter/Notifier shall take the waste back if the shipment or the recovery or disposal has not been completed as intended or if it has been effected as an illegal shipment, in accordance with *[Art. (8) and (9) of the BC] [Chapt. II D.(3) and (4) of the OECD-Decision] [Art. 22 and 24(2) of the EU-WSR]**.

3. The Importer/Consignee shall recover or dispose the waste if it has been effected as an illegal shipment, in accordance with *[Art. 9 of the BC] [Chapt. II D. (3) of the OECD-Decision] [Article 24(3) of the EU-WSR]**.

---

*   insert applicable law or regulation
*   chose applicable stipulations

4. The Importer/Facility shall, *[within 3 days of receipt of the waste,]#)* provide confirmation in writing that the waste has been received in accordance with *[Art. 6 (9) of the BC] [Chapt. II D.(2) k) of the OECD-Decision] [Article 16(d) of the EU-WSR]•)*. The Facility shall complete box 18 of the movement document and send a copy of the form to the Exporter/Notifier and to all competent authorities concerned.

5. The Facility shall provide, according to *[Art. 6 (9) of the BC] [Chapt. II D.(2) l) of the OECD-Decision] [Article 5(3)(c) and Article 16(e) of the EU-WSR]•)*, a certificate that the waste has been recovered or disposed of, in accordance with the notification and the conditions specified therein and the requirements of the *[BC] [OECD-Decision] [EU-WSR]•)*. The Facility shall, as soon as possible, *[but no later than 30 days after completion of recovery or disposal operation, and no later than 12 months following receipt of the waste,]#)* complete box 19 of the movement document and send a copy of the form to the Importer/Notifier and to all competent authorities concerned (certificate of recovery/disposal).

6. If the Facility issues a certificate of recovery or disposal in such a way as to result in an illegal shipment, with the consequence that the financial guarantee is released, *[Art. 9 (3) of the BC] [Art. 24(3) and 25(2) of the EU-WSR]•)* shall apply. The Importer/Consignee shall bear the costs arising from the disposal or recovery in an environmentally sound manner, including possible transport and storage costs.

This contract covers the shipments of waste under the above-mentioned notification and is effective at the time of the notification and for the duration of the shipments until a certificate for the completion of the *[recovery/disposal]* operation is issued in accordance with *[Art. 6 (9) of the BC] [Chapt. II D.(2) of the OECD-Decision] [Article 16(e) of the EU-WSR]•)*.

Should there be any contradictions between this contract and any other contract concluded directly or indirectly between the Contractors, the provisions of this contract shall prevail.

Date: *[date of signature]*

For the Exporter/Notifier

Date: *[date of signature]*

For the Importer/Consignee/Facility

____

Name and title

____

Name and title

Signature:_______________________

Signature:_______________________

____

#)   requirement of the OECD-Decision and the EU-WSR

•)   chose applicable stipulations

*)   insert applicable law or regulation

II.12 Contract for Shipments of Waste Subject to the Information Requirements *(EU specific)*

Example template[1] for the contract concluded according to Article 18 of Regulation (EC) No 1013/2006.

> **Contract for Shipments of Waste Subject to the Information Requirements of Article 18 of Regulation (EC) No 1013/2006 on Shipments of Waste**

between **"The Person who arranges the shipment/Exporter"[2]**

[Company name]

[Contact information]

and **"The Consignee/Importer"**

[Company name]

[Contact information]

Concerning shipments for recovery of the following waste(s) and the recovery operation(s) corresponding to each waste[3]:

[usual description(s) of the waste; waste identification(s) according to box 10 of Annex VII; operation(s) R...]

The parties to this agreement, being the Person who arranges the shipment and the Consignee, shall comply with the requirements of Regulation (EC) No 1013/2006 in respect of the shipment of waste referred to in Art. 3(2) of this Regulation. Shipments shall be carried out in accordance with Art. 18 and the information provided in the Annex VII document and under the terms of this contract.

The Person who arranges the shipment agrees to deliver the waste to the Consignee and/or recovery facility for the recovery thereof and the Consignee agrees, in case it is also the recovery facility, to recover it according to Regulation (EC) No 1013/2006 on shipments of waste.

It is hereby agreed between the parties to this agreement that the following legal duties and obligations will be observed, as required by Regulation (EC) No

---

[1] This annex includes the following example template for the wording of the contract referred to in Art. 18(2) and block 12 of Annex VII to the EU-WSR providing information that is consistent with that provided in the Annex VII document

[2] The person who arranges the shipment must fall under the jurisdiction of the country of dispatch.

[3] In case of multiple wastes, provide a list of the wastes with their corresponding recovery operation(s). In each Annex VII document corresponding to this contract, only one recovery operation is to be indicated according to Appendix I, paragraph 16 of Correspondents" Guidelines No. 10.

1013/2006:

(a) The Person who arranges the shipment shall ensure that the waste is accompanied by an Annex VII document.

(b) The Annex VII document shall be signed by the Person who arranges the shipment before the shipment starts, and by the recovery facility and the Consignee when the waste is received.

(c) This contract between the Person who arranges the shipment and the Consignee shall be effective when the shipment starts.

(d) Where a shipment of waste or its recovery cannot be completed as intended, or where it has been effected as an illegal shipment, the Person who arranges the shipment or, where this person is not in a position to complete the shipment of waste or its recovery (for example, due to insolvency) the Consignee, shall take the waste back or ensure its recovery in an alternative way, and provide, if necessary, for its storage in the meantime.

(e) This contract remains valid for the duration of the shipment effected on the Annex VII document until the recovery operation at the facility has been completed.

(f) The Person who arranges the shipment or the Consignee shall provide a copy of the contract upon request by the authority involved in inspections.

For the Person who arranges the shipment:

Name and title: [*Name and title*]

____

Date: [Date of signature]

_____________________________Signature:_______________________

For the Consignee:

Name and title: [Name and title]

____

Date: [Date of signature]

_____________________________Signature:_______________________

II.13 Correspondents' guidelines *(EU-specific)*

Correspondents' guidelines represent the common understanding of all Member States on how Regulation (EC) No 1013/2006 (Waste Shipment Regulation – EU-WSR) should be interpreted. The guidelines were agreed by the correspondents at a meeting organised pursuant to Article 57 of EU-WSR. They are not legally binding. The binding interpretation of Community law is the exclusive competence of the European Court of Justice.

The guidelines agreed by the focal points (correspondents) shall apply in each case, as shown in the text of the relevant guideline, from the date of adoption and shall be reviewed no later than five years after adoption.

The following correspondents' guidelines have been agreed:

Correspondents' Guidelines No 1 on Shipments of Waste Electrical and Electronic Equipment (WEEE) and of used Electrical and Electronic Equipment (EEE) suspected to be WEEE

Correspondents' Guidelines No 2 concerning information on imports into the Community of waste generated by armed forces or relief organisations according to Article 1(3)(g) of Regulation (EC) No 1013/2006 on shipments of waste

Correspondents' Guidelines No 3 on a certificate for subsequent non-interim recovery or disposal according to Article 15(e) of Regulation (EC) No 1013/2006 on shipments of waste

Correspondents' Guidelines No 4 on classification of waste electrical and electronic equipment and fly ash from coal-fired power plants according to Annex IV, part I, note (c) to Regulation (EC) No 1013/2006 on shipments of waste

Correspondents' Guidelines No 5 on classification of wood waste under entries B3050 or AC170

Correspondents' Guidelines No 6 on classification of slags from processing of copper alloys under entries GB040 and B1100

Correspondents' Guidelines No 7 on classification of glass waste originating from cathode ray tubes (CRT) under entries B2020 or A2010

Correspondents' Guidelines No 8 on classification of waste cartridges containing toner or ink, according to Regulation (EC) No 1013/2006 on shipments of waste

Correspondents' Guidelines No 9 on shipment of waste vehicles

Correspondents' Guidelines No 10 on shipments of waste pursuant to Article 18 of Regulation (EC) No 1013/2006 on shipments of waste

Correspondents' Guidelines No 11 on the specification of a data model for the electronic data interchange under Regulation (EC) No 1013/2006 on shipments of waste.

Correspondents' Guidelines No 12 on the classification of plastic waste.

They can be downloaded via:

https://ec.europa.eu/environment/topics/waste-and-recycling/waste-shipments/waste-shipments-correspondents-guidelines_en
https://waste-move.eu/?p=539&lang=en

II.14 Bank Guarantee

Sample template of a bank guarantee

Guarantee for waste shipments

The company [applicant/exporter/notifier]

> **[Company name]**

> **[Contact information]**

has entered into an agreement with the company [importer/consignee of the waste]

> **[Company name]**

> **[Contact information]**

for the recovery/disposal of [quantity] [name of waste] - waste code: [waste identification code according to the [BC/OECD-Decision/EU-WSR]

with the company [importer/consignee of the waste].

For the shipment of waste according to the above-mentioned contract and under notification no.: [indicate the notification number of the notification document], the deposit of a financial guarantee is required in accordance with [Art. 6 of the BC] [Chapt. II D.(1) b) of the OECD Decision] [Art. 6 of the EU-WSR], which covers the costs to be borne by the notifier in accordance with [Art. 9 of the BC] [Chapt. II D.(3) and (4) of the OECD Decision] [Art. 23 and 25 of the EU-WSR in the cases of Art. 22 and 24 of the EU-WSR].

Having said this, we hereby undertake

> **[address of the bank].**

vis-à-vis the authority

> **[address of competent authority of export/dispatch]**

the directly enforceable guarantee for all claims of the company [applicant/exporter/notifier] on the above-mentioned basis up to the amount of

> [€] [$][1] ....

> in words: [€] [$][1] ....

with the proviso that we can only be held liable for the payment of money upon first written demand.

The guarantee is unlimited in time. It shall expire as soon as this document has been returned to us also via third parties.

---

[1]   insert applicable currency

ABOUT THE AUTHOR

Dr. Joachim Wuttke studied chemistry at the Ruhr University of Bochum, where he received his PhD in organic photochemistry in 1985. From November 1985 to April 2018, he was a scientific employee at the German Federal Environment Agency. From 1993 to 2018, he was head of the section "Municipal Waste Management, Hazardous Waste, Focal Point to the Basel Convention". From 1990 to 2005, Dr Joachim Wuttke was a lecturer for "Packaging and Environment" at the University of Applied Science in Berlin. He has worked as a short-term expert in EU programmes such as "Twinning" and TAIEX in Bulgaria, Estonia, Israel, the Slovak Republic, Poland, Turkey and Egypt. He has also worked as a consultant for the OSCE and other organisations.

He chaired the project group "Collection and Transboundary Movement of Used and End-of-life Mobile Phones" of the Mobile Phone Partnership Initiative of the Basel Convention. Within the Partnership for Action on Computer Equipment (PACE) of the Basel Convention, he co-chaired the project groups "Transboundary Movement of Used and End-of-Life Computing Equipment" and "Environmentally Sound Material Recovery/Recycling of End-of-Life Computing Equip-ment".

In the Framework of Basel Convention he developed the Biomedical and Health-Care Waste Guideline of the Basel Convention, including inter-governmental negotiations. Furthermore he developed and proposed amendments of the Waste Lists of the Basel Convention, including inter-governmental negotiations of these amendments.

From May 2018, he has been working as a consultant (e.g. for the Basel Convention Secretariat) and is, among other things, a lecturer/speaker in particular on issues related to transfrontier waste shipments and waste classification.

He has published more than 75 articles in the field of waste management, operates a website and is the author of a practical handbook on transfrontier waste shipments (in German).

wuttke@waste-move.eu

www.waste-move.eu